Spin Glasses and Other Frustrated Systems

Princeton Series in Physics

edited by Arthur S. Wightman
and Philip W. Anderson

Quantum Mechanics for Hamiltonians Defined as Quadratic Forms
by Barry Simon

Lectures on Current Algebra and Its Applications
by Sam B. Treiman, Roman Jackiw, and David J. Gross

Physical Cosmology *by P.J.E. Peebles*

The Many-Worlds Interpretation of Quantum Mechanics
edited by B. S. DeWitt and N. Graham

Homogeneous Relativistic Cosmologies *by Michael P. Ryan, Jr. and Lawrence C. Shepley*

The $P(\Phi)_2$ Euclidean (Quantum) Field Theory *by Barry Simon*

Studies in Mathematical Physics: *Essays in Honor of Valentine Bargmann edited by Elliott H. Lieb. B. Simon, and A.S. Wightman*

Convexity in the Theory of Lattice Gases *by Robert B. Israel*

Works on the Foundations of Statistical Physics
by N.S. Krylov

The Large-Scale Structure of the Universe
by P.J.E. Peebles

Surprises in Theoretical Physics *by Rudolf Peierls*

Statistical Physics and the Atomic Theory of Matter,
From Boyle and Newton to Landau and Onsager *by Stephen G. Brush*

Quantum Theory and Measurement *edited by*
John Archibald Wheeler and Wojciech Hubert Zurek

Current Algebra and Anomalies *by Sam B. Treiman, Roman Jackiw, Bruno Zumino, and Edward Witten*

Spin Glasses and Other Frustrated Systems
by Debashish Chowdhury (*Spin Glasses and Other Frustrated Systems* is published in co-operation with World Scientific Publishing Co. Pte. Ltd., Singapore.)

Spin Glasses and Other Frustrated Systems

by
Debashish Chowdhury

Department of Physics
Temple University, Philadelphia
USA

and

Institut für Theoretische Physik
Universität zu Köln, Köln
West Germany

Princeton Series in Physics

Princeton University Press
Princeton, New Jersey

Published by

Princeton University Press
41 William Street
Princeton, New Jersey 08540

Library of Congress Cataloging-in-Publication Data are available.

SPIN GLASSES AND OTHER FRUSTRATED SYSTEMS

ISBN 0-691-08461-0

Printed in Singapore by Kim Hup Lee Printing Co. Pte. Ltd.

TO MY

PARENTS

Acknowledgements

I am indebted to A. Aharony, D.J. Amit, P.W. Anderson, K. Baberschke, K. Binder, J.R.L. de Almeida, B. Derrida, G. Dreyfus, J. Ferre, G.S. Grest, H. Gutfreund, C.L. Henley, J.J. Hopfield, R.M. Hornreich, D.L. Huber, J. Kertesz, M.A. Klenin, T.C. Lubensky, A.P. Malozemoff, H. Meyer, M. Mezard, Y. Miyako, G.J. Nieuwenhuys, A.T. Ogielsky, J.P. Provost, J.P. Renard, T.M. Rice, W. Saslow, J.P. Sethna, J. Souletie, D. Stauffer, N.S. Sullivan, J.L. Tholence, Y. Yeshurun, D. Wuertz, G. Vallee, J.L. van Hemmen, J. Vannimenus for crtical reading of the various chapters of the preliminary versions of the manuscript, for their comments, and suggestions. However, I am responsible for all the error(s), if any, in this book. My sincere thanks go to K. Banerjee, J.K. Bhattacharjee, B.K. Chakrabarti, S.F. Edwards, D. Forster, E.T. Gawlinsky, S.K. Ghatak, M. Grant, J.D. Gunton, K. Kaski, D.C. Khan, D. Kumar, A.K. Majumdar, A. Mookerjee, G. Parisi, M. Schreckenberg, K.D. Usadel, E.P. Wohlfarth for useful discussions and/or correspondences over the last few years. I would like to thank C.L. Henley for sending a copy of his Ph.D. thesis. However, the contents of this book do not necessarily reflect the opinions of the experts named above.

It is my great pleasure to thank Jim Gunton and Dietrich Stauffer without whose encouragements and support this book could never have been completed.

I would like to thank Mrs. A. Schneider for drawing some of the figures in this book and Mrs. L. Tucker for typing the references. My friends T. Ala-Nissila, S. Kumar, and especially, A. Saxena corrected several typographical errors in the earlier versions of the manuscript. The first version of the manuscript was typed using the software written by Martin Schulte, a high school student in Köln, for the text editing system of a PDP11/34 computer. Being the author as well as the typist, I cannot blame anyone else for the typographical error(s) in this book.

My stay at Köln has been supported by the Alexander von Humboldt Foundation in the form of a Humboldt Fellowship. I also acknowledge financial support from the Office of Naval Research, USA, in the form of a postdoctoral fellowship from the grant N00014-83-K-0382. I would also like to thank the International Center for Theoretical Physics, Trieste, Italy, for excellent hospitality.

Finally, I am happy that the evenings and the weekends which I devoted during the last year learning the physics of the frustrated systems, instead of visiting the German pubs or the American beaches, have not been frustrating. I shall consider those long hours as "sacrifice for a good cause" if the reader finds this book useful.

A note for the readers:

In the limited space of a few hundred pages it is not possible to introduce the reader to the various techniques and methodologies before applying them to the spin glasses. However, I have tried my best to provide references of the introductory as well as advanced level texts and reviews so that a beginner can master the 'tricks of the trade' before applying them to spin glasses. Most often I have followed notations compatible with the original papers for the convenience of those readers who would like to look into the latter. Therefore, it has not been possible to use unique notation system in this book. However, I believe the meaning of the various notations and symbols will be clear in all the relevant contexts.

CONTENTS

CHAPTER 1

REAL SPIN GLASS (SG) MATERIALS, SG-LIKE MATERIALS AND SG MODELS

Magnetic systems exhibit various different types of ordering depending on the temperature T, external magnetic field H, etc. (see Hurd 1982 for an elementary introduction). An experimentalist usually identifies a magnetic material as a spin glass (SG) if it exhibits the following characteristic properties:
(i) the low-field, low-frequency a.c. susceptibility $\chi_{a.c.}(T)$ exhibits a cusp at a temperature T_g, the cusp gets flattened in as small a field (H) as 50 Gauss, (a better criterion is the divergence of the nonlinear susceptibility, as we shall see in chapter 16),
(ii) no sharp anomaly appears in the specific heat,
(iii) below T_g, the magnetic response is history-dependent; viz. the susceptibility measured in a field-cooled sample is higher than that when cooled in zero-field,
(iv) below T_g, the remanent magnetization decays very slowly with time,
(v) below T_g, a hysteresis curve, laterally shifted from the origin, appears,
(vi) below T_g, no magnetic Bragg scattering, chracteristic of long-range order (LRO), is observed in neutron scattering experiments, thereby demonstrating the absence of LRO
(vii) susceptibility begins to deviate from the Curie law at temperatures $T >> T_g$.
(see appendix A for a list of the SG materials and appendix B for a summary of the general features of the experimental results). There are several other systems (not necessarily magnetic) which share at least some of the features of SG and hence we shall call these materials SG-like systems (see chapter 27).

From the theoretical point of view, the SG materials have two common features: (a) the disorder is "quenched" and (b) the interaction between the spins are in conflict with each other leading to "frustration" (don't worry if you are not familiar with the two terms within quotes, we shall explain these terms in sections 1.2 and 1.3, respectively.

1.1 SG Models:

The simplest model for a d-dimensional magnetic system is the Heisenberg Hamiltonian (see Mattis 1981, 1985)

$$\mathcal{H} = - \Sigma J_{ij} \vec{S}_i \cdot \vec{S}_j \qquad (1.1)$$

where $\vec{S}_i$, $\vec{S}_j$, etc. are the m-component spin vectors (the models corresponding to m=1 and m=2 are called the Ising model and the XY model, respectively) at the positions $\vec{r}_i$, $\vec{r}_j$, etc. in a d-dimensional space interacting with one another, the strength of the interaction being J_{ij}. In the general case J_{ij} depends on $\vec{r}_i - \vec{r}_j$. (The nature of J_{ij} in some of the real SG materials has been listed in appendix A.) Most of the theoretical works so far have focussed attention only on classical spins. Such a description, in terms of "spin vectors" is valid provided there exist "good" local moments in the system. There exists a class of random magnetic alloys, e.g., AuCo, RhFe, RhCo, etc. (Murani and Coles 1970), whose SG transition temperature T_g is much lower than the Kondo temperature T_K. Such systems, called Stoner glass (Hertz 1979, 1980), are described better by the Stoner model (Shimizu 1981, Gautier 1982) rather than by the Heisenberg Hamiltonian (1.1). However, we shall not discuss Stoner glasses further in this book.

The exchange interaction J_{ij} on the right hand side of (1.1) is called short-ranged or long-ranged depending on (Mattis 1985) whether $\sum_j |J(r_{ij})| < \infty$ or $\sum_j |J(r_{ij})| = \infty$.

Disorder can be introduced into magnetic models in two different ways - either through "bond disorder" or "site disorder". In the bond-random models of SG one assumes the exchange bonds J_{ij} to be independent random variables distributed as

$$P(J_{ij}) = c_b P_c(J_{ij}) + (1-c_b)\ \delta(J_{ij}) \qquad (1.2)$$

where P_c is some suitable continuous or discrete distribution and the presence of the second term incorporates dilution. In other words, only a finite fraction c_b of the exchange bonds are nonzero, the strength of the latter are determined by the distribution P_c. In the Gaussian model

$$P_c(J_{ij}) = (2\pi J^2)^{-1/2}\ \exp[\,-\,(J_{ij} - J_0)^2/2J^2) \qquad (1.3)$$

In the nondilute Gaussian model, most often used in the literature (Edwards and Anderson 1975), $c_b = 1$. In the Sherrington-Kirkpatrick (SK) model each spin is assumed to interact with every other spin in the system (i.e., the lattice coordination number $z \to \infty$) with an exchange interaction distributed as in (1.3). However, the requirement that the free energy is an extensive thermodynamic quantity is fulfilled provided

$$J_0 = \tilde{J}_0/N \quad \text{and} \quad J = \tilde{J}/N^{1/2}$$

where $\tilde{J}_0$ and $\tilde{J}$ are both intensive. Thus, SK model is truly long-ranged.

In the generalized ±J model, $P_c(J_{ij})$ is assumed to be discrete, having only two values +J and -J with probabilities c_{bf} and c_{ba} respectively, i.e.,

$$P_c(J_{ij}) = c_{bf}\ \delta(J_{ij} - J) + c_{ba}\ \delta(J_{ij} + J) \qquad (1.4)$$

where $c_{bf} + c_{ba} = c_b$. Notice that for $c_{ba} = 0$ (1.4) represents a dilute ferromagnet and for $c_b = c_{bf} + c_{ba} = 1$ (1.4) reduces to the nondilute ±J model (Toulouse 1977). Another possible generalization of the nondilute ±J model is to have a variable strength of one of the interactions (say, the antiferromagnetic) by replacing J by aJ where a is a variable parameter (Wolff and

Zittartz 1985). So far most of the theoretical attention has been focussed on the Gaussian model and the ±J model.

Now let us consider the site-disorder models where only a finite nonzero fraction c of the lattice sites are occupied (randomly) by the spins; the remaining sites are occupied by nonmagnetic atoms or molecules. One such model is defined by the Hamiltonian (Binder et al.1979)

$$\mathcal{H} = - J_1 \Sigma c_i c_j S_i S_j - J_2 \Sigma c_i c_j S_i S_j \qquad (1.5)$$

where $J_1 > 0$ and $J_2 < 0$, the first summation is carried over nearest neighbor pairs and the second summation is carried over next-nearest-neighbor pairs; c_i is the occupation probability of the i-th site ($c_i = 1$ if the i-th site is occupied by a spin, and $c_i = 0$ otherwise, so that $(1/\mathcal{N})\Sigma c_i = c$, where $\mathcal{N}$ is the total number of lattice sites). In another site-disordered model the interaction J_{ij} between any two sites $\vec{r}_i$ and $\vec{r}_j$ (not necessarily nearest-neighbors) is given by

$$J_{ij} = J_{RKKY} (\vec{r}_i - \vec{r}_j)\, c_i\, c_j \qquad (1.6)$$

where J_{RKKY} is the Ruderman-Kittel-Kasuya-Yoshida (RKKY) exchange interaction (see Mattis 1981, 1985). The latter interaction arises from the indirect exchange interaction between the localized moments mediated via the conduction electrons, the conduction electron-spin interacts with a localized impurity spin through the s-d exchange interaction J_{sd}. The expression for J_{RKKY} in a d-dimensional nonrandom system is given by (Larsen 1981)

$$J_{RKKY}(r)=\{CJ_{sd}^2/(d-1)\}\{r^{2-2d}/(2\pi)^d\}(k_F r)^d$$
$$[J_{d/2-1}(k_F r)Y_{d/2-1}(k_F r) + J_{d/2}(k_F r)Y_{d/2}(k_F r)] \qquad (1.7)$$

where C is a constant depending on the characteristic parameters of the system, k_F is the Fermi wave vector, J_{sd} is the strength of the s-d exchange interaction and J_ν and Y_ν are the Bessel and the Neumann functions, respectively. The expression (1.7) reduces to the appropriate forms for J_{RKKY} in d=2 (Fischer and Klein 1975) and in d=3 (Ruderman and Kittel 1954, Kasuya 1956, Yoshida 1957); the latter is given by

$$J^{d=3}_{RKKY} = V_0 \; \{\cos\,(2k_F r_{ij} + \phi)\}/r^3_{ij} \qquad (1.7a)$$

where V_0 determines the "amplitude" of the oscillating factor of the interaction and ϕ is a phase factor. Is the RKKY interaction long-ranged or short-ranged? The answer to this question will be presented in chapter 17. In random magnetic systems the RKKY interaction gets damped by the disorder (deGennes 1962, Kaneyoshi 1975, 1979, Poon 1978, deChatel 1981); for weak disorder the latter interaction in d=3 (see Larsen 1985 for the general expression in d dimension) is given by

$$J^r_{RKKY} = J_{RKKY} \; \exp(-|\vec{r}_i - \vec{r}_j|/\delta) \qquad (1.8)$$

where δ is the mean free path of the conduction electrons.

In all the models (1.2)-(1.8) $P(J_{ij})$ is independent of temperature and hence the quenched nature of the disorder. As we shall see in the next section, the random sign of the interaction leads to frustration. However, randomness in the sign of the interaction does not necessarily guarantee that it is frustrated. For example, let us consider the Mattis model (1976) which is defined by (1.1) together with

$$J_{ij} = \xi_i \, \xi_j \qquad (1.9)$$

where ξ_i and ξ_j are independent random variables assuming the values +1 and -1 with equal probability. Mattis model is a special case ($J_1 = J_3 = 0$) of the Luttinger model (1976) which is defined by

$$J_{ij} = (1/N)\; [J_1 + J_2\, \xi_i \xi_j + J_3(\xi_i + \xi_j) \qquad (1.10)$$

where J_1, J_2 and J_3 are constants. However, as we shall in section 1.3, the Mattis model as well as the Luttinger model do not contain frustration. Therefore, these two models fail to capture one of the essential features of SG.

The van Hemmen model (1982) is defined by

$$\mathcal{H} = -(J_0/N)\,\Sigma\, S_i S_j - \Sigma\, J_{ij}\, S_i S_j \qquad (1.11)$$

where J_0 is the nonrandom ferromagnetic exchange and randomness is incorporated through J_{ij} which is given by

$$J_{ij} = (J/N)\,(\xi_i \eta_j + \xi_j \eta_i)$$

where ξ_i and η_i are identically distributed independent random variables with zero mean.

Provost and Vallee (1983) generalized the van Hemmen model as follows:

$$\mathcal{H} = -(J_0/N)\,\Sigma S_i S_j - (1/N)\,\Sigma\Sigma\, J_{\mu\nu}\, \xi_{i\mu}\, \xi_{j\nu}\, S_i S_j \qquad (1.12)$$

where ξ_i are N independent identically distributed random p-vectors ($\mu = 1,\ldots,p$) with mean zero. The special case p=2, $J_{11} = J_{22} = 0$, $J_{12} = J_{21} = J$, $\xi_{i1} = \xi_i$ and $\xi_{i2} = \eta_i$ is identical with the van Hemmen model.

The models described so far can be divided into two classes: separable and non-separable models. The Mattis model, the van Hemmen model and the generalized van Hemmen model (Provost-Vallee model) belong to the former class whereas the SK model belongs to the latter. An elegant way of writing all these separable models of random magnetic systems is as follows:

$$\mathcal{H} = -(1/N)\,\Sigma\,(\xi_i, J\xi_j)\, S_i S_j$$

where J is a p X p symmetric matrix and ξ_i are N independent identically distributed Gaussian p-vectors with zero mean. For example, the Mattis model corresponds to a scalar J, the van Hemmen model corresponds to

$$J = \begin{pmatrix} 0 & J \\ J & 0 \end{pmatrix}$$

and so on. All the separable models discussed so far have a common characteristic, viz., the dimension of the matrix J is finite. The close correspondence between the SK model and the separable model with infinite value of p (Benamira et al.1985) will be examined in chapter 14.

A model with only a long-ranged interaction or with only a short-ranged interaction is an ideal situation whereas the Hamiltonian of a real SG material is expected to have both long-ranged and short-ranged parts. Bowman and Halley (1982) studied a model

$$\mathcal{H} = \mathcal{H}_S + \mathcal{H}_L \qquad (1.13)$$

where the short-ranged part $\mathcal{H}_S$ is nearest-neighbor ferromagnetic or antiferromagnetic exchange Hamiltonian whereas the long-ranged part $\mathcal{H}_L$ is the SK Hamiltonian. Notice that the short-ranged parts in the van Hemmen as well as in the Bowman-Halley model are nonrandom. Morgenstern and van Hemmen (1984) generalized the model (1.11) so as to incorporate randomness also in the short-ranged part of the Hamiltonian, the latter was assumed to be Gaussian-distributed.

All the Hamiltonians (1.1)-(1.13) assume the spins to be "hard", i.e., the spins can have only fixed finite values. For example, an Ising spin is "hard" because it can take only two values +1 and -1. In other words, the spin weighting function for Ising spins consists of two delta functions:

$$\exp[-W_s(S_i)] = \delta(S_i + 1) + \delta(S_i - 1) \qquad \text{for all } i$$

Now, one can generalize this idea to introduce "soft" spins where each component of the spin is allowed to vary between $-\infty$ and $+\infty$ and the distribution $\exp[-W_s(S_i)]$ is usually assumed to be

$$\exp[-W_s(S)] = \exp[- r_i |S|^2 - u_i |S|^4] \quad (u>0)$$

in the so-called Landau-Ginzburg-Wilson model. The partition function is given by (see, for example, Fisher 1982, Sherrington 1981)

$$Z = \mathrm{Tr} \exp(- H_{eff})$$

where

$$H_{eff} = - \Sigma K_{ij} S_i S_j - \Sigma W_s(S_i) \quad (1.14)$$

with $K_{ij} = J_{ij}/k_B T$, k_B is the Boltzmann constant. One must remember that r_i and u_i are quenched variables in SG. One recovers the hard spin model (1.1) in the limit $r_i \to -\infty$, $u_i \to +\infty$, so that $(-2r_i/u_i) \to S_i^2$. The competing nature of the random interaction in SG is taken into account by assuming a distribution of K_{ij} in (1.14) with, say, zero mean (Hertz and Klemm 1979).

All the models of random magnets listed above, including (1.14) are defined on a discrete lattice. However, continuum version of (1.1), viz.,

$$H = (1/2) \int d^dx\, [r_0|\phi|^2 + (u/4)|\phi|^4 + |\{\nabla - iQ(x)\}\phi(x)|^2] \quad (1.15)$$

with a gauge-invariant derivative of the field ϕ, has also been studied (Hertz 1978). For a given Q, (1.15) describes a spin density wave with the wave vector Q. The effective Hamiltonian (1.15) for SG differs from the corresponding expressions for random ferromagnets (RF) and random antiferromagnets (RAF) by the fact that Q is x-dependent in SG in contrast to x-independent Q for RF and RAF. However, we shall see in chapter 20 that just arbitrary x-dependence of Q does not necessarily guarantee that (1.15) is indeed a good model for SG. The continuum analogue of the discrete Mattis model (1.9) will be identified in chapter 20.

In an attempt to classify the various SG systems into different universality classes, Proykova and Rivier (1981) studied the effect of varying the range of the interaction given by

$$J(r) = \mu^2 r^{-\sigma} \exp(-r/l) \qquad (1.16)$$

which reduces to the RKKY interaction (1.7) for d=3 and σ=3. Since almost all attempts in solving the models discussed above (except, of course, the SK model) exactly have failed so far, several simpler models have been proposed in the recent years. For example, Kotliar et al.(1983) proposed a one-dimensional Ising model

$$\mathcal{H} = \Sigma\epsilon_{ij} S_i S_j/[a(i-j)]^{\sigma}, \; 1/2 < \sigma < 3/2 \qquad (1.17)$$

where a is the lattice constant and ϵ_{ij} are independent, Gaussian-distributed, random variables. The model (1.17) can be easily generalized to d dimensions as (Kotliar et al.1983, Chang and Sak 1984)

$$\mathcal{H} = \Sigma\, \epsilon_{ij} S_i S_j/[a(i-j)]^{\sigma}, \quad (d/2) < \sigma < (d+2)/2 \qquad (1.18)$$

which contains rich physics in the various domains of the values of the parameters d and σ. Feigelman (1983), Feigelman and Ioffe (1983) and Ioffe and Feigelman (1983a,b) introduced a model where

$$J(x) = J_0 \cos(x/l_0) \exp(-!x!/l) \quad \text{in } d=1 \qquad (1.19)$$

$$J(r) = \Sigma\, [\text{const.} \sin(r/l_0) \exp(-r/l)] \quad \text{in } d=3 \qquad (1.20)$$

All the models introduced above consist of two-spin interactions. One can generalize these models to write a p-spin interaction of the form

$$\mathcal{H}_p = \Sigma A_{i_1,i_2,\ldots,i_p} S_{i_1} S_{i_2} \ldots S_{i_p} \qquad (1.21)$$

where $A_{i_1,i_2,\ldots,i_p}$ are the exchange interactions. In analogy with the SK model, one can introduce infinite-ranged p-spin interaction model, where the distribution of the exchange interaction is given by

$$P(A_{i_1,i_2,\ldots,i_p})=\{N^{p-1}/(\pi J^2 p!)\}^{1/2}\exp[-A^2_{i_1,\ldots,i_p} N^{p-1}/(J^2 p!)] \qquad (1.22)$$

The SK model corresponds to p=2 and the limit p=∞ is called the Random Energy Model (REM). The REM (Derrida 1980a,b, 1981) is defined in chapter 10.

The Hamiltonian (1.1) is the model in the absence of external magnetic fields. In the presence of external fields the Hamiltonian is given by

$$\mathcal{H} = -\Sigma J_{ij} \vec{S}_i\cdot\vec{S}_j - \Sigma \vec{H}_i\cdot\vec{S}_i$$

where $\vec{H}_i$ is a site-dependent external magnetic field. In most of the real laboratory experiments the external field is spatially uniform, i.e., $\vec{H}_i = \vec{H}$ for all i. Moreover, the Hamiltonian (1.1) has O(n) symmetry in spin space. However, in most real SG materials there is also an anisotropic contribution to the Hamiltonian. We shall define the different types of anisotropies in chapter 15 and survey the corresponding consequences.

For earlier reviews on general as well as specialized topics on the SG models, solutions of these models and some closely related topics see Aharony 1978, 1983, Anderson 1977, 1978a, 1979, Blandin 1978, Binder 1977, 1978, 1979, 1980a, b, Binder and Kinzel 1981, 1983, Chowdhury and Mookerjee 1984a, De Dominicis 1979, 1983, 1984, Fischer 1983, 1985a, Hertz 1985, Kirkpatrick 1979, Krey 1983, Lubensky 1979, Morgenstern 1983a, Palmer 1982, 1983, Parisi 1981, Rammal and Souletie 1982, Sherrington 1981, 1983, Sompolinsky and Zippelius 1983a, Toulouse 1981, 1982a, b, 1983, van Hemmen 1983, Walker and Walstedt 1983,

Walstedt 1982, Walstedt and Walker 1982, Wolff and Zittartz 1983 Young 1979, 1983a, 1985, Zittartz 1984).

1.2 Annealed versus Quenched Disorder; Replica Trick

In this section we shall compare and contrast the meaning and the implications of the annealed and quenched disorders in random magnetic systems. While dealing with random systems, one usually assumes the system to be "spatially ergodic" which is the spatial counterpart of Boltzmann's assumption of "temporal ergodicity". According to the hypothesis of spatial ergodicity, the physical properties (e.g., the free energy) of macroscopically large system is identical with the same property averaged over all possible spatial configurations. In other words, the bulk of a macroscopically large system can be divided into a large number of sub volumes each with a configuration corresponding to a member of the ensemble of systems characterized by the distribution of random variables I (for example, in random AuFe alloys, the configurations of the Fe atoms correspond to the "impurity" configuration denoted by {I}). We shall critically examine the validity of temporal as well as spatial ergodicity in SG in chapter 9.

So far as configuration averaging is concerned, one has to carry out averages over the spin configurations, denoted by {S}, as well as over the impurity configurations {I}. In annealed systems the impurity degrees of freedom I are in thermal equilibrium with the spin degrees of freedom S. Therefore, the partition function

$$Z = \mathrm{Tr}_{\{S\},\{I\}} \exp(-\beta H[\{S\},\{I\}]) \qquad (\beta = 1/k_B T)$$

for an annealed system can be written as

$$Z_{\mathrm{annealed}} = \mathrm{Tr}_{\{S\}} \exp(-\beta H_{\mathrm{eff}}[\{S\}])$$

where

$$\exp(-\beta H_{\mathrm{eff}}[\{S\}]) = \mathrm{Tr}_{\{I\}} \exp(-\beta H[\{S\},\{I\}])$$

or, equivalently,

$$Z_{annealed} = Tr_{\{I\}} \exp(-\beta\mathcal{H}'_{eff}[\{I\}]) \qquad (1.23)$$

where

$$\exp(-\beta\mathcal{H}'_{eff}[\{I\}]) = Tr_{\{S\}} \exp(-\beta\mathcal{H}[\{S\},\{I\}])$$

Therefore, for annealed systems, one can perform the configuration average of the partition function.

The impurity degrees of freedom {I} are frozen rigidly in quenched systems. In reality, if the impurities are frozen for time intervals over which the experiments are carried out, the disorder can be assumed to be quenched. Therefore, the impurity variables {I} cannot be treated on the same footing as the spin variables {S}. In this case one must first calculate the free energy corresponding to a given configuration {I} as

$$F[\{I\}] = -k_BT \ln Z[\{I\}]$$

where

$$Z[\{I\}] = Tr_{\{S\}} \exp(-\beta\mathcal{H}[\{I\},\{S\}]).$$

Then the free energy is averaged over all possible impurity configurations to yield

$$F_{quenched} = [F]_{av} = \int F[\{I\}]\, P[\{I\}]\, d[\{I\}] \qquad (1.24)$$

where P[{I}] is the probability distribution of the impurity configuration.

The original arguments of Brout (1959) leading to (1.24) goes as follows: suppose, we divide a large sample into a large number of smaller subunits (each of which is statistically large) in such a way that each of these subunits can be regarded as an independent member of an ensemble of systems characterized by the random distribution {I}. Since each of the subunits is statistically large, surface interaction will be assumed to be negligible. If the original system is annealed, each of the subunits will soon acquire all possible configurations in them because positions of the impurities are not frozen. But, if the original system is quenched, neglect of surface interaction leads to factorization of the total partition function into those corresponding to each of the subunits, i.e.,

$$\ln Z = \Sigma \ln Z_{su},$$

where Z_{su} is the partition function of a subunit. Division of both sides of the above relation by the number of subunits leads to (1.24).

In all the models introduced in section 1.1 the disorder is assumed to be quenched and hence require configuration average of the free energy, rather than that of the partition function. But, in practice, averaging Z over all possible configurations is much easier than averaging the free energy. However, using the identity

$$\ln Z = \lim_{n\to 0} (1/n)\ (Z^n - 1) \qquad (1.25)$$

the problem is simplified enormously, because one needs to average Z^n only. For integral n, $Z^n = \Pi\ Z_\mu$ can be interpreted as the partition function of a composite system consisting of n noninteracting replica of the system and hence the above technique is popularly called "replica trick" (Edwards and Anderson 1975, Edwards 1970, Grinstein and Luther 1976, Emery 1975). This trick will be used extensively in some of the later chapters in this book.

1.3 An Elementary Introduction to Frustration

Let us consider the ±J model. Now, we focus our attention on Fig.1.1. Such a unit of a lattice (here square) will be called a plaquette. In this figure + corresponds to $J > 0$ and − to $J < 0$. The individual bond energies are minimized if the two spins connected by any arbitrary bond <ij> are parallel to each other for $J > 0$ and antiparallel for $J < 0$. Notice that all the bond

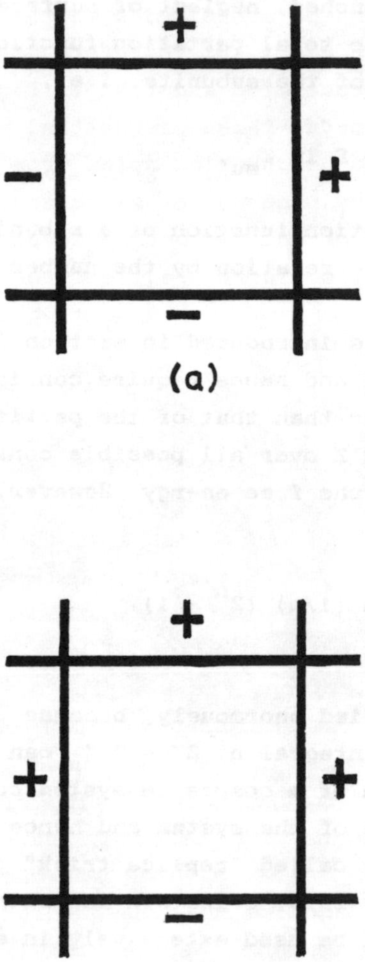

Fig. 1.1 (a) Unfrustrated plaquette, (b) frustrated plaquette on a square lattice. The frustration function $\phi = 1$ for the unfrustrated plaquette (a) and $\phi = -1$ for the frustrated plaquette (b).

energies in Fig.1.1a can be minimized simultaneously whereas the same is not possible in Fig.1.1b. Therefore, the plaquette in Fig.1.1a is called unfrustrated and that in Fig.1.1b is called frustrated. In other words, those plaquettes, where topological constraints prevent the neighboring spins from adopting a configuration with every bond energy minimized, are called frustrated. The number of antiferromagnetic bonds in Fig.1.1a is even, but odd in Fig.1.1b. Therefore, the frustration function (Toulouse 1977)

$$\Phi = \prod_{(C)} \text{sign}(J_{ij}) \qquad (1.26)$$

defined over a closed contour (C) of connected bonds, is a measure of the frustration within the contour. A plaquette is frustrated or unfrustrated depending on whether $\Phi = -1$ or $+1$.

It is easy to see that competition between the ferromagnetic and antiferromagnetic exchange interactions is often sufficient (except in situations like the Mattis model) but not necessary for frustration. For example, a triangular plaquette with only nearest-neighbor antiferromagnetic interaction is frustrated. It should be noted that although the quantitative measure of frustration in terms of the frustration function is a rather recent concept (Toulouse 1977) the impossibility of satisfying all the antiferromagnetic bonds on a triangular lattice and some of its consequences was realized as early as 1950 (see, for example, Wannier 1950).

The importance of frustration lies in the fact that it can distinguish "trivial disorder" from "nontrivial disorder". As an example, let us consider the Mattis model (1.9). By a simple change of variable $S_i \rightarrow S_i' = \xi_i S_i$ (i.e., by reversing an appropriate subset of spins) we recover a nonrandom Hamiltonian. The implications of this transformation are subtle and will be discussed in detail in chapter 20. However, we emphasize that since $\Phi = +1$ in the Mattis model, the latter is not a true model of SG.

So far we have discussed the elementary implications of

frustration based on the definition (1.26) of the frustration function. The above definition assumes nearest-neighbor exchange interaction on a Bravais lattice and randomness only in the sign of J. However, this definition is too restricted. For example, there are no closed 'plaquettes' on a Bethe lattice, but it can be frustrated (Bowman and Levin 1982, Thouless 1986). Moreover, presence of frustration is believed to be a necessary condition for a system to behave like a SG. Therefore, a definition of frustration, much more general than (1.26), is possible (Anderson 1978b). Suppose the energies of two blocks of spins, each of size L^d on a d-dimensional hypercubic lattice, have been minimized separately. Let the two blocks be then glued together without changing their internal states. Then the total surface energy is given by

$$\Delta E = \Sigma J_{ij} S_{i_1} S_{j_2}$$

where the summation is to be carried over the interacting pairs on the two sides of the surface. In a ferromagnet with only nearest-neighbour interactions, where $J_{ij} = J$ for each pair (ij),

$$[\Delta E]_{av} \sim L^{d-1}$$

$$\{[(\Delta E)^2]_{av}\}^{1/2} \sim L^{d-1}$$

so that the interface energy is an extensive quantity. On the other hand, in a short-ranged SG

$$[\Delta E]_{av} \simeq 0$$

and

$$\{[(\Delta E)^2]_{av}\}^{1/2} \sim L^{(d-1)/2}$$

The latter difference between the interface energies can be utilized to define frustration. Suppose, A be the number of bonds intersecting the interface. The spin system is called frustrated if

$$\lim_{A\to\infty} \{[(\Delta E)^2]_{av}\}^{1/2}/A = 0$$

The most important consequence of frustration is that it leads to high degeneracy of the ground state of the system. We shall discuss the more formal aspects of frustration in chapter 20.

CHAPTER 2

A BRIEF HISTORY OF THE EARLY THEORIES OF SG

"To understand science, it is necessary to know its history"
- Auguste Comte

It was Blandin (1961) who first exploited the power law behaviour of the exchange interaction to derive some universal properties of canonical SG alloys. The dominant exchange interaction in these alloys is the RKKY interaction (1.7a). Suppose, the cosine factor in (1.7a) can be dropped and its effect can be simulated by supplying +1 and -1 randomly. Note that dividing the Hamiltonian $\mathcal{H}$ and the temperature T in the partition function

$$Z = \mathrm{Tr}\,\exp(-\mathcal{H}/k_B T)$$

by the same constant leaves Z, and hence the thermodynamic properties, unaltered. Suppose, this constant is the concentration of the magnetic impurities (e.g., Fe in AuFe), c. There is a characteristic length scale R_c associated with the given concentration c, viz., $R_c^3 \sim c^{-1}$ in d = 3 so that a volume R_c^3 contains, on the average, a constant number cR_c^3 of the impurity atoms. Since $J(r) \sim r^{-3}$ in RKKY SG, dividing the Hamiltonian (1.7a) by c we get $\mathcal{H}/c$ which is a function of $(r/R_c)^3$ and hence independent of concentration. Moreover, if an external field H is applied, $\mathcal{H}/c$ becomes a function also of H/c. Therefore, the free energy F obeys the scaling relation

$$F \sim c^2 f(T/c, H/c) \qquad (2.1)$$

where f is an universal function of its arguments. Since all the thermodynamical properties of SG can be derived from the free energy taking the appropriate derivatives, these derived quantities should also follow universal scaling laws. The magnetization and the specific heat of dilute SG alloys were shown (Souletie and Tournier 1969) to follow scaling. Thus, one can calculate the thermodynamic properties for all concentration if the corresponding property for a single concentration is known. One can show (see, Rammal and Souletie 1982) that for a system where the interaction obeys a power law of the form $r^{-\lambda}$ the free energy has the scaling form

$$F \sim c^{1+\lambda/d} f(T/c^{\lambda/d}, H/c^{\lambda/d}) \qquad (2.2)$$

Since, $\lambda = d$ for RKKY SG, one can compute the thermodynamic properties of all canonical SG for all concentrations provided the corresponding properties for at least one concentration of one system is known. It follows from these scaling laws that the SG transition temperature $T_g \propto c$. The dependence of T_g on c in real SG alloys is weaker than linear (see appendix B). This limitation of the scaling theory in predicting the c-dependence of T_g arises from the fact that the latter theory is valid only for $c \to 0$.

The scaling functions (2.1) and (2.2) for the free energy were derived from the geometrical properties of the interaction without explicitly calculating the free energy. The scaling theory is not a microscopic theory of SG alloys; it correlates the properties of all the dilute canonical SG alloys reflecting their common universal features. But it needs the free energy of at least one system at a concentration as the input; the latter has to be calculated from a true microscopic theory beginning with the Hamiltonian (1.1).

The first attempt to calculate the specific heat of canonical SG alloys from a microscopic approach, supplemented by intuitive arguments, was made by Marshall (1960). The latter theory was put on a more formal basis by Klein and coworkers (Klein and Brout 1963, Klein 1964, 1969, Klein and Shen 1972, also Liu 1967). The basic idea was to evaluate the free energy from

$$[F]_{av} = \int P(\{h_i\})\, F(\{H_i\})\, d\{h_i\} \qquad (2.3)$$

where $P(\{h_i\})$ is the distribution of the local random magnetic field. The evaluation of the latter distribution, even within the mean-field approximation, is quite difficult problem, as we shall show more explicitly later in chapter 26.1. Although the $P(\{h_i\})$ for the Ising SG yields the low temperature specific heat in closer agreement with the experimental data, the agreement turned poorer for the vector SG (Held and Klein 1975)! (See Corbelli and Morandi (1979) for a comparison of the form of $P(\{h_i\})$ and the corresponding prediction of the scaling theory). Generalization of the latter theory did not improve the theory significantly (Kinzel and Fischer 1977a).

The linear T-dependence of the specific heat at very low temperatures is a common feature of SG as well as glasses. A phenomenological model was suggested by Anderson, Halperin and Varma (AHV) (1972) and by Phillips (1972) (see Phillips 1981 for a review) to explain the latter feature. AHV's idea consists mainly of two parts:

(i) in a disordered solid, e.g., a glass (or a SG), certain atoms or groups of atoms (or spins) can have two available positions (or orientations) which correspond to two minima of the free energy separated by a finite barrier (see Fig. 2.1),

(ii) due to disorder the splitting of the two minima, Δ, is a random function whose distribution ρ_{TLS} is assumed to be constant $\rho_{TLS}(0)$ (also see Harris 1983). For obvious reasons, this theory is called the theory of two-level systems (TLS). Suppose, the barriers are such that thermal equilibrium can be attained within the time scale of specific heat measurement. Therefore, the energy

$$E = \int \rho_{TLS}(\Delta E)\, f(\Delta E)\, d(\Delta E)$$

$$= \int \rho_{TLS}(\Delta E)\, [\,\Delta E/\{1 + \exp(\Delta E/k_B T)\}]\, d(\Delta E)$$

leads to the specific heat $C(T) \sim T\rho_{TLS}(0)$ in qualitative

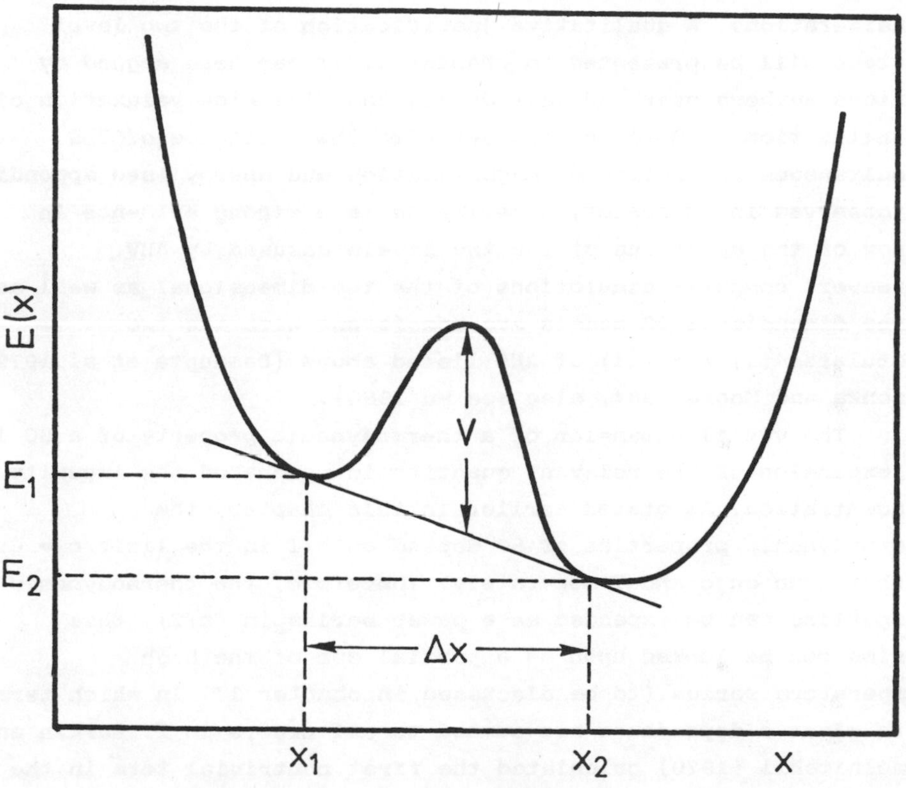

Fig. 2.1 A schematic representation of a two-level system. The energy E of the system is plotted as a function of a generalized coordinate x, measuring position along a line connecting two nearby local minima of E (after Anderson et al. 1972).

agreement with experiments. It should be emphasized that AHV's theory of TLS is a phenomenological theory; it assumes the existence of the two levels structure of the energy levels without any quantitative justification from microscopic considerations. A qualitative justification of the two level systems will be presented in chapter 5. It has been argued by various authors over the last decade that the slow relaxation of magnetization in SG is consistent with the existence of TLS. Simultaneous relaxation of magnetization and energy (see appendix B) observed in laboratory experiments is a strong evidence in favor of the splitting of the two levels assumed by AHV. Moreover, computer simulations of the two-dimensional as well as three dimensional SG models are consistent with the two postulates (i) and (ii) of AHV listed above (DasGupta et al.1979, Grzonka and Moore 1984, also see Ma 1980).

The virial expansion of a thermodynamic property of a SG is an expansion of the relevant quantity in powers of the impurity concentration. As stated earlier in this chapter, the thermodynamic properties of SG depend on c/T in the limit $c \to 0$ rather than on c and T separately. Therefore, the thermodynamic properties can be expanded as a power series in (c/T), this series can be looked upon as a partial sum of the high temperature series (to be discussed in chapter 17) in which terms with given c-dependence are summed to all orders in T. Larkin and Khmelnitskii (1970) calculated the first nontrivial term in the virial expansion for the specific heat and magnetic susceptibility. Owen (1983) computed the first two nontrivial terms for the specific heat, susceptibility and nonlinear susceptibility. However, this approach to the understanding of the SG transition does not seem to be efficient.

The experimental observation of a sharp cusp in χ_{ac} (Cannella and Mydosh 1972) and its interpretation as a phase transition in the light of the Edwards-Anderson mean-field theory (Edwards and Anderson 1975) mark the beginning of the "modern era" in SG physics. During the last decade experimentalists have not only discovered SG behaviour in diverse classes of systems,

they have also come across several new surprises while studying the SG over wider ranges of frequencies, external fields, etc. At the same time, because of the subtleties of the SG models, theoreticians took almost a decade to develope a satisfactory mean-field theory. However, the concepts developed for the study of the SG have helped in understanding the behaviour of several other systems which I call "SG-like". The stories of these exciting developments will be narrated in the remainder of this book. (See Toulouse 1983, Coles 1985 for the summaries of the main historical developments).

CHAPTER 3

SG "PHASE TRANSITION": ORDER PARAMETERS AND MEAN-FIELD THEORY

The spontaneous magnetization is a measure of the long-range orientational (spin) order in space, and is the order parameter that distinguishes ferromagnetic phase from the paramagnetic phase (for an excellent introduction to the concepts of order parameter and broken symmetry see Anderson 1984). There is no long-ranged order (LRO) in SG, as demonstrated by neutron scattering experiments (see appendix B). However, the cusp in $\chi_{a.c.}$ at T_g suggets the onset of some kind of orientational ordering. In order to describe this new kind of ordering, Edwards and Anderson (EA) (1975) introduced an order parameter, defined as

$$q_{EA} = [\langle S_i \rangle^2]_{av} \qquad (3.1)$$

(Our discussion in chapters 3-11 will be confined to the case of Ising spins, if not stated otherwise). The dynamics of the system, whose equilibrium ordering is characterized by the order parameter q_{EA}, is described by the dynamic order parameter (Edwards and Anderson 1976)

$$q_{EA}(t) = [\langle S_i(t) S_i(0) \rangle]_{av} \qquad (3.2)$$

which is a measure of the long-ranged auto-correlation in time. The symbols q and q_{EA} will be used interchangeably in this book. For ergodic systems, one would expect (3.2) to be identical with (3.1) in the long-time limit $t \to \infty$.

In order to get an intuitive feeling for the differences between the three types of orientational order, namely, paramagnetic (P), SG and ferromagnetic (F), let us perform a "gedanken experiment". Suppose the snapshots of the spin configurations in each of the three phases are taken at successive intervals of time. If snapshots of the three phases were available for only one particular instant of time one would not be able to distinguish between the P and the SG phases, because the <u>mutual</u> orientations of different spins in both these phases are random. But, if a sequence of the snapshots at the successive intervals of time are avilable, one can easily distinguish the SG from the P phase. The successive snapshots of the SG would be identical with each other (in reality, there will be thermal fluctuations even at $T \to 0$), i.e., each spin retains its orientation over very long periods of time. On the other hand, in the P phase not only the relative orientation between the different spins is random, but even the orientation of the same spin at successive instants of time is random.

Note that the EA order parameter gives no information about the short-ranged order (SRO) among the different spins in the system. SRO can be taken into account by a simple generalization (Medvedev 1979a) of the EA order parameter

$$[\langle S_i\rangle\langle S_j\rangle]_{av} = q\,\phi_{ij} \qquad (3.3)$$

where $\phi_{ij} \to 0$ as $|\vec{r}_i-\vec{r}_j| \to \infty$. The EA order parameter is recovered in the limit $\phi_{ij} = \delta_{ij}$.

Unfortunately, as we shall see later, q_{EA} is inadequate for the mean-field theoretic description of the SG phase. Even the simplest model, viz. the SK model needs an order parameter function $q(x)$ (Parisi 1979a,b, 1980a, 1981, Sompolinsky 1981a) where the parameter x is defined over the unit interval, i.e., $0 \le x \le 1$. The reason for the introduction of such an unusual order parameter and its physical interpretation will be explained in the chapters 4-9.

While simulating the SG models with computers, Binder and co-workers realized the importance of the study of the "similarity" (or overlap) between the various ground states of the system. Let the set of the phase factors $\{\phi_1^{(l)},\phi_2^{(l)},\ldots,\phi_N^{(l)}\}$ defined by the directions of the spins $\{S_1, S_2,\ldots, S_N\}$ in the l-th ground state, define an N-dimensional space with unit vectors

$$\vec{\phi}^{(l)} = (1/\sqrt{N})\{\phi_1^{(l)}, \ldots, \phi_N^{(l)}\}$$

The order parameter is defined by

$$\vec{\Psi}^{(l)} = \vec{\phi}^{(l)} \cdot \vec{X} \tag{3.4}$$

which is just the projection of the state

$$\vec{X} = (1/\sqrt{N})\ \{S_1, \ldots, S_N\}$$

onto the l-th ground state.

The definition (3.1) should be written as

$$q_{EA} = \lim_{H \to 0} \lim_{N \to \infty} [\langle S_i \rangle^2]_{av}$$

where the limits must be taken in the order shown in order to break the symmetry properly. In other words, the thermodynamic limit must be taken before taking the limit $H \to 0$; otherwise, the thermal average $\langle S_i \rangle$ will vanish. But, if the system is finite, e.g., systems studied by computer simulation, the thermodynamic limit cannot be taken. Under such circumstances, the smallest field required to break the time-reversal symmetry is $T/N^{1/2}$ in contrast to T/N required in the case of ferromagnets (Young and Kirkpatrick 1982). Then, one cannot use the EA order parameter as defined above; but it is more convenient to use the Morgenstern-Binder order parameter (Morgenstern and Binder 1979, 1980a)

$$q_{MB} = \lim_{r_{ij} \to \infty} [\langle S_i S_j \rangle^2]_{av}$$

where contributions from the short-range order vanishes because of the limit $r_{ij} \to \infty$. In the case of the infinite-ranged SG all the distinct pairs of spins are equivalent and the order parameter appropriate for studying finite systems is (Young and Kirkpatrick 1982, Kirkpatrick and Young 1981)

$$q^{(2)} = [\langle S_i S_j \rangle^2]_{av} \qquad (i \neq j) \qquad (3.5)$$

where the superscript 2 implies that $q^{(2)}$ is "something like q^2 but not exactly equal to this". For the numerical study of a system consisting of N spins interacting with each other via long-ranged RKKY interaction, Ariosa et al. (1982) used the order parameter

$$q_{ADM} = [1/\{N(N-1)\}] \; [\Sigma \langle S_i S_j \rangle^2]_{av}$$

which is different from $q^{(2)}$.

Suppose, $\langle i | \lambda \rangle$ is the i-th element of the eigenstate of J corresponding to the eigenvalue J_λ. Then, the staggered magnetization in the λ-th mode is given by

$$m_\lambda = \Sigma \langle i | \lambda \rangle \langle S_i \rangle \qquad (3.6)$$

Sompolinsky (1981b) (compare with Anderson 1970, 1979) suggested that m_λ should be used as the order parameter for SG. The relation between the latter approach and some of the other more familiar approaches will be clarified in chapter 23. (See Suzuki and Miyashita 1981 for comparing the various order parameters introduced before the Parisi order parameter q(x)).

In an attempt to develope a simple theory of SG that would capture the essential features of the real SG materials, EA introduced the model (1.1) where the distribution of the exchange interactions is given by (1.3). Unfortunately, even such a simple model cannot be solved exactly. The simplest approximation that yields a (approximate) solution is the mean-field approximation (MFA). The mean-field theory (MFT) of the paramagnetic-to-ferromagnetic phase transition is quite well understood (see Mattis 1985). The basic philosophy of the latter theory is to neglect spatial fluctuations, e.g.,

$$\langle (S_i - \langle S_i \rangle)(S_j - \langle S_j \rangle) \rangle \simeq 0$$

EA developed the MFT of the Ising SG using q_{EA} as the relevant order parameter and calculated the configuration-averaged free energy by applying the replica trick. In order to work within the replica formalism, they assumed that

$$M = \lim_{n\to 0} M_\alpha = \lim_{n\to 0} [\langle S_i^\alpha \rangle]_{av} \qquad (3.7)$$

and, more crucially, that

$$q = \lim_{n\to 0} q_{\alpha\beta} = \lim_{n\to 0} [\langle S_i^\alpha \rangle \langle S_i^\beta \rangle]_{av} \quad (\alpha \neq \beta) \qquad (3.8)$$

(where $\alpha,\beta=1,\ldots,n$ label the replicas)
are the only relevant order parameters, and neglected all other combinations including the higher order ones. EA showed, in the MFA, that for the Gaussian distributed exchange interactions, there exists a temperature T_g, such that $q_{EA} = 0$ for $T > T_g$. T_g is called the SG transition temperature (we shall study the detailed techniques of the calculation by the replica method in the context of the SK model in the next chapter). The zero-field susceptibility χ is given by

$$\chi = \chi_p \, (1 - q_{EA}) \qquad (3.9)$$

where

$$\chi_p = N\, S\, (S + 1)\, (g\mu_B)^2/3k_B T$$

is the susceptibility of free spins. $q_{EA} \to 0$ continuously as $T \to T_g$ but dq_{EA}/dT is discontinuous at T_g and hence the qualitative agreement with the experimentally observed cusp at T_g (see appendix B). Note that the expression (3.9) is consistent with the linear response theory (LRT) because the standard result

$$\chi = \Sigma\{[\langle S_i S_j \rangle]_{av} - [\langle S_i \rangle \langle S_j \rangle]_{av}\}\chi_p/NS(S+1) \qquad (3.10)$$

and the definition (3.2) of q_{EA} leads to (3.9) for classical spins and to the expression

$$\chi = \chi_p \,[1 - q_{EA}/\{S(S+1)\}] \qquad (3.11)$$

for quantum spins (Fischer 1975, Sherrington and Southern 1975). The relation (3.9) has been derived also by Sherrington (1975) and Southern (1975) without using the replica trick (for an excellent comparison of the various types of MFA for disordered magnets, including SG, see Southern 1976 and for χ in the presence of external magnetic field H see Fischer 1976). Although the EA MFT explains the qualitative behaviour of the susceptibility, it fails to account for the smooth variation of the specific heat with T. EA MFT predicts a cusp at T_g also in the specific heat, no such cusp has been observed so far in any experiment.

EA predict that $T_g \propto c^x$ where x=1/2 in the MFA. This agrees neither with the corresponding prediction of the scaling theory (discussed in chapter 2) nor with the experimental results (appendix B); x seems to be system-dependent in real SG materials. It has been suggested for a long time (Larsen 1977, Kinzel and Fischer 1977a) that the deviation of T_g from the linearity arises from the mean-free path effect, i.e., from the damping of the RKKY interaction (see equation 1.8), which is neglected in the scaling theory. Moreover, attempts have already been made (Henley 1985) to explain the experimentally observed result (appendix B)

$$T_g\,(c_1, c_2, \rho) = T_g(c_1, 0, \rho) + T_g(0, c_2, \rho) \qquad (3.12)$$

for alloys with two magnetic impurities, with "properly formulated cutoffs for the damped RKKY interaction". However, mean-free path effect alone is insufficient in expalining the c-dependence of T_g. As stated in appendix B, addition of nonmagnetic impurities can not only decrease T_g, but can, occassionally, also increase T_g. The latter observation cannot be justified as a finite mean-free path effect. For possible interpretation of the latter observation see Levy and Zhang

(1986).

As a historical note I would like to mention that the mean-field treatment of the random-bond models by the Japanese groups led to the discovery of the so-called glass-like phase (GLP) (Matsubara and Sakata 1976, Katsura 1976a) and the random ordered phase (ROP) (Ueno and Oguchi 1976a,b, Oguchi and Ueno 1977, Ono 1976a,b). Both GLP and ROP are essentially identical with the SG phase (see Takano 1980, Sakata et al.1977 for comparisons of the GLP and ROP with the SG).

Let us now develope a phenomenological Ginzburg-Landau (GL) theory for the SG transition which will be the starting point for further studies in chapters 13 and 16. The GL phenomenological free energy for Ising SG can be written as (Suzuki 1977)

$$F = F_0 + (aM^2 + bM^4 + \ldots) + (cq_{EA}^2 + dq_{EA}^3 + \ldots)$$
$$+ (eq_{EA}M^2 + \ldots) - MH \qquad (3.13)$$

where the coefficients a, b, c, d, e etc. are functions of temperature. Note that odd powers of M are not allowed by symmetry whereas the odd powers of q_{EA} are not forbidden. The term $eq_{EA}M^2$ in (3.13) denotes the coupling between M and q_{EA}. Minimizing F with respect to M and q_{EA}, we get

$$2(a + eq_{EA} + \ldots)M + 4bM^3 + \ldots = H \qquad (3.14)$$

and

$$2cq_{EA} + 3dq_{EA}^2 + \ldots + eM^2 + \ldots = 0 \qquad (3.15)$$

For H=0, (3.14) has two possible solutions, viz., M=0 and M≠0. Using these two solutions in (3.15), we ultimately get the following sets of solutions:

$$M = 0 \text{ and } q_{EA} = 0 \quad (P),$$
$$M = 0 \text{ and } q_{EA} = -2c/(3d) + \ldots \quad (SG),$$
$$M \neq 0 \text{ and } q_{EA} \neq 0 \quad (F).$$

Therefore, the SG transition temperature T_g is given by

$$c(T_g) = 0 \qquad (3.16)$$

Now, let us expand the magnetization M as

$$M = \chi_0 H + \chi_2 H^3 + \ldots \qquad (3.17)$$

where χ_0 is the linear susceptibility and χ_2 is a nonlinear susceptibility. From equation (3.14), we get

$$\chi_0 = 1/[2\{a + eq_{EA}(T)\}] \simeq [1 - (e/a)\, q_{EA}(T)]/\{2a(T)\}$$

which reduces to (3.9) for $2a = \chi_p^{-1}$ and $e = a$, and leads to discontinuity in $d\chi_0/dT$ at T_g. Thus, the GL theory is consistent with the EA MFT. Some other predictions of the GL phenomenological theory will be summarized in chapter 16.

We shall now investigate the behaviour of the κ-dependent susceptibility $\chi(\kappa)$. In principle, one can calculate it as follows: suppose, the externally applied magnetic field is $H = H(\kappa) \exp(i\, \vec{\kappa}\cdot\vec{r}_i)$. Then, the susceptibility $\chi(\kappa) = \partial M/\partial H(\kappa)$. More generally, suppose that the external magnetic field is
$H = \Sigma\, H(\kappa) \exp(i\, \vec{\kappa}\cdot\vec{r}_i)$
and the corresponding magnetization
$M = \Sigma\, M(\kappa) \exp(i\, \vec{\kappa}\cdot\vec{r}_i)$, and
$\chi(\kappa) = \partial M(\kappa)/\partial H(\kappa)$. Using EA mean-field equations for M one can calculate $\chi(\kappa)$. Such a calculation (Bhargava and Kumar 1977) lead to a κ-independent T_g, in contrast to the κ-dependent T_g observed experimentally.

An alternative approach to the $\chi(\kappa)$ is to start with the GL free energy (3.13). Expressing the free energy as a function of the Fourier transformations of M and H, and defining $\chi(\kappa)$ as $\chi(\kappa) = \partial M(\kappa)/\partial H(\kappa)$, Ghatak (1979) observed qualitative agreement with the experimental results.

We know that the Bethe-Peirls-Weiss (BPW) approximation is an improvement over the naive MFA, because the latter treats the interaction between the nearest-neighbors excatly, while treating the effect of the farther neighbors in the MFA (see Ghatak and

Moorjani 1976, Klein et al.1979 for the earlier results for SG in the BPW approximation). Levin and coworkers (Soukoulis and Levin 1977, 1978, Soukoulis 1978, Soukoulis et al. 1978, Levin et al.1979) developed a cluster MFT (CMFT) along this line. The Hamiltonian is divided into two parts as

$$\mathcal{H} = -\Sigma J_{\nu\lambda}\vec{S}_{\nu}\cdot\vec{S}_{\lambda} + \Sigma\Sigma J^{0}_{ij}\vec{S}_{i\nu}\cdot\vec{S}_{j\nu} - g\mu_{B}\Sigma\vec{H}\cdot\vec{S}_{i\nu} \tag{3.18}$$

where $\vec{S}_{i\nu}$ represents the spin at the i-th site in the ν-th cluster, J^{0}_{ij} and $J_{\nu\lambda}$ are the intracluster and intercluster exchange interactions, respectively. The intercluster interaction is assumed to be independent of the spin indices i and j; the latter assumption is justified in the dilute limit because the clusters are then far apart. In order to make the concept of clusters meaningful, the intracluster interaction is assumed to be stronger than the intercluster interaction. The intercluster exchange is assumed to be Gaussian-distributed and treated in the MFA whereas the intracluster exchange is treated exactly. The contribution to the specific heat from the intracluster interaction has a rounded maximum at a temperature T_0 whereas the corresponding intercluster contribution exhibits a cusp at T_g ($T_g < T_0$). For sufficiently large clusters, the intracluster contribution dominates and hence the qualitative agreement with the experimental results is better (see Fig. 3.1) as compared to that with the EA MFT.

The total neutron scattering intensity is given by

$$I(\kappa,T) = k_B T\chi(\kappa,T) + I_B(\kappa,T)$$

where I_B is the Bragg term and the first term corresponds to critical scattering. Although $I(\kappa,T)$ exhibits a maximum at a κ-dependent temperature, the temperature corresponding to the maximum of $\chi(\kappa,T)$ is argued to be independent of T in the CMFT. In order to check this prediction the experimental data was reanalyzed. Since I_B is proportional to an order parameter it was assumed to be a monotonically decreasing function of temperature and was assumed to vanish at a sharply defined temperature independent of κ. Subtracting such a term from the experimentally measured total intensity Soukoulis et al.(1978) argued that the temperature corresponding to the maximum of $\chi(\kappa)$ is indeed independent of κ. Some objections against the latter analysis will be examined in chapter 21. The CMFT of Soukoulis, Levin and Grest assumes identical size of all the clusters in the system. This theory can be easily extended to incorporate a distribution of the cluster size (Bien and Usadel 1986).

The quantum effect on the SG ordering would be strongest in systems with $S = 1/2$. The MFT of the latter system has been developed (Fischer 1975, Klemm 1979, Bray and Moore 1980b, Sommers 1981) within the replica formalism as well as along the TAP-like approach (TAP theory will be explained in chapter 5). The partition function, in the replica approach, is given by

$$[\ln Z]_{av} = \lim (1/n) \{[Z^n]_{av} - 1\}$$

with

$$Z^n = \mathrm{Tr}\ T \exp\{\int d\tau\ \Sigma \beta J_{ij}\ \Sigma\ S_i^\alpha(\tau) S_j^\alpha(\tau)\}$$

where τ is the "Matsubara time" (see Ramond 1981) and T represents the time ordering operator. The transition temperature T_g is depressed by the quantum fluctuations. Nevertheless, $T_g > 0$ for all S in the MFA (Bray and Moore 1980b, Sommers 1981). Recently, a replica-symmetric MFT of the quantum Heisenberg SG has been developed (Theumann and Gusmao 1984, Theumann 1986) by representing the spin operators as bilinear combinations of two fermion fields:

$$S_i^z = (1/2)(a_{i+}^* a_{i+} - a_{i-}^* a_{i-})$$

$$S_i^+ = a_{i+}^* a_{i-} = (a_i^-)^*$$

where the operators a_{is}^* and a_{js} (s = + or −) satisfy the usual fermion anticommutation relations. (See Xu et al.1985 for an alternative approach).

It might seem surprising to a beginner that a model as simple as (1.1) with (1.3) can capture the main features of the real SG materials. But, indeed this model does reproduce several qualitative features of the SG materials. Monte Carlo simulation of the latter model (see chapter 24 for an introduction) yield the susceptibility (Fig.3.2), specific heat (Fig.3.3), IRM and TRM (Fig.3.4), etc. in qualitative agreement with the corresponding experimental data (Binder and Schroeder 1976, see Binder 1977, 1980a,b, Binder and Kinzel 1981, 1983 for reviews).

The MFT discussed in this chapter treat all the replicas on the same footing and therefore, these are called replica-symmetric MFT. The shortcomings of the replica-symmetric MFT at low temperatures and the possible remedies will be explained in detail in the chapters 4-10, 13 and 15.

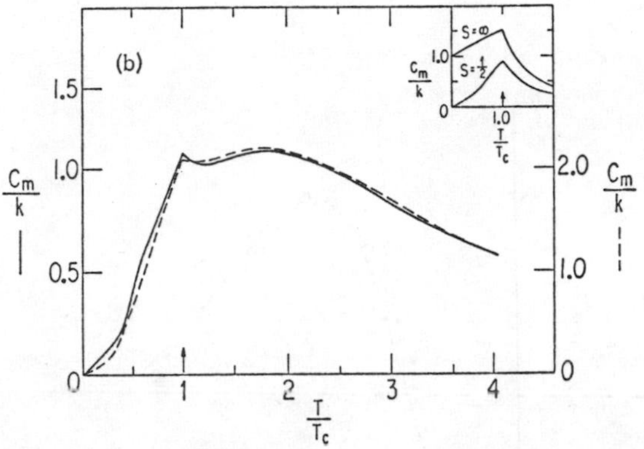

Fig. 3.1. Temperature dependence of the specific heat per cluster in the cluster mean-field theory of Soukoulis and Levin assuming antiferromagnetic intracluster exchange interactions for three-spin (solid curve) and six-spin (dashed curve) clusters (after Soukoulis and Levin 1977).

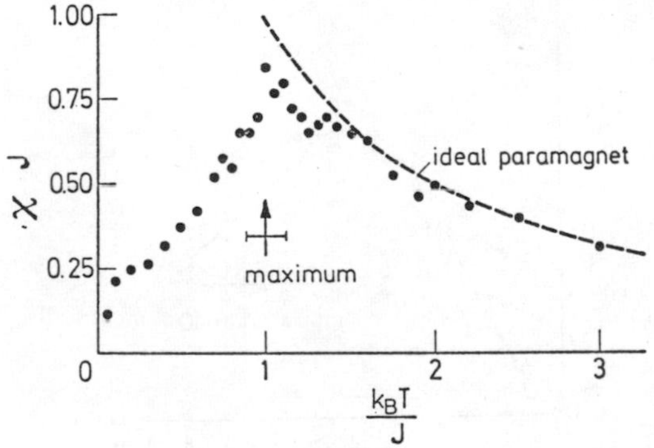

Fig. 3.2. Susceptibility in the two-dimensional Ising model with Gaussian-distributed nearest-neighbour exchange interactions. The system sizes are 34 X 35 (after Binder 1977a).

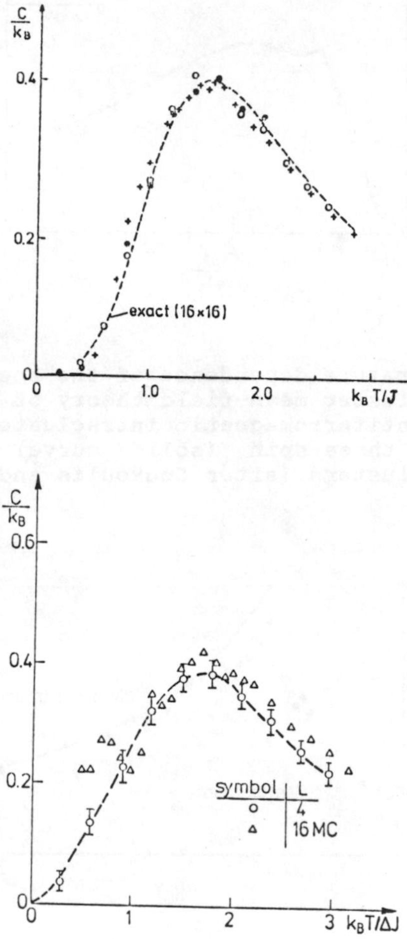

Fig. 3.3. (a) Specific heat for the ±J model in d=2 for two different system sizes. (b) Specific heat for the Gaussian model in d=3 (after Binder and Kinzel 1981).

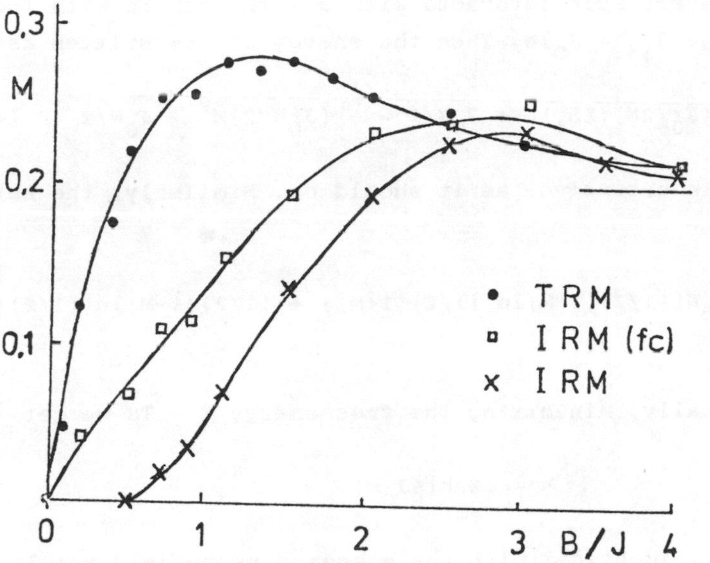

Fig. 3.4. Isothermal remanent magnetization (IRM), thermo remanent magnetization (TRM) for the two-dimensional Gaussian model. The IRM (fc) is IRM after fast cooling of the sample (after Kinzel 1979).

CHAPTER 4

SHERRINGTON-KIRKPATRICK (SK) MODEL AND THE SK SOLUTION

We know that the MFT for nn ferromagnets becomes exact in the infinite-range limit. As an example, consider an N-spin system where every spin interacts with all the others with the same exchange $J_{ij} = J_0/N$. Then the energy can be written as

$$E = -(J_0/2N)(\Sigma S_i)^2 + J_0N/2 = -(J_0N/2)M^2 + J_0N/2 \qquad (4.1)$$

which is extensive, as it should be. Similarly, the entropy is given by

$$S = -k_BN[(1/2)(1+M)\ln\{(1/2)(1+M)\} + (1/2)(1-M)\ln\{(1/2)(1-M)\}] \qquad (4.2)$$

and finally, minimizing the free energy E - TS we get

$$M = \tanh(\beta J_0 M)$$

which is identical with the standard mean-field result.

What happens when the latter treatment is extended to EA model of SG? In order to examine whether the EA MFT becomes exact in the infinite range limit, Sherrington and Kirkpatrick (SK) (1975) proposed a model, now called the SK model (defined in chapter 1), and solved it by applying the replica trick (1.25).

The free energy, averaged over all possible configurations, is given by

$$[F]_{av} = -k_BT \lim_{N\to\infty} \lim_{n\to 0} (1/n)\{[Z^n]_{av} - 1\} \qquad (4.3)$$

where

$$[Z^n(\{J_{ij}\})]_{av} = \int \Pi\, dJ_{ij}\, P(\{J_{ij}\})\, Z^n(\{J_{ij}\}) \qquad (4.4)$$

The partition function

$$Z^n(\{J_{ij}\}) = \mathrm{Tr}_{\{S^\alpha\}} \exp[-(1/k_BT)\, \Sigma\, \mathcal{H}^\alpha] \tag{4.5}$$

corresponds to the replica Hamiltonian

$$\mathcal{H}^\alpha = -\, \Sigma\, J_{ij}\, S_i^\alpha\, S_j^\alpha - H\, \Sigma S_i^\alpha \tag{4.6}$$

where $\alpha(\alpha = 1,\ldots,n)$ labels the replicas and i,j $(i,j = 1,\ldots,N)$ denote the spins. Now, substituting (1.3), (4.5) and (4.6) into (4.4) and integrating over J_{ij} we get

$$[Z^n\{J_{ij}\}]_{av} = \exp[(J^2N^2n)/\{4(k_BT)^2\}$$

$$\mathrm{Tr}_{\{S^\alpha\}} \exp[\Sigma J^2/\{2(k_BT)^2\}(\Sigma S_i^\alpha S_i^\beta)^2$$

$$+ \{J_0/(2k_BT)\}\Sigma(\Sigma S_i^\alpha)^2] + (H/T)\Sigma(\Sigma S_i^\alpha) \tag{4.7}$$

Using the identity

$$\exp(\lambda a_\mu^2) = \int (1/\sqrt{\pi})\, dt_\mu\, \exp\{-(1/2)t_\mu^2 + \sqrt{2\lambda}\, a_\mu t_\mu\}; \quad \mu = (\alpha\beta) \text{ or } \alpha$$

we get

$$[F]_{av} = -T \lim_{N\to\infty} \lim_{n\to 0} (1/n)[\exp\{(\tilde{J}^2Nn)/(4k_B^2T^2)\}$$

$$\int\{\Pi(N/2\pi)^{1/2}\, dx_\alpha\}\{\Pi(N/2\pi)^{1/2}\, dy_{\alpha\beta}\}$$

$$\{\exp(NG) - 1\}] \tag{4.8}$$

where

$$G = (1/2)\Sigma y_{\alpha\beta}^2 - (1/2)\Sigma x_\alpha^2 +$$

$$\ln \mathrm{Tr}_{\{S^\alpha\}} \exp[\{\tilde{J}/(k_BT)\}\Sigma y_{\alpha\beta} S^\alpha S^\beta +$$

$$\sqrt{\{\tilde{J}_0/(k_BT)\}}\Sigma x_\alpha S^\alpha + \{H/(k_BT)\}\Sigma S^\alpha] \tag{4.9}$$

SK assumed that the thermodynamic limit ($N \rightarrow \infty$) and the replica limit ($n \rightarrow 0$) can be interchanged. Then, carrying out the integrals in (4.8) by the method of steepest descent, we get

$$(F/k_BTN) = -\{\tilde{J}^2(1-q)^2/4k_B^2T^2\} + \{(\tilde{J}_0M^2)/(2k_BT)\} -$$

$$(1/\sqrt{2\pi})\int dz \exp(-z^2/2)$$

$$\ln 2\cosh[\{\tilde{J}_0M)/(k_BT)\} + \{\tilde{J}z\Sigma q^{1/2}/(k_BT)\} + \{H/(k_BT)\}] \qquad (4.10)$$

where

$$M = \lim_{n\to 0} M_\alpha = (2\pi)^{-1/2}\int dz \exp(-z^2/2)$$
$$\tanh[\{\tilde{J}_0M/(k_BT)\}+\{\tilde{J}q^{1/2}z/(k_BT)\}+\{H/(k_BT)\}] \qquad (4.11)$$

and

$$q = \lim_{n\to 0} q_{\alpha\beta} = (2\pi)^{-1/2}\int dz \exp(-z^2/2)$$
$$\tanh^2[\{\tilde{J}_0M/(k_BT)\}+\{\tilde{J}q^{1/2}z/(k_BT)\}+\{H/(k_BT)\}] \qquad (4.12)$$

Equations (4.10)-(4.12) constitute the SK solution (also see Klein 1976). Three phases appear on the phase diagram of the SK model (Fig.4.1):

(a) the paramagnetic (P) phase, characterized by $M = 0$ and $q = 0$,
(b) the SG phase ($M = 0$ and $q \neq 0$); and
(c) the ferromagnetic (F) phase ($M \neq 0$ and $q \neq 0$).

Equations for M and q, similar to (4.11) and (4.12) respectively, have been derived also for the site-disordered SG model with the RKKY interaction (1.7a) using Klein-Brout-like arguments (Mookerjee 1978). The parameters $\tilde{J}_0/k_B$ and $\tilde{J}/k_B$ in equations (4.11) and (4.12) are replaced by T_c and T_g, respectively, for the SG system with the given concentration c of the magnetic constituent. For a sample that exhibits the SG transition at $T_g > 0$, T_c is hypothetical and is determined by extrapolating high temperature susceptibility. Self-consistent

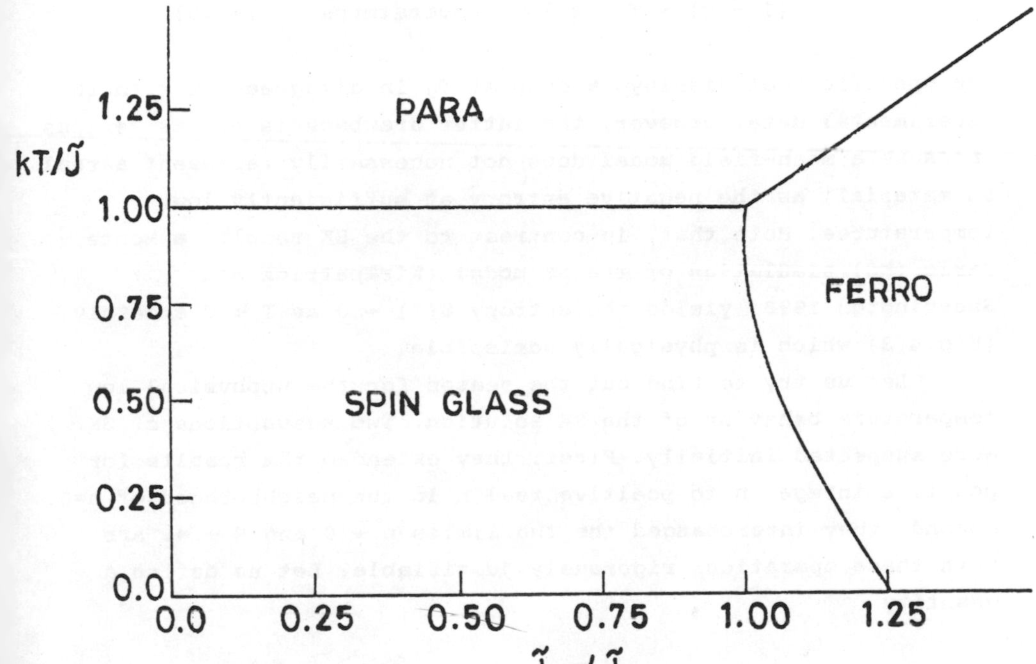

Fig. 4.1. The phase diagram for the SK model predicted by the replica-symmetric SK solution (after Sherrington and Kirpatrick 1975).

solutions of these SK-like equations have been computed (Mookerjee and Chowdhury 1983a) numerically as shown in Fig.4.2 (also see Nieuwenhuys et al.1978).

From the SK solution we get

$$(1 - q) \sim T \quad \text{at low temperatures} \qquad (4.13)$$

The specific heat displays a cusp at T_g in disagreement with the experimental data. However, the latter drawback is not as serious (because a mean-field model does not necessarily represent a real SG material) as the negative entropy at sufficiently low temperatures. Note that, in contrast to the SK result, a Monte Carlo (MC) simulation of the SK model (Kirkpatrick and Sherrington 1978) yields the entropy $S(T) \to 0$ as $T \to 0$ smoothly (Fig.4.3) which is physically admissible.

Let us try to find out the reason for the unphysical low temperature behavior of the SK solution. Two assumptions of SK were suspected initially. First, they extended the results for positive integer n to positive real n in the neighborhood of n=0; second, they interchanged the two limits $n \to 0$ and $N \to \infty$. Are both these operations rigorously justifiable? Let us define a quantity

$$\psi(n) = (1/n) \ln [Z^n]_{av}. \qquad (4.14)$$

Van Hemmen and Palmer (1979) argued that the interchange of the two limits is allowed provided $\psi(n)$ is analytic and bounded for n in a neighborhood of n = 0. Moreover, for the SK model, $\psi(n)$, considered as a function of real n, is convex and hence continuous and differentiable at all n including n=0. Since the simple minded continuation of n from integer to real n led SK to the wrong branch of the free energy, there must be at least another branch which would lead naturally to the stable solution. We shall analyze these two branches in chapter 7. One must also realize that the replica trick is not really any "trick" in the sense that it is a sensible method that yields, if used properly,

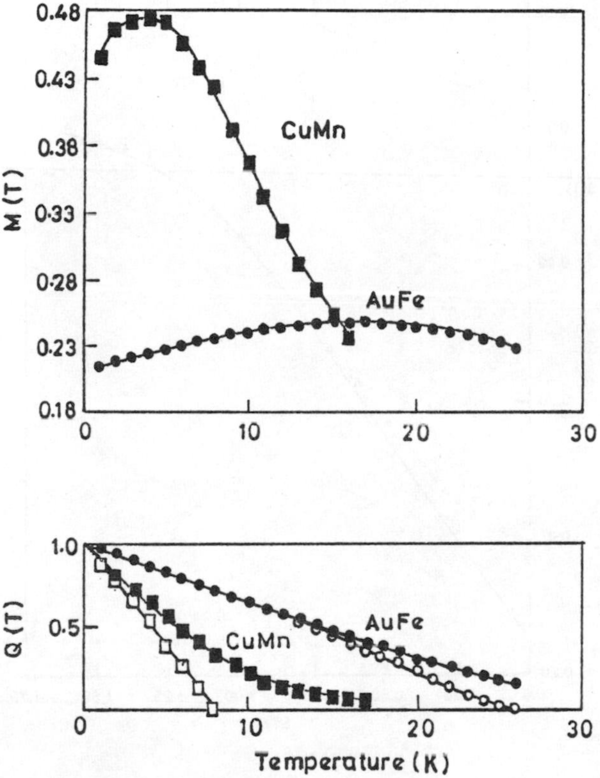

Fig. 4.2. The temperature dependence of M and q for AuFe (6.6 at.%) and CuMn (0.7 at.%) alloys for two different values of the external magnetic field H. The open and the closed symbols correspond to the H = 0 and H = 16kG, respectively (after Mookerjee and Chowdhury 1983a).

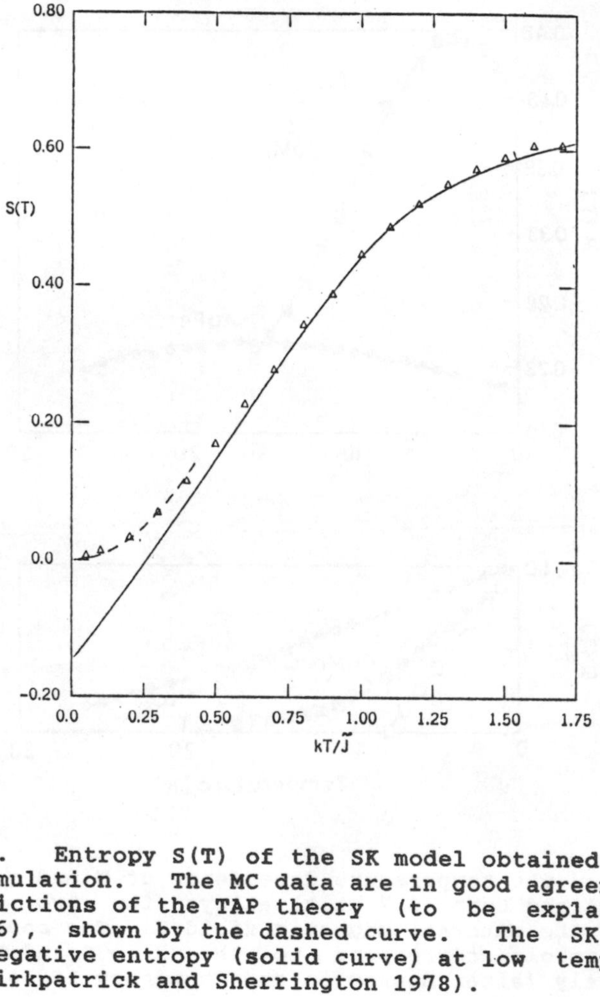

Fig. 4.3. Entropy S(T) of the SK model obtained by Monte Carlo simulation. The MC data are in good agreement with the predictions of the TAP theory (to be explained in chapter 6), shown by the dashed curve. The SK solution yields negative entropy (solid curve) at low temperatures (after Kirkpatrick and Sherrington 1978).

physically admissible solution of the problem. For example, in the case of the spherical model (chapter 12) this method gives the same results that could be obtained by an alternative approach without using the replica method. The unphysical low temperature behavior of the SK solution is a consequence of its instability (de Almeida and Thouless 1978), as we shall see in the next chapter. This instability arises from the breakdown of the permutation symmetry between the replicas at low temperatures.

CHAPTER 5

INSTABILITY OF THE SK SOLUTION

The unphysical low temperature behaviour of the SK solution is a consequence of its instability, as was shown subsequently (de Almeida and Thouless 1978). In order to test the stability of the SK solution, let us introduce the variables

$$x^{\alpha} = x + \epsilon^{\alpha}, \quad \text{and} \quad y^{(\alpha\beta)} = y + \eta^{(\alpha\beta)},$$

where

$x^{\alpha}=(J_0/k_BT)^{1/2}\langle S^{\alpha}\rangle=(J_0/k_BT)^{1/2}M$ and $y^{(\alpha\beta)}=(J/k_BT)\langle S^{\alpha}S^{\beta}\rangle=Jq/k_BT$, x and y are the equilibrium values given by the SK solution. Substituting for x^{α} and $y^{(\alpha\beta)}$ in the SK free energy expression and keeping terms up to the second order in ϵ^{α} and $\eta^{(\alpha\beta)}$, the free energy difference was given by
$-(1/2)\Delta$ where

$$\Delta = \Sigma[\delta_{\alpha\beta} - (J_0/k_BT)(\langle S^{\alpha}S^{\beta}\rangle-\langle S^{\alpha}\rangle\langle S^{\beta}\rangle)]\epsilon^{\alpha}\epsilon^{\beta}$$
$$+ \{2JJ_0^{1/2}/(k_BT)^{3/2}\}\ \Sigma\ [(S^{\delta}\rangle\langle S^{\alpha}S^{\beta}\rangle-\langle S^{\alpha}S^{\beta}S^{\delta}\rangle)]\epsilon^{\delta}\eta^{(\alpha\beta)}$$
$$+ \Sigma\ [\delta_{(\alpha\beta)(\gamma\delta)} - (J/k_BT)^2(\langle S^{\alpha}S^{\beta}S^{\gamma}S^{\delta}\rangle-\langle S^{\alpha}S^{\beta}\rangle\langle S^{\gamma}S^{\delta}\rangle]\eta^{(\alpha\beta)}\eta^{(\gamma\delta)} \qquad (5.1)$$

This quadratic form must be positive definite for a stable solution of the problem. AT diagonalized the matrix G associated with this quadratic form and concluded that the above condition is violated by terms of the order q^2 at temperatures T very close to T_g.

Let us find the eigenvalues of the matrix G utilizing the symmetry properties. The eigenvectors of G are of the form

$$\mu = \begin{pmatrix} \{\epsilon^{(\alpha)}\} \\ \\ \{\eta^{(\alpha\beta)}\} \end{pmatrix} \quad \alpha,\beta = 1,2,..,n \qquad (5.2)$$

where $\{\epsilon^{(\alpha)}\}$ and $\{\eta^{(\alpha\beta)}\}$ are column vectors with n and n(n-1)/2 elements respectively. We are interested in the solutions of the eigenvalue equation

$$G\ \mu = \lambda\ \mu \tag{5.3}$$

Let us begin with the vector μ_1 with elements

$$\epsilon^{(\alpha)} = a \text{ for all } \alpha \text{ and } \eta^{(\alpha\beta)} = b \text{ for all } (\alpha\beta) \tag{5.4}$$

The latter have the same structure as $q^{(\alpha\beta)}$. These eigenvectors are called "longitudinal" eigenvectors in the field theoretic language (DeDominicis 1984). Substituting (5.4) into the eigenvalue equation (5.3) leads to the eigenvalues

$$\lambda=(1/2)\{(A-B+P-4Q+3R)\pm[A-B-P+4Q-3R)^2-8(C-D)^2]^{1/2}\} \text{ for } n=0 \tag{5.5}$$

where

$$A = G_{\alpha\alpha}=1 - (J_0/k_BT)(1-\langle S^{\alpha}\rangle^2) \tag{5.6}$$

$$B = G_{\alpha\beta}=-(J_0/k_BT)(\langle S^{\alpha}S^{\beta}\rangle - \langle S^{\alpha}\rangle^2) \tag{5.7}$$

$$C = G_{\alpha(\alpha\beta)}=G_{(\alpha\beta)\alpha}=JJ_0^{1/2}(k_BT)^{-3/2}(\langle S^{\alpha}\rangle\langle S^{\alpha}S^{\beta}\rangle - \langle S^{\beta}\rangle) \tag{5.8}$$

$$D = G_{\gamma(\alpha\beta)}=G_{(\alpha\beta)\gamma}=J_0^{1/2}(k_BT)^{-3/2}(\langle S^{\gamma}\rangle\langle S^{\alpha}S^{\beta}\rangle-\langle S^{\alpha}S^{\beta}S^{\gamma}\rangle) \tag{5.9}$$

$$P = G_{(\alpha\beta)(\alpha\beta)}=1 - (J/k_BT)^2(1 - \langle S^{\alpha}S^{\beta}\rangle^2) \tag{5.10}$$

$$Q = G_{(\alpha\beta)(\alpha\gamma)}=1 - (J/k_BT)^2(\langle S^{\beta}S^{\gamma}\rangle - \langle S^{\alpha}S^{\beta}\rangle^2) \tag{5.11}$$

$$R = G_{(\alpha\beta)(\gamma\delta)}=-(J/k_BT)^2(\langle S^{\alpha}S^{\beta}S^{\gamma}S^{\delta}\rangle - \langle S^{\alpha}S^{\beta}\rangle^2) \tag{5.12}$$

Similarly, for the vectors μ_2 of the form

$$\epsilon^{(\alpha)} = a \text{ for } \alpha = \theta, \text{ and } \epsilon^{(\alpha)} = b \text{ for } \alpha \neq \theta$$

and (5.13)

$$\eta^{(\alpha\beta)} = c \text{ for } \alpha \text{ or } \beta = \theta \text{ and } \eta^{(\alpha\beta)} = d \text{ for } \alpha,\beta \neq \theta$$

(these eigenvectors are called "anomalous" eigenvectors (DeDominicis 1984)) the eigenvalues reduce to the same form (5.5) for n=0. Finally, for the vectors of the form

$$\epsilon^{(\alpha)} = a \text{ for } \alpha = \theta \text{ or } \nu, \quad \epsilon^{(\alpha)} = b \text{ for } \alpha \neq \theta, \nu$$

and (5.14)

$$\eta^{(\theta\nu)} = c, \; \eta^{(\theta\alpha)} = \eta^{(\nu\alpha)} = d \text{ for } \alpha\neq\theta, \nu, \; \eta^{(\alpha\beta)} = e \text{ for } \alpha,\beta \neq\theta,\nu$$

are called the "replicon" eigenstates (DeDominicis 1984) and the corresponding eigenvalue is given by

$$\lambda = P - 2Q + R \qquad \text{for } n = 0 \qquad (5.15)$$

Using the definitions (5.10)-(5.12) for P, Q and R one can show that the condition that the eigenvalue (5.15) be positive reduces to the condition

$$(k_B T/J)^2 > \{1/(2\pi)^{1/2}\} \int dz \exp(-z^2/2) \operatorname{sech}^4[\{Jq^{1/2} z/(k_B T)\} + J_0 M/(k_B T)] \qquad (5.16)$$

The latter condition is satisfied at all $T > J/k_B$, i.e., in the paramagnetic phase but violated for all T in the SG phase. By expanding near T_g, one can also show that the condition (5.16) is violated by terms of the order of q^2 near T_g. In the field-theoretic language, the negative mass of the replicon mode expresses the instability of the SK's replica-symmetric solution. Pytte and Rudnick (1979) and Bray and Moore (1978, 1979b) arrived at the same conclusion starting from the Landau-Ginzburg-Wilson effective (replica) Hamiltonian. They also identified that the term in the effective Hamiltonian responsible for the latter instability is proportional to $\Sigma(q^{\alpha\beta})^4$; this term is often referred to as the "multiple-loop" term as opposed to the single loop terms of the form $q^{\alpha_1\alpha_2} q^{\alpha_2\alpha_3} q^{\alpha_3\alpha_4} \ldots q^{\alpha_i\alpha_1}$ (the reason for the latter terminology is obvious in the diagrammatic formulation of the field theory). This observation also explains why the replica-symmetric theory for the spherical model does not suffer from the diseases of the SK solution of the SK model; the quartic term is absent in the spherical model. AT suggested that the permutation symmetry between the replicas has to be broken at low temperatures in order to get a stable solution of the SK model. At least three different prescriptions for the replica symmetry breaking have been propsed so far (Bray and Moore 1978, 1979a, 1980c, Blandin et al.1980, Parisi 1979a,b, 1980a), among which only Parisi's scheme yields a stable (marginally stable!) solution (see chapter 7).

Following Brout's (1965) diagrammatic approach, Khurana and Hertz (1980) calculated the EA order parameter susceptibility χ_{EA}, defined as

$$\chi_{EA} = (\partial q/\partial \tilde{h}^2) \qquad (5.17)$$

where $\tilde{h}^2$ is the mean-square value of the local (random) field, without using the replica trick. χ_{EA} becomes negative for $T \le T_g$ therby signalling the instability of the mean-field solution. This approach also indicates the insufficiency of the single EA order parameter to describe the SG phase. We shall see in chapter 7 that both the suggestions, namely, the insufficiency of the single order parameter (Khurana and Hertz 1980) and breaking of the replica symmetry (Almeida and Thouless 1978) are consistent with each other; breaking the permutation symmetry between the replicas leads to (effectively) infinite number of order parameters!

Let us carry out the stability analysis in the presence of the external field H. For convenience let us assume $J_0 = 0$. The stability condition (5.16) is now modified to

$$(k_BT/J)^2 > (2\pi)^{-1/2}\int dz \, \exp(-z^2/2)\operatorname{sech}^4[\{(Jq^{1/2}z)/(k_BT)\}+\{H/(k_BT)\}] \qquad (5.18)$$

where

$$q = (2\pi)^{-1/2} \int dz \exp(-z^2/2) \tanh^2[\{Jq^{1/2}z/(k_BT)\}+\{H/(k_BT)\}]$$

The stability condition is simplified to

$$H^2 > (4J^2/3)\{1 - (T/T_g)\}^3 \quad \text{for small } H \qquad (5.19)$$

and

$$k_BT > (4/3)(2\pi)^{-1/2} J \exp(-H^2/2J^2) \quad \text{for large } H \qquad (5.20)$$

(see Fig.5.1). The expression (5.19) leads to the so-called AT line

$$\tilde{t} \sim H^{2/3} \text{ for small } H \qquad (5.21)$$

where $\tilde{t} = (1 - T/T_g)$. The physical significance of the AT line will be explored in chapter 7. For the experimental determination of the "AT-line" for SG materials see appendix B.

Bray (1982) examined the effect of including a site-dependent magnetic field proportional to the eigenvector with the largest eigenvalue of the random matrix J in the SK Hamiltonian. The motivation was to investigate if such an apparently natural "ordering field" would suppress the AT instability by projecting out one valley in the generalized phase space. However, it turned out that such an ordering field can strongly suppress, but cannot remove, the AT instability in the SK model.

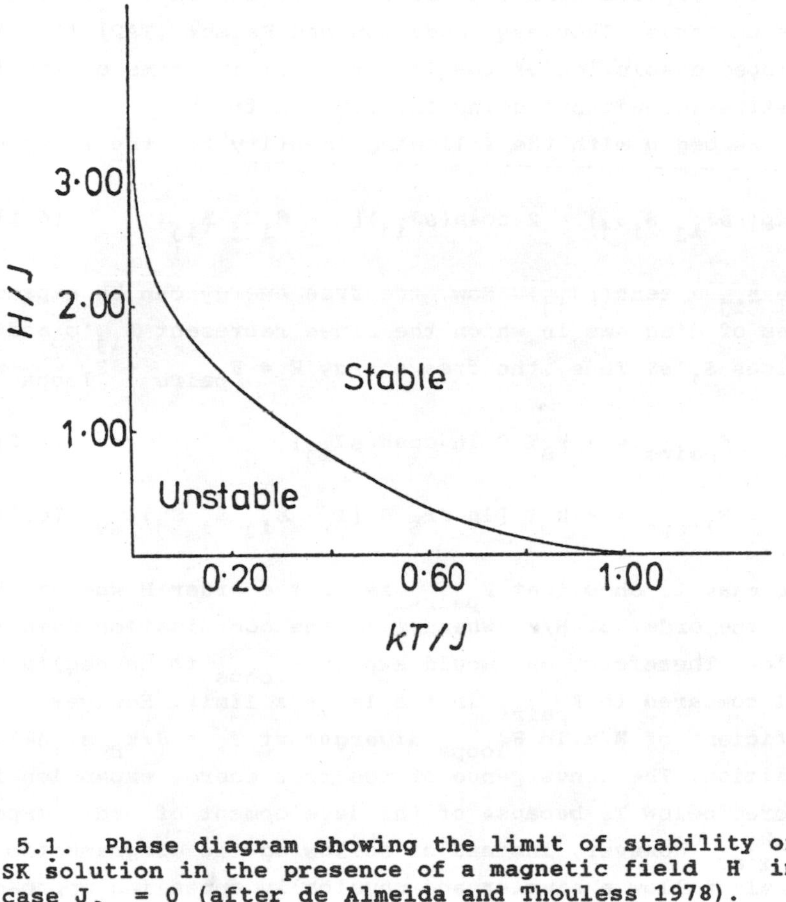

Fig. 5.1. Phase diagram showing the limit of stability of the SK solution in the presence of a magnetic field H in the case $J_0 = 0$ (after de Almeida and Thouless 1978).

CHAPTER 6

THOULESS-ANDERSON-PALMER (TAP) SOLUTION OF THE SK MODEL

Since the replica-symmetric solution of the SK model turned out to be unstable, Thouless, Anderson and Palmer (TAP) (1977) developed a solution of the latter model in terms of the local magnetization without using the replica trick.

We begin with the following identity for the Ising spins

$$\exp(-\beta J_{ij}\ S_i S_j) = 2\cosh(\beta J_{ij})[1 - S_i S_j\ \mathcal{B}_{ij}] \qquad (6.1)$$

where $\mathcal{B}_{ij} = \tanh(\beta J_{ij})$. Now, the free energy can be expanded as a series of diagrams in which the lines represent $\mathcal{B}_{ij}$'s and the vertices S_i's. Thus, the free energy $F = F_{pairs} + F_{loops}$ where

$$F_{pairs} = -\ k_B T\ \Pi\ \ln\cosh(\beta J_{ij}) \qquad (6.2)$$

and

$$F_{loops} = -\ k_B T\ [\ln \mathrm{Tr}_S\ \Pi\ (1 - \mathcal{B}_{ij}\ S_i\ S_j)]_{av} \qquad (6.3)$$

It is easy to show that F_{pairs} is of the order N whereas F_{loops} is of the order of N/z, where z is the coordination number of the lattice. Therefore, one would expect F_{loops} to be negligibly small compared to F_{pairs} in the large z limit. However, the coefficient of N/z in F_{loops} diverges at $T_g = J/k_B$ signalling the transition. The convergence of the free energy expansion is restored below T_g because of the development of order expressed by $[M_i^2]_{av}$. However, instead of following the diagrammatic method we shall follow a simpler and physically appealing approach (Anderson 1979).

Let us divide the Hamiltonian as $\mathcal{H} = \mathcal{H}_0 + \mathcal{H}_{fluc}$ where

$$\mathcal{H}_0 = \Sigma\ J_{ij}\ (M_i M_j - M_i S_j - M_j S_i) \qquad (6.4)$$

is a soluble mean-field Hamiltonian and

$$\mathcal{H}_{fluc} = - \Sigma J_{ij} (S_i - M_i)(S_j - M_j) \quad (6.5)$$

is the fluctuating part that is neglected in the conventional MFT. However, TAP theory does not drop $\mathcal{H}_{fluc}$. The free energy is given by $F = F_0 + F_{fluc}$ where

$$F_0 = U_0 - T S_0, \quad (6.6)$$

with

$$U_0 = -\Sigma J_{ij} M_i M_j \quad (6.7)$$

and

$$S_0 = k_B \Sigma [\{(1+M_i)/2\} \ln \{(1+M_i)/2\} + \{1-M_i)/2\} \ln \{(1-M_i)/2\}] \quad (6.8)$$

Note that U_0 is the energetic contribution and TS_0 is the entropic contribution to the free energy of a set of independent spins constrained to have the value M_i. Also note that the equations (6.7) and (6.8) for the random systems are the analogs of the equations (4.1) and (4.2) for nonrandom systems. In the TAP theory ($z \rightarrow \infty$)

$$F_{fluc} = - \{1/(2k_B T)\} \Sigma (1 - M_i^2)(1 - M_j^2) J_{ij}^2 \quad (6.9)$$

Then from the extremization condition $\partial F/\partial M_i = 0$ we get the so-called TAP equation

$$M_i = \tanh(\beta h_i) \quad (6.10a)$$

where

$$h_i = \Sigma J_{ij} M_j - \beta \Sigma J_{ij}^2 (1 - M_j^2) M_i \quad (6.10b)$$

(see Southern and Young 1977a for an alternative derivation). The physical interpretation of (6.10b) is very interesting. The field produced at the j-th site by the i-th spin is $J_{ij}S_i$. This field induces a magnetization $J_{ij}S_i\chi_j$ at the j-th site, the latter magnetization producing a field $J_{ij}^2 S_i \chi_j$ and hence the latter is called the reaction field. In the SK model,

$\chi_j = \beta(1 - M_j^2)$ and thus finally we identify the last term in (6.10b) as the reaction field in the SK model. Therefore, equations (6.10) implies that the local magnetization at an arbitrary site i is determined by the cavity field which is the field in the absence of the i-th spin. In the m-vector SG (Palmer and Pond 1979, Bray and Moore 1981, Sommers 1981) equation (6.10b) generalizes to

$$h_i = \Sigma\, J_{ij}\, M_j - \Sigma\, J_{ij}^2\, \chi_j M_j \qquad (6.11)$$

Nakanishi (1981) developed two-spin and three-spin cluster theories of SG. The two-spin cluster theory yields the equation (6.10b) because of its mathematical equivalence with the TAP theory. The corresponding result for the three-spin cluster theory is

$$h_i = \Sigma J_{ij} M_j - \Sigma\, J_{ij}^2 \chi_j - \Sigma J_{ij} J_{jk} J_{ki}\, \chi_j \chi_k M_i \qquad (6.12)$$

Why is the reaction field so important in SG but can be ignored in ferromagnets? Let us estimate the orders of magnitude of the three fields- the Weiss field, the reaction field and the cavity field in nonrandom ferromagnets and in SG. In a nonrandom nn ferromagnet $J_{ij} = J$ for all the nearest-neighbour pairs <ij> on a lattice with the coordination number z. Besides, the susceptibility near T_c (Curie temperature) is $\chi \simeq (1/T_c)$ and $T_c \simeq zJ$. On the other hand, in a random system let us assume $J_{ij} = \pm J$ with equal probability. The susceptibility in such a system is given by $\chi \simeq (1/T_g)$ and $T_g \simeq z^{1/2} J$. Therefore, in a nonrandom system the reaction field R and the Weiss field W are given by

$$R \simeq zJ^2 \chi M \quad \text{and} \quad W = zJM$$

where we have neglected the site dependence of χ and M. For $T \sim T_c$

$$R/W \sim (1/z) << 1$$

and hence the reaction field is negligible in comparison with the Weiss field. On the other hand, in a SG (Cyrot 1979)

$$R \simeq z^{1/2}JM \text{ and } W \simeq z^{1/2}JM \text{ for } T \sim T_g$$

and hence R is of the same order of magnitude as W. Therefore, in SG the reaction field is not negligible.

Since the magnitude of the reaction field in SG is comparable to that of the Weiss field, it can account for the existence of two-level systems (introduced in chapter 2). There will be a finite fraction of sites where the magnitude of the reaction field is larger than that of the cavity field and consequently the Weiss field at those sites is parallel to that of the reaction field and antiparallel to the cavity field. Thus, depending on the direction of the cavity field there are two energy levels available- one with energy $-\langle S_i \rangle H_i^{cav} - \lambda_i$ (spin parallel to the cavity field) and the other with energy $\langle S_i \rangle H_i^{cav} - \lambda_i$ (spin antiparallel to the cavity field but parallel to the Weiss field), the first being stable and the second is metastable, where $\lambda_i = \Sigma J_{ij}^2 \chi_j$. However, the extension of the latter argument to the vector SG models is nontrivial. It has been argued (Nozieres 1982) that the TLS in vector SG arises from the anisotropic nature of the local susceptibility χ_j. An alternative argument will be presented in chapter 20.

Let us now solve the TAP equations (6.10). The latter are a set of coupled N equations which have to be solved self-consistently. Suppose μ_i is the eigenvector corresponding to the largest eigenvalue $(J_\lambda)_{max} = 2\tilde{J}$ of the exchane matrix J, i.e., $\Sigma J_{ij}\mu_j = 2\tilde{J}\mu_i$. Now linearizing the TAP equation and defining $M_i = \mu_i + \delta M_i$, together with $q = [\mu_i^2]_{av}$, we get (Thouless et al.1977)

$$q = 1 - (T/T_g) \text{ near } T_g = \tilde{J} \text{ (of course, } T < T_g). \quad (6.13)$$

Since $M_i = \text{sign}(h_i)$ at T = 0, one can estimate q at $T \ll T_g$ provided the behaviour of the local field distribution $P(h_i)$ is known. From numerical investigation, TAP concluded that $P(h_i)$ is linear in h at low temperatures (see chapter 26.1). Therefore,

$$q = 1 - 1.665\ (T/T_g)^2 \quad \text{for } T \ll T_g \qquad (6.14)$$

instead of the linear T-dependence (equation (4.13)) predicted by the SK solution (also see Plefka 1982a). The latter prediction is in good agreement with the corresponding Monte Carlo result (Kirkpatrick and Sherrington 1978). Unfortunately, because of the complicated structute of the TAP equations, it has not been possible to find analytically any general solution valid at all temperatures. Bray and Moore (1979a) solved the TAP equations at all temperatures T below T_g numerically for finite systems of size N = 40, 100, 250 following the iterative scheme

$$M_i^{(n+1)} = \tanh[\beta\Sigma J_{ij} M_j^{(n)} - \beta^2 \Sigma J_{ij}^2 \{1-(M_j^{(n)2}\} M_i^{(n)}] \qquad (6.15)$$

where $M_i^{(n)}$ is the value of M_i after n iterations. Several sets of the Gaussian-distributed random exchange J_{ij} were used. However, the rate of convergence of the iteration (6.15) to the solutions was very slow for most of the sets of $\{J_{ij}\}$ used (see Nemto and Takayama 1985 for a faster convergence). Later, more efficient numerical works have conclusively proved the existence of a large number of solutions of the TAP equations.

The most severe difficulty faced by the theorists in the seventies was not to get a solution of the SK model but to get one that is <u>stable</u>. Using the numerical solutions of the TAP equations, TAP computed the eigenvalue spectrum of the Hessian A. One can easily check that $(A^{-1})_{ij} = \beta^{-1}\ \chi_{ij}$ so that $(1/N)\ Tr(A^{-1}) = (1/N)\Sigma(1/\lambda) = \int\{\rho(\lambda)/\lambda\}\ d\lambda = (1-q)$. Since 1-q is finite at all remperatures below T_g the distribution of the eigenvalues of A, $\rho(\lambda)$ must vanish at λ=0. Bray and Moore (1979a) showed that $\rho(\lambda)$ indeed, vanishes at λ=0 but the eigenvalue spectrum of A extends down to zero. The latter result implies that the TAP solutions are marginally stable! The latter result

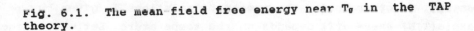

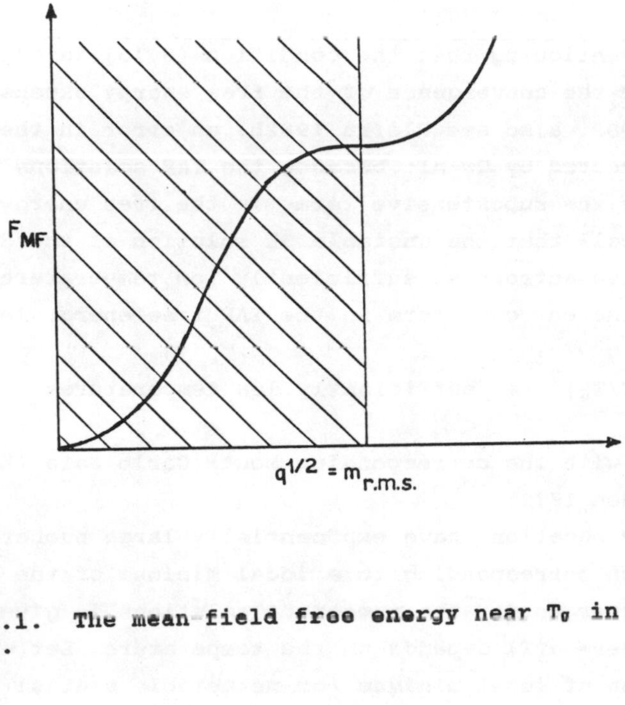

Fig. 6.1. The mean-field free energy near T_g in the TAP theory.

was also anticipated in the original paper of TAP (1977) (see Fig.6.1). Moreover, from a locator expansion analysis Bray and Moore concluded that the vanishing of $\rho(\lambda)$ at $\lambda=0$ is assured provided

$$(\beta J^{-2}) > (1 - 2q_2 + q_4) \qquad (6.16)$$

where

$$q_\nu = (1/N) \Sigma M_i^\nu. \qquad (6.17)$$

It is worth mentioning that the condition (6.16) is identical with that for the convergence of the free energy expansion (6.2)-6.3) (Owen 1982, also see Plefka 1982b, an error in the latter has been corrected by Owen), because the TAP solutions cannot be stable unless the subextensive terms in the free energy expansion converge. Recall that the unstable SK solution of the SK model led to negative entropy at sufficiently low temperatures. On the other hand, the entropic term in the TAP free energy leads to

$$S(T) \sim (T/T_g)^2 \text{ at sufficiently low temperatures} \qquad (6.18)$$

in agreement with the corresponding Monte Carlo data (Kirkpatrick and Sherrington 1978).

The TAP equations have exponentially large number of solutions each corresponding to a local minimum of the free energy. In other words, the number of solutions is given by $\exp\{\alpha(T)N\}$ where $\alpha(T)$ depends on the temperature. Let us outline the definition of local minimum (or metastable states) and the methods of estimating their number. The local change of exchange energy as a consequence of the flipping of, say, the Ising spin at the i-th site is $\Delta\epsilon_i = \epsilon_i' - \epsilon_i = 2 S_i h_i$ where h_i is the local magnetic field. If the initial state stable against such a single-spin-flip event, then $\Delta\epsilon_i = 2S_i h_i \geq 0$. A state where the latter relation is valid for all i is called a metastable state. Defining $h_i = \lambda_i S_i$, the condition for metastability reduces to the condition that each spin must be parallel to the corresponding local field. The average number of metastable states in an Ising SG is given by

$$\langle g_0 \rangle = (1/2) \int \Pi \, d\lambda_i \, Tr_{\{S\}} \, [\Pi \, \delta(\Sigma J_{ij} S_j - \lambda_i S_i)]_{av} \qquad (6.19)$$

where the prefactor 1/2 ensures exclusion of the trivial degeneracy that arises from the simultaneous flipping of all the spins in the system. Evaluation of the R.H.S. of (6.19) for the SK model yields $\alpha \simeq 0.2$ for Ising spin system (Tanaka and Edwards 1980a, DeDominicis et al.1980, Bray and Moore 1980a). In the investigation of the number of metastable states in the SK model by Monte Carlo simulations (Kirkpatrick and Sherrington 1978) one begins with an arbitrary random spin configuration for a given random bond configuration and lets the system evolve according to the Metropolis algorithm (to be explained in chapter 24) so as to arrive at a (local) minimum-energy configuration. Then the system is heated up and cooled down again thereby reaching a new metastable configuration. This alternate heating and cooling procedure is repeated to visit as many new metastable states as possible. Such simulations strongly suggested the existence of very large number of metastable states for the SK model (see also Kaplan 1981, 1986, Tanaka and Edwards 1980b).

The distribution of the metastable states $g_s(E)$ is no less interesting. Since it is anticipated that $g_s(E) \sim \exp(\alpha' N)$ one should average the extensive quantity $\ln g_s(E)$ over all possible configurations rather than averaging $g_s(E)$. However, configuration averaging of $g_s(E)$ needs the use of replica trick for the same reason as for the evaluation of the free energy in the EA MFT. It turns out that there exits a critical value E_c of E below which $[\ln g_s(E)]_{av}$ differs from $\ln[g_s(E)]_{av}$ whereas both are equal for $E > E_c$. The latter is a consequence of the onset of correlations between the metastable states below E_c (see Bray and Moore 1980a, 1981a,b,d, Roberts 1981).

The third law of thermodynamics states that "as the temperature tends to zero, the magnitude of the entropy change in any reversible process tends to zero" (Pippard 1966). It might appear at first sight (Wohlfarth 1977) that since

$$H(\partial\chi/\partial T)_H = (\partial M/\partial T)_H = (\partial S/\partial H)_T \qquad (6.20)$$

slopes of M-T and χ-T curves must approach zero as $T \to 0$ if the third law of thermodynamics is to hold. Since for the TAP solution of the SK model relation (6.20) is violated Wohlfarth (1977) claimed that the TAP solution is erroneous. However, that is not true. In fact, there are several examples for which the Nernst law, $\lim S(T,H) = 0$, holds although $\partial\chi/\partial T$ is nonzero at $T = 0$ (Anderson et al.1977, Sherrington and Fernandez 1977). Thus the TAP does not necessarily violate the third law of thermodynamics. Besides, one must also keep in mind that the third law of thermodynamics is not a law on the same footing as the first and the second laws of thermodynamics.

As stated earlier, all the diagrams with closed loops in the free energy expansion were neglected by TAP because $z \to \infty$ in the SK model. However, for large but finite z, corrections from the simplest loop diagrams are not negligible. Suppose, T_L is the temperature ($T_L > T_g$) at which the correction to the self-field from the loops becomes comparable to the Bethe lattice reaction field. It can be shown (Morgan-Pond 1982) that

$$T_L(z)/T_g(z) = 1 + (1/2z) + \ldots$$

One interesting question is "how large z should be in order to be a good approximation for the mean-field ($z=\infty$) result"? Klein et al.(1979) observed that for $z \geq 8$ the phase diagram differs very little from the corresponding phase diagram for $z = \infty$.

The TAP approach to the SK model hints that the SG transition is a kind of "blocking" transition; the spins are blocked by frustration from going to the lower free energy state that these spins would have if allowed to adjust independently (Anderson 1979, also see Garel 1980).

Using a diagrammatic expansion of the free energy, Sommers (1978, 1979) showed that (i) the free energy resulting from averaging over only the ring diagrams is identical with the TAP free energy, (ii) averaging over both ring as well as tree diagrams indicates the existence of two solutions, one of which is the SK solution and the other is called the Sommers solution. The Sommers solution is characterized by two order parameters, q and Δ. On the other hand, the SK solution corresponds to $\Delta = 0$. Unlike the SK solution, the entropy vanishes at $T = 0$ for the Sommers solution. However, entropy vanishes exponentially as $T \neq 0$. It was realized very soon that the latter behaviour of the entropy would imply an unique ground state separated by a gap from the first excited state. This contradicts our earlier assertion in this chapter that there are exponentially large number of (TAP) solutions of the SK model. Indeed, the Sommers solution turned out to be unstable (DeDominicis and Garel 1979).

CHAPTER 7

PARISI SOLUTION OF THE SK MODEL AND ITS STABILITY

Almeida and Thouless (1978) (AT) suggested that breaking the permutation symmetry between the replicas might yield stable solution(s) of the SK model. Suppose P_n is the group of permutation of n elements. At first it might appear that a pattern

$$P_n \rightarrow P_{n/m} \otimes (P_m)^{n/m}$$

i.e.,

$$P_0 \rightarrow P_0 \otimes (P_m)^0 \text{ in the limit } n \rightarrow 0$$

would be sufficient (Blandin et al.1980). Unfortunately, such a weak symmetry breaking is not sufficient and Parisi (1979a,b, 1980a) suggested the scheme

$$P_0 \rightarrow (P_{m_1})^0 \otimes (P_{m_r})^0 \otimes \dots \otimes P_0$$

so that one can have infinitely broken symmetry. However, parametrizing an n X n matrix in the limit n=0 is very difficult. Fortunately, the following physical requirements provide some guidelines:

(a) $[\lim_{n\to 0} (1/n)\Sigma q_{\alpha\beta}^2] < \infty$,

so that the configuration-averaged free energy is finite,

(b) $\Sigma q_{\alpha\beta} = \Sigma q_{\gamma\beta}, \quad \alpha \neq \gamma$,

so that the free energy is an extremum with respect to q and

(c) $-\lim_{n\to 0} (1/n)\, \Sigma q_{\alpha\beta}^2 \geq 0$,

so that at high temperatures the minimum of the free energy is located at $q_{\alpha\beta} = 0$.

Parisi's schme of replica symmetry breaking goes as follows:

first, one begins with the SK replica-symmetric matrix

$$q_{\alpha\beta}^{(0)} = \overbrace{\begin{array}{|c|}\hline \quad q_0 \quad \\ \hline\end{array}}^{n} \longrightarrow \qquad q_{\alpha\beta}^{(1)} = \overbrace{\begin{array}{|c|c|c|}\hline q_1 & q_0 & q_0 \\ \hline q_0 & q_1 & q_0 \\ \hline q_0 & q_0 & q_1 \\ \hline\end{array}}^{n} \Big\}\, m_1$$

Thus, after the first stage, the matrix q consists of $(n/m_1) X (n/m_1)$ constant blocks each of size $m_1 X m_1$. At the next stage, the off-diagonal blocks q_0 are left unmodified whereas the diagonal blocks q_1 are further subdivided as

$$\overbrace{\begin{array}{|c|}\hline \quad q_1 \quad \\ \hline\end{array}}^{m_1} \rightarrow \overbrace{\begin{array}{|c|c|c|}\hline q_2 & q_1 & q_1 \\ \hline q_1 & q_2 & q_1 \\ \hline q_1 & q_1 & q_2 \\ \hline\end{array}}^{m_1} \Big\}\, m_2$$

Continuing the above procedure we subdivide the diagonal blocks q_2 further into smaller blocks and so on. Finally, the latter process is terminated after R such iterations, with the smallest diagonal block q_R of size $m_R X m_R$.

By construction, the successive sizes of the blocks are

$$n \geq m_1 \geq m_2 \ \ldots \geq m_R \geq 1$$

which in the limit $n \rightarrow 0$ become

$$0 \leq m_1 \leq m_2 \ \ldots \leq m_R \leq 1$$

In the limit $R \rightarrow \infty$ we have

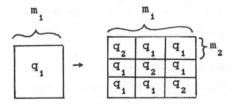

$$m_k/m_{k+1} \to 1 - (dx/x)$$

and $q_k \to q(x)$ where x is defined over the unit interval, i.e., $0 \le x \le 1$. Thus, the order parameter in the Parisi's theory is a function! The functional form of q(x) is shown in Fig.7.1. The physical significance of the parameter x will be explained in chapter 9. (Jonsson's (1982) theory involves an order parameter q(x,y) which is a function of the two parameters x and y both defined over the unit interval. For possible physical interpretation see Chowdhury and Banerjee 1984).

The free energy in Parisi's theory can be expressed as the solution of the differential equation (Parisi 1980b, Duplantier 1981)

$$(\partial F/\partial x) = -(1/2)\ (dq/dx)[(\partial^2 F/\partial H^2) + x(\partial F/\partial H)^2] \quad (7.1)$$

where F(x,H) is a generalized free energy such that F(0,H) gives the actual free energy. The corresponding boundary condition is $F(1,H) = \ln [2 \cosh (\beta H)]$.

By a generalization of the AT analysis for the SK solution, the stability of the Parisi solution has also been analysed (Thouless et al.1980, DeDominicis and Kondor 1983). In the limit $R \to \infty$, all the eigenvalues turn out to be nonnegative. However, the spectrum of the eigenvalues of the Hessian extends down to zero thereby implying marginal stability of the Parisi solution (see also Khurana 1983).

Let us summarize the scenario without going through the technical details. We have stated in chapter 5 that the three types of eigenvectors of the Hessian exhibit degeneracies in the case of the SK solution in the limit $n = 0$. The replica symmetry breaking acts like switching on a "perturbation". The latter perturbation not only lifts the degeneracy but also stabilizes the system. The resulting structure of the spectrum of the Hessian corresponding to finite as well as for infinite R have been shown in Fig.7.2. The importance of the limit $R \to \infty$ is demonstrated by the fact that $S(T=0) \to 0$ only as $R \to \infty$ (Parisi

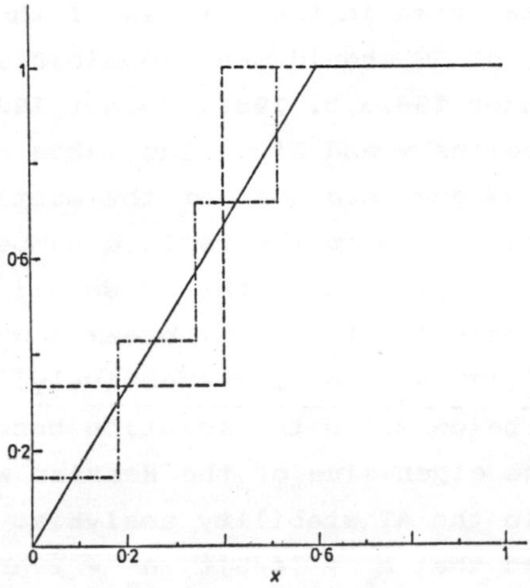

Fig. 7.1. The Parisi order parameter (function) q(x) for three values or R; the dashed curve, the dashed-dotted curve and the full curve, respectively, correspond to R = 1, R = 4 and R = ∞ (after Parisi 1980a).

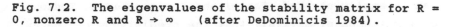

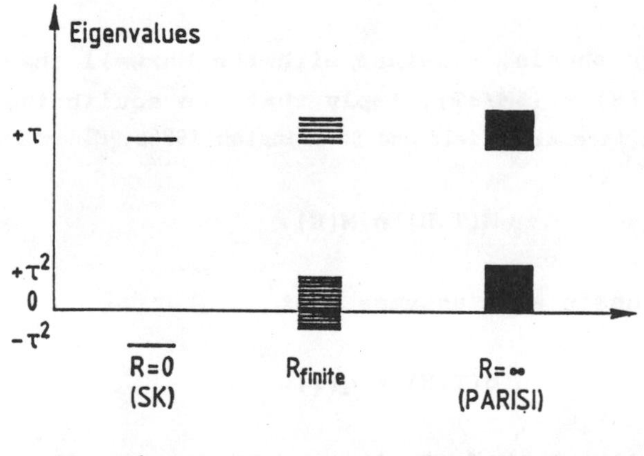

Fig. 7.2. The eigenvalues of the stability matrix for R = 0, nonzero R and R → ∞ (after DeDominicis 1984).

1980a). Readers interested in the details of the recent progress in the field theory of SG should see DeDominicis 1984, DeDominicis and Kondor 1984a,b, 1985, Kondor 1985, Goltsev 1984a,b, 1986, Sompolinsky and Zippelius 1983a,b.

In chapter 4 we examined some of the subtleties related to the analytical continuation in the replica number n. Rammal (1981) showed that in the case of the SK solution below T_g, the convexity and monotonicity of of $\psi(n)$ break down at the critical values $n_c(T)$ and $n_m(T)$ ($n_m > n_c$), respectively. Suppose $n_s(T)$ be the replica number below which the solution becomes unstable, i.e., if $\lambda(n)$ is the eigenvalue of the Hessian which becomes negative at n = 0 in the AT stability analysis, then $\lambda(n_s) = 0$. Kondor (1983) showed that $n_s = (4/3)\tilde{t}$, $n_m = \tilde{t}$ and $n_c = (2/3)\tilde{t}$, where $\tilde{t} = (T_g-T)/T_g$ and therefore, $n_s(T) > n_m(T) > n_c(T)$. In other words, the SK solution becomes unstable before the breakdown of convexity and monotonicity takes place.

Parisi-Toulouse (PaT) (1980) hypothesis states that in the SG phase the entropy does not vary with the magnetic field, i.e.,

$$S(T,H) = S(T)$$

The latter hypothesis, together with the Maxwell thermodynamic relation $(\partial S/\partial H) = (\partial M/\partial T)$, imply that the equlibrium magnetization (see Elderfield and Sherrington 1982b, Elderfield 1983)

$$M(T,H) = M(H).$$

The PaT hypothesis also assumes that

$$q(T,H) = q(T)$$

It can be easily checked (Parisi and Toulouse 1980, Toulouse 1980, Vannimenus et al.1981) that the predictions of the PaT hypothesis, e.g., the low-temperatute behaviour of q and S, the stability criterion of TAP, the field-dependence of the susceptibility, are in good agreement with the corresponding results of TAP and Parisi. Moreover, Ma and Payne (1981) computed the entropy by numerical method on finite SG systems using the coincidence counting method (Ma 1981) and found approximate agreement between the result extrapolated to infinite system-size and the PaT prediction. For temperatures $T > T_g$, the entropy S is a decreasing function of the field H. The instability line, the so-called AT line (5.21), corresponds to the locus of the points where entropy S exhibits a maximum. The latter observation, together with the relation $(\partial\chi/\partial T)_{H=0} = (\partial^2 S/\partial H^2)_{H=0}$, implies that the locus of the temperature $T_g(H)$ where the magnetization M(T,H) attains its maximum also follows the AT line. Since the PaT hypothesis identifies the latter curve $T_g(H)$ as the SG transition temperature, the AT line is believed to be the SG transition line in external field. (Also see Toulouse et al. 1982). Finally, I would like to mention that the PaT hypothesis was introduced for computing the relevant physical properties of the SK model starting from a simple hypothesis which yields results in agreement with the Parisi formalism without the necessity of going through the tedious algebra.

CHAPTER 8

SOMPOLINSKY'S DYNAMICAL SOLUTION OF THE SK MODEL AND ITS STABILITY

Sompolinsky (1981a) proposed a dynamical approach to the statics of the SK model based on the Sompolinsky-Zippelius (1981, 1982) dynamical theory (to be discussed in chapter 19). As mentioned in chapter 3, Sompolinsky (1981a) introduced an order parameter (function) q(x) which is defined as

$$q(x) = [\langle S_i(0)S_i(t_x)\rangle]_{av} \tag{8.1}$$

where x varies between 0 and 1. q(x) is a measure of the correlation that has not decayed at the time scale t_x. Sompolinsky also introduced another order parameter

$$\Delta(x) = T\chi(x) - (1 - q_{EA}) \tag{8.2}$$

where $\chi(x)$ is the local susceptibility measured at the frequency $\omega_x = t_x^{-1}$. Both q(x) and $\Delta(x)$ are assumed to be continuous functions of x.

First we shall briefly sketch Sompolinsky's derivation of the free energy functional and the susceptibility. Then we shall show how the latter results can be derived from replica formalism. Then, using the latter formulation of Sompolinsky's solution we shall investigate the stability of the latter solution.

In order to extract the statics of the SK model from the dynamics, Sompolinsky (1981a) wrote the dynamic autocorrelation and the local response as

$$C(\omega) = [\langle S_i(\omega)S_i(-\omega)\rangle] = \tilde{C}(\omega) + q\ \delta(\omega) \tag{8.3}$$

and

$$\chi(\omega) = [\partial\langle S_i(\omega)\rangle/\partial h_i(\omega)] = \tilde{\chi}(\omega) + \Delta\ \delta_{\omega,0}, \tag{8.4}$$

respectively, where $\tilde{C}$ and $\tilde{\chi}$ are the finite-time parts of C and χ. q and Δ represent the time-persistent parts of C and χ, respectively. Note that $\delta(\omega)$ and $\delta_{\omega,0}$ are, respectively, the Dirac delta and Kronecker delta functions and are formally related by the fluctuation-dissipation theorem (FDT):

$$\tilde{C}(\omega) = 2\,\omega^{-1}\,\mathrm{Im}\,\tilde{\chi}(\omega). \qquad (8.5)$$

Some important implications of the relation (8.5) will be investigated in more detail in the next chapter.

Let us assume that there exist a large number of ground states separated by free-energy barriers. The distribution of the barrier heights gives rise to the distribution of relaxation times τ. Therefore, one can write

$$q\,\delta(\omega) = \sum_{i=1}^{R} q_i'\,\delta^i(\omega) \qquad (8.6)$$

and

$$\Delta\,\delta_{\omega,0} = \sum_{i=1}^{R} \Delta_i'\,\delta^i_{\omega,0} \qquad (8.7)$$

where $\delta^i(\omega)$ and $\delta^i_{\omega,0}$ are normalized Dirac delta function and Kronecker delta, respectively, only in the thermodynamic limit $N \to \infty$, but, have widths $\omega_i \sim \tau_i^{-1}$ for finite system sizes. Thus, in Sompolinsky's theory all the relaxation times are assumed to diverge in the thermodynamic limit. The latter assumption has been supported by Monte Carlo simulation of the SK model (Mackenzie and Young 1982). Then, in the limit $N, R \to \infty$, defining a continuous parameter $x = i/R$, the partial sums

$$q(x) = \int_0^x q'(y)\,dy \qquad (8.8)$$

and

$$\Delta(x) = - \int_x^1 \Delta'(y)\, dy \qquad (8.9)$$

serve as the two order parameters in Sompolinsky's theory. But, because of the arbitraryness of the scale of x, the latter theory does not determine q(x) and Δ(x) independently; these two order parameters are related by a gauge transformation, as we shall see in the next chapter. The equlibrium susceptibility is given by

$$\chi = \beta\{1 - q_{EA} + \Delta\}. \qquad (8.10)$$

The latter must be contrasted with the quasiequilibrium susceptibility (susceptibility after a time large compared to microscopic time scales but small compared to macroscopic equlibration time scales)

$$\chi = \beta\{1 - q_{EA}\} \qquad (8.11)$$

(see Sompolinsky and Zippelius 1982 and Sommers 1983 for the details of the dynamical derivation). The physical significances of these two forms for χ will be explained in the light of the FDT in chapter 9. The free energy functional is given by

$$- \beta F\{q,\Delta\} = (1/4)\beta^2\tilde{J}^2\{[1-q(1)]^2 + 2\int \Delta'(x)q(x)\, dx\}$$

$$+ \int \Pi\ [dz(x)/\{2\pi q'(x)\}^{1/2}]\ \exp[-(1/2)\int\{z^2(x)/q'(x)\}dx]$$

$$[\ln 2\cosh(h\{z\}) + (1/2)\beta^2\tilde{J}^2\int\Delta'(x)m_x^2\{z\}dx] \qquad (8.12)$$

where

$$h\{z\} = \beta\tilde{J}\ \int z(x)\, dx - \beta^2\tilde{J}^2\ \int\Delta'(x)m_x\{z\}\, dx + \beta H$$

and

$$m_x\{z\}=\int\Pi[dz(y)/\{2\pi q'(y)\}^{1/2}]\exp[-(1/2)\int dy\{z^2(y)/q'(y)\}]\tanh(h\{z\})$$

A replica derivation of Sompolinsky's free energy functional (8.12) was proposed by DeDominicis et al.(1981). In this approach one begins with the order parameter matrix

$$q_{\alpha\beta}^{(o)} = \overbrace{\begin{array}{|c|c|c|}\hline q_0 & r_0 & r_0 \\ \hline r_0 & q_0 & r_0 \\ \hline r_0 & r_0 & q_0 \\ \hline \end{array}}^{n} \;\} p_0$$

where the matrix consists of $(n/p_0)X(n/p_0)$ block matrices q_0 (the diagonal blocks) and p_0Xp_0 block matrices r_0 (off-diagonal blocks). Note that the Parisi scheme begins with the SK order parameter matrix whereas the Sompolinky scheme does not. Moreover, unlike the Parisi scheme, at the first step,

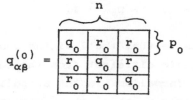

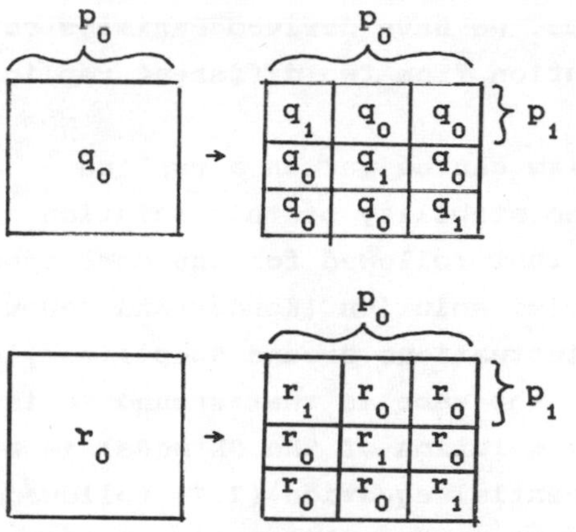

so that both the diagonal and off-diagonal blocks are split into subblocks of smaller sizes. At the second step, the off-diagonal blocks remain unchanged whereas the diagonal blocks are subjected to further subdivision into smaller subblocks. The latter procedure is continued until the R-th step, when the smallest diagonal blocks q_R and r_R are of size p_RXp_R. Thus,

$$p_0 >> p_1 >> \ldots >> p_R >> 1.$$

I would like to emphasize that the limiting procedure in the latter formalism is quite diffetent from the corresponding procedure in Parisi's formalism. In the Sompolinsky's formalism, one first takes the limit $p_0 >> .. >> p_R \to \infty$ <u>in that order</u> <u>followed by</u> the limit $n \to 0$. Then, in addition to the q_k, we also have the so-called "anomaly" parameters

$$-\Delta_k^{'} = p_k \, (q_k - r_k)$$

As $R \to \infty$, the above-mentioned parametrization leads to the free energy functional (8.12). Thus, we have derived Parisi's solution as well as Sompolinsky's solution from two different replica formalisms.

Since Sompolinsky's formalism can be put on a replica formalism, as shown above, the stability of this solution can be analysed in a way similar to that followed for the corresponding stability analysis of the Parisi solution (Kondor and DeDominicis 1983). The spectrum of the fluctuations around Sompolinsky's solution, thus calculated, is the same as that around Parisi's solution. Thus, Sompolinsky's solution of the SK model is also marginally stable. The differential equation (7.1) followed by Parisi's solution is also obeyed by Sompolinsky's solution provided (DeDominicis et al. 1982)

$$\Delta'(x) = -\, xq'(x)$$

It is the latter relation between q and Δ that leaves only one of these as the independent mean-field order parameter for SG, as was noted in Sompolinsky's original paper (1981a). (Also see Elderfield 1984b).

CHAPTER 9

ERGODICITY, PURE STATES, ULTRAMETRICITY, AND FLUCTUATION-DISSIPATION THEOREM

In the formalism of equilibrium thermodynamics, a system is called ergodic if its properties observed experimentally are equal to the corresponding ensemble average. To put the ideas on a more precise footing, let us describe the dynamical evolution by T_t, so that if w is a point in the phase space Ω at t = 0 then $T_t w$ is the position in the phase space at time t. Suppose $\mu(w)$ is the probability measure in the phase space. The phase space average of a quantity f is given by

$$\langle f \rangle = \int d\mu(w)\ f(w)$$

A system is called ergodic with respect to T_t if the null set ϕ and the set Ω are the only sets in Ω invariant under T_t. On the other hand, a system is called nonergodic if we can split the phase space Ω into disjoint subsets Ω_i each of which is invariant with respect to T_t, i.e., $\Omega = \cup\, \Omega_i$, $\Omega_i \cap \Omega_j = \phi$, $T_t\ \Omega_i \subseteq \Omega_i$. As an example of ergodicity breaking, consider the two-dimensional Ising model with nearest-neighbour exchange interaction. At the phase transition temperature T_c the phase space decomposes into two disconnected parts- corresponding to "up" and "down" spontaneous magnetizations, respectively. In the thermodynamic limit $N \to \infty$, the free energy barrier separating these two "pure states" diverges and the system remains "confined" within one minimum for all finite times of observation. Similarly, because of the divergence of the corresponding free energy barriers in the SK model (Mackenzie and Young 1982), the latter model also exhibits "absolutely broken ergodicity" (Young 1981, Young and Kirkpatrick 1982) and hence the differences between the corresponding time-averages and the ensemble-averages. What happens if at least some of the barriers do not diverge even in the thermodynamic limit? As an example, the system remains confined within one of the two components in the phase space of a finite-sized 2-d nearest-neighbour Ising model for time scales smaller than $\exp(\Delta F)$ where ΔF is the free energy barrier. Similarly, the short-ranged SG, which has finite free energy barriers in the phase space (Morgenstern and Horner 1982), also exhibits "effective broken ergodicity" for all times smaller than the smallest barrier hopping time (see Palmer 1982, 1983 for an excellent introduction to the concept of ergodicity).

We assume (van Enter and van Hemmen 1984a) that the Gibbsian-average of a variable O

$$\langle O\rangle = [\Sigma\ O \exp\{-\beta\mathcal{H}\}]/[\exp\{-\beta\mathcal{H}\}]$$

can be decomposed as a sum over the "pure states", viz.,

$$\langle O\rangle = \Sigma\ P_\alpha \langle O\rangle_\alpha$$

where P_α is the statistical weight for the α-th pure state and $\langle\rangle_\alpha$ denotes the restricted average in the α-th pure state. Both "canonical weight" (DeDominicis and Young 1983a, DeDominicis 1983) of the form

$$P_\alpha = \exp[-\ \beta F\{m^\alpha\}]/\Sigma\ \exp[-\beta F\{m^\delta\}]$$

and "white weights" (DeDominicis et al.1983, DeDominicis 1983) of the form

$$P_\alpha = \text{constant within an appropriate window of states}$$
$$= 0 \text{ otherwise,}$$

have been proposed. The magnetization at the i-th site in the α-th pure state is given by

$$m_i^\alpha = \langle S_i\rangle_\alpha$$

The overlap of the magnetization in any two arbitrary states α and β is given by

$$q^{\alpha\beta} = (1/N) \Sigma m_i^{\alpha} m_i^{\beta} \qquad (9.1)$$

For a given configuration {J}, we define the probability distribution of the overlap by

$$P_J(q) = \Sigma P_{\alpha} P_{\beta} \delta(q - q^{\alpha\beta}) \qquad (9.2)$$

Let us denote the average of $P_J(q)$ over all possible distributions of the couplings J by

$$P(q) = [P_J(q)]_{av} \qquad (9.3)$$

Let us introduce a function

$$x(q) = \int dq' \, P(q'). \qquad (9.4)$$

Now consider two real physical copies of the same system and label these copies by the superscript a (a=1,2). Also define a Hamiltonian $\mathcal{H}_2 = \Sigma \mathcal{H}(\{S^a\})$. Then in the limit $N \to \infty$, we have

$$g(z) = \langle\exp(z\Sigma S_i^1 S_i^2/N)\rangle_2 \simeq \Sigma\Sigma P_{\alpha} P_{\beta} \exp(zq^{\alpha\beta}) = \int dx \, \exp\{zq(x)\} \qquad (9.5)$$

where $\langle\rangle_2$ denotes statistical expectation value with respect to H_2. On the other hand, following Parisi's ansatz for breaking the replica symmetry in the replica space (as explained in chapter 7), we get

$$g(z) = [1/\{n(n-1)\}] \, \Sigma \, \Sigma \, \exp(zq_{ab}) \qquad (9.6)$$

where q_{ab} is Parisi's order parameter matrix in the replica space. In the appropriate replica limit $n \to 0$,

$$g(z) = \int dx \, \exp[zQ(x)] \qquad (9.7)$$

where $Q(x)$ is the Parisi order parameter introduced in chapter 7. Comparing (9.5) and (9.7) we identify $q(x)$ with $Q(x)$ and thereby establish the physical interpretation of the parameter x (Parisi 1983a,b, Houghton et al.1983a,b). More conveniently,

$$y(q) = 1 - x(q) = \int dq' \; P(q') \qquad (9.8)$$

is the probability that two pure states have an overlap larger than q. Thus, $q(0)$ is the minimum overlap between two states From the above observation one might be tempted to use $P(q)$ as the order parameter for SK model (a probability law as a measure of the order !). But, we shall see soon in this chapter that the appropriate measure of the order for the SK model is the probability distribution of $P_J(q)$ (the probability law of a probability law as a measure of the order!!).

Suppose, $P(q_1,q_2,q_3)$ is the probability for any three pure states to have overlaps q_1, q_2 and q_3. Then, generalizing Parisi's method (1983a) for the computation of $P(q)$, we get (Mezard et al.1984a,b)

$$g(z_1,z_2,z_3) = \int \Pi \, dq_i \; \exp(\Sigma z_i q_i) \; P(q_1,q_2,q_3)$$

$$= [1/\{n(n-1)(n-2)\}] \; \Sigma \; \exp[z_1 Q_{ab} + z_2 Q_{bc} + z_3 Q_{ca}] \qquad (9.9)$$

Note that the latter expression is an appropriate generalization of (9.6). This leads to

$$P(q_1,q_2,q_3) = (1/2)P(q_1)x(q_1)\delta(q_1-q_2)\delta(q_1-q_3)$$

$$+ (1/2)[P(q_1)P(q_2)\Theta(q_1-q_2)\delta(q_2-q_3) + \text{permutations}] \qquad (9.10)$$

The expression (9.10) implies that the pure states are such that for any three given states with mutual overlaps q_1, q_2 and q_3 at least two are equal to each other and the third is either larger than or equal to these two. This result can be interpreted geometrically as shown in Fig.9.1). "Spaces" with such geometrical structure are called ultrametric and had been studied earlier by mathematicians. The problem of diffusion in such ultrametric spaces has become fashionable and will be reviewed in chapter 27.6. (See Sibani and Hertz 1985 for the nature of P(q) in two simpler models, viz. the tapeworm model and the random Dyson model). The overlap function P(q) has been computed numerically for the SK model (Young 1983c) and is shown in Fig.9.2. In principle, one could calculate P(q) (= dx/dq) provided the function q(x) is known. The function q(x) computed first numerically by Parisi (1980a) is shown in Fig. 7.1. No analytical closed form expression for q(x) valid over the whole domain $0 \le x \le 1$ is available. However, extending an earlier work (Sommers and Dupont 1984), Sommers (1985) calculated the function q(x) analytically near T_g. q(x) is constant for $x > x_1$ where x_1 is the break point shown in Fig.7.1. For $x < x_1$,

$$q(x) = (1/2)\{1 + 3\tilde{t}\}x - (1/8)\, x^3 + O(\tau^4) \qquad (9.11)$$

where $\tilde{t} = (T_g - T)/T_g$ is small.
The latter result, as well as independent approach from the scaling behaviour of q(x) (Bray et al.1984), suggest that

$$P(q) = (1 - x_1)\, \delta(q - q(1)) + \tilde{P}(q) \qquad (9.12)$$

where $\tilde{P}(q)$ is a smooth "background" contribution. It has been conjectured (Bray et al.1984) that the first term in (9.12) arises from self-overlap only, i.e.,

$$1 - x_1 = \Sigma\, P_\alpha^2$$

Let us go one step further and define the overlaps q_1 and q_2 as those between two pure states α_1 and α_2 and between α_3 and

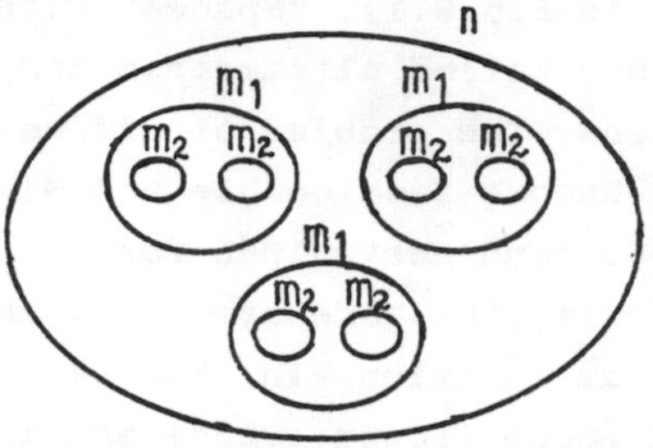

Fig. 9.1. Representation of the ultrametric structure of replica space.

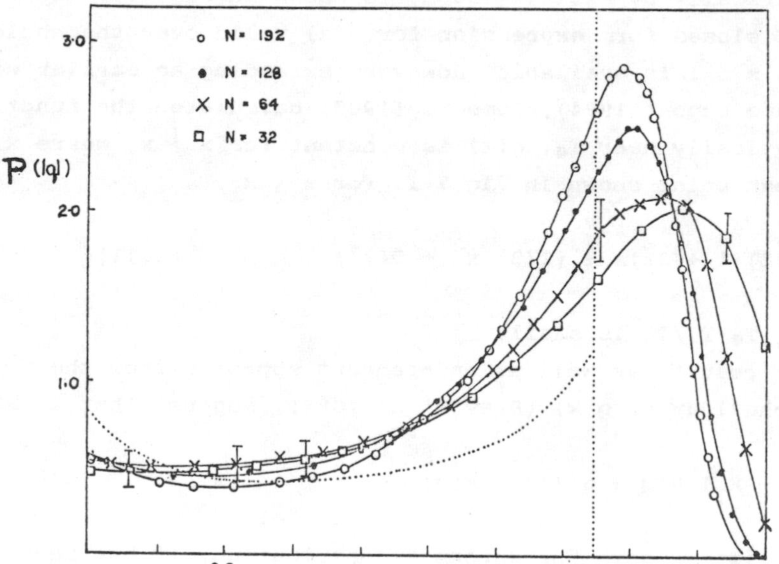

Fig. 9.2. P(q) for T = 0.4 T_g, H = 0 (after Young 1983c).

α_4 respectively. Then,

$$g(z_1,z_2) = \int dq_1\, dq_2\, \exp[z_1q_1+z_2q_2]\, P(q_1,q_2)$$

$$= \int dq_1\, dq_2\, \exp[z_1q_1+z_2q_2]\, [P_J(q_1)P_J(q_2)]_{av}$$

where the summations are to be carried over replicas such that a,b,c,d are all different. In the limit $n \to 0$, one gets (Mezard et al.1984a,b)

$$[P_J(q_1)P_J(q_2)]_{av} = (1/3)P(q_1)\delta(q_1-q_2)+(2/3)P(q_1)P(q_2) \neq P(q_1)P(q_2) \quad (9.13)$$

The expression (9.13) implies that $P_J(q)$ fluctuates with J even in the thermodynamic limit. In other words, $P_J(q)$ is not self-averaging! (Recall that in chapter 1 we invoked the idea of spatial-ergodicity which is based on the assumption of self-averaging property of disordered systems). The fluctuation of $P_J(q)$ from sample to sample have been observed in numerical investigations of the SK model (Parga et al.1984). Does it imply that the relevant physical properties of the infinite-ranged SG are also non-self-averaging? Among the physically observable properties, let us consider magnetization M, and susceptibility χ. Let us define

$$M = [M]_J = \lim_{n\to 0} (1/n) \sum_\alpha M^\alpha$$

and similarly for χ. Now, following replica technique one can show that (Young et al.1984 , see also Elderfield 1984a)

$$[M^2]_J = \lim [1/\{n(n-1)\}] \sum M^\alpha M^\beta = [M]_J^2, \quad (9.14a)$$

because in the Parisi formalism any quantity that depend only on one replica is the same for all the replicas. On the other hand,

$$[\bar{\chi}^2]_J - [\bar{\chi}]_J^2 = (1/T^2)[\int q^2(x)\, dx-\{\int q(x)\, dx\}^2] \quad (9.14b)$$

and hence χ is not self-averaging when replica symmetry is broken (i.e., when $q(x)$ depends on x). At first sight the two results (9.14a) and (9.14b) may appear to contradict each other. But, it has been argued (Young et al.1984) that although M is self-averaging, χ can be non-self-averaging because of the "hopping" between the pure states when the external field is varied to probe the susceptibility. The deeper implications of such "hoppings" in the context of the fluctuation-dissipation theorem (FDT) will be examined later in this chapter. It should be emphasized that the physically <u>observable</u> properties of the SK model turn out to be self-averaging and the fact that some of the quantities, e.g., $q(x)$, which are not directly observable, are non-self-averaging, is irrelevant from an experimentalist's point of view.

It has been argued (Mezard et al.1985) that the sample-to-sample fluctuations in the SK model are consequence of the fact that the free energies of the pure states are independent random variables with an exponential distribution. A deeper relation between this result and the corresponding result for the random energy model (REM) will be explored in the next chapter.

Since $P_J(q)$ fluctuates, one would like to know the statistics of $P_J(q)$. The direct way would be to compute the quantities $[P_J(q_1,q_2,\ldots,P_J(q_k)]_{av}$. But for technical reasons it is much easier to calculate the moments $[Y_J^k(q)]_{av}$ where

$$Y_J(q) = \int dq'\ P_J(q')$$

with q_{max} as the maximum possible overlap between two pure states at the given temperature T (Mezard et al.1984a,b). For example,

$$[Y_J^2(q)]_{av}=\int dq_1 \int dq_2\ [P_J(q_1)P_J(q_2)]_{av}=(1/3)y(q)+(2/3)y(q)^2,$$

where $y(q) =[Y_J(q)]_{av}$. (See Parga et al.1984 for the corresponding numerical results).

Let us now investigate the statistical weights of the pure

states in some more detail. Given an overlap q, one can organize into "groups" all the pure states that have an overlap larger than q. Let us define the weight of the I-th group as

$$W_I = \Sigma P_\alpha \text{ such that } \Sigma W_I = 1$$

Suppose $f_J(W,y)$ is the number of the groups that have weights between W and W + dW for a given configuration of the exchange couplings {J}, i.e.,

$$f_J(W,y) = \Sigma\delta(W-W_I).$$

Then,

(i) $[f_J(W,y)]_{av} = [W^{y-2}(1-W)^{-y}/[\Gamma(y)\Gamma(1-y)]$

(ii) $[f_J(W_1,y)f_J(W_2,y)]_{av} = \delta(W_1-W_2)[f_J(W_1,y)]_{av}$

$$+ [\sum_{I\neq I'} \delta(W_I-W_1)\delta(W_{I'}-W_2)]_{av} \qquad (9.15a)$$

where

$$[\sum_{I\neq I'} \delta(W_I-W_1)\delta(W_{I'}-W_2)]_{av} =$$

$$[(1-y)\Theta(1-W_1-W_2)(W_1W_2)^{y-2}(1-W_1-W_2)^{1-2y}]/[\Gamma(y)\Gamma(y)\Gamma(2-2y)] \qquad (9.15b)$$

where Θ is the standard step function.

(iii) Indeed, most generally,

$$[\sum_{I_1,\ldots,I_k}' \prod_{p=1}^{k} \delta(W_{I_p}-W_p) = [\{(1-y)^{k-1}\delta(k)\}/\{\delta(y)^k\delta(k-ky)\}]$$

$$\Theta(1-\sum_{p=1}^{k} W_p)^{y-2}(1-\sum_{p=1}^{k} W_p)^{k-ky-1}$$

where Σ' means that all the indices $I_1, \ldots, I_k$ are different. Several physical consequences emerge from these expressions:
(a) there are infinitely large number of groups with $W = 0$,
(b) the total number of groups

$$\lim_{\epsilon \to 0} \int_{\epsilon}^{1} [f_J(W,y)]_{av} \, dW \to \infty$$

whereas

$$\lim_{\epsilon \to 0} \int_{\epsilon}^{1} W[f_J(W,y)]_{av} \, dW \to 1,$$

(c) so far as the relative weights are concerned, only one isolated group dominates as $y \to 1$ for a given distribution $\{J\}$. (for more details about the ultrametric structure of the pure states see Mezard and Virasoro 1985, Rammal et al.1986).

We have presented several interesting features of the SK model which are related to the existence of a large number of metastable states. The basins of attraction of these states is a quantitative measure of the "spread" compared to the corresponding "depth". Therefore, Parga and Parisi (1986) investigated the basins of attraction of the SK model in the following way: starting from a random intial configuration the state of the spins were updated by the T=0 algorithm of Metropolis, i.e., a spin is flipped if and only if the energy decreases. The algorithm was terminated when in a sweep through the system none of the spins is flipped. Then the basin of attraction was calculated as the ratio between the number of states in a given energy interval and the number of trials. The structure of the basins of attraction for the different system sizes are shown in Fig.9.3.

It is well known that the usual form of the FDT is

$$C(\omega) = 2\, \omega^{-1} \, \mathrm{Im}\, \chi(\omega) \qquad (9.16)$$

where the <u>full</u> response function is related to the <u>full</u>

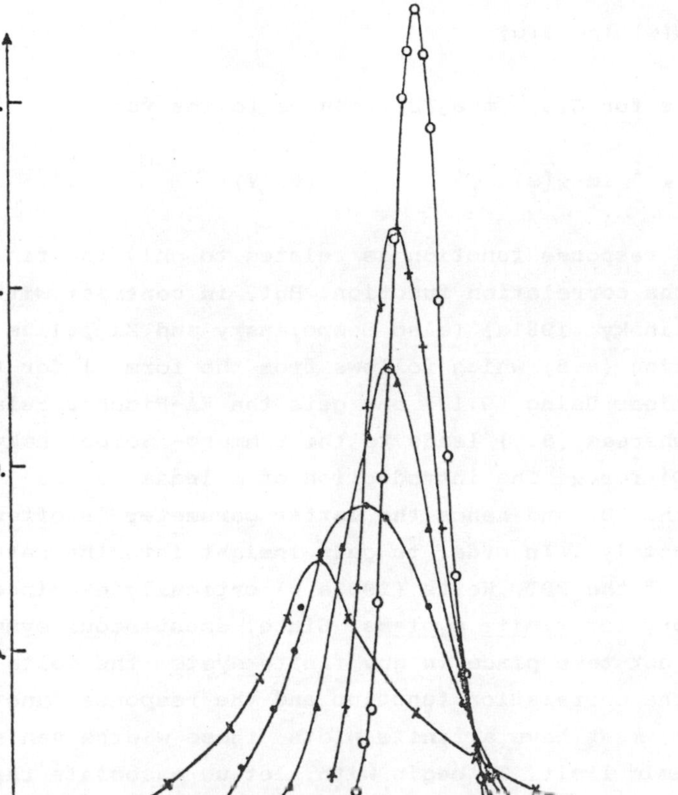

Fig. 9.3. The basin of attraction for different system sizes: x corresponds to N=32, the dots to N=64, the triangles to N=128, ⍭ to N=256 and the open circles to N=512 (after Parga and Parisi).

correlation function. In order to construct the MFT of SG we would like to have a nonzero time-persistent part q in the correlation function for $T < T_g$. This is achieved by writing

$$C(\omega) = \tilde{C}(\omega) + q\ \delta(\omega)$$

Using this form for $C(\omega)$, the FDT reduces to the form

$$\tilde{C}(\omega) = 2\ \omega^{-1}\ \mathrm{Im}\ \chi(\omega) \qquad (9.17)$$

i.e., the full response function is related to only the finite-time part of the correlation function. But, in contrast with (9.17), Sompolinsky (1981a) (also Sompolinsky and Zippelius 1981) used the relation (8.5) which follows from the form () for the response function. Using (9.17) one gets the EA-Fischer relation (3.9) for χ, whereas (8.5) leads to the Sommers- Sompolinsky relation (8.10) for χ. The introduction of Δ leads to the breakdown of the FDT and hence the latter parameter is often called the "anomaly". In order to gain insight into the nature of the breakdown of the FDT, Hertz (1983a,b) crtically examined the dynamical theory for finite systems. Since, spontaneous symmetry breaking does not take place in any finite system the delta functions in the correlation function and the response function (8.6) and (8.7) must have a finite width; these widths vanish in the thermodynamic limit. To begin with, let us associate the widths ϵ and ϵ' with C and χ, respectively. The FDT ensures $\epsilon = \epsilon'$. The spin correlations die out on time scales $t \gg \epsilon^{-1}$ and beyond this time scale the susceptibility is just that of free spins (i.e., paramagnetic susceptibility). Therefore, a truly infinite system is only in dynamic equilibrium.

Thus, the breakdown of the FDT in Sompolinsky's theory is closely related to the broken ergodicity. In such systems, the relative order in which the thermodynamic limit $N \to \infty$ and the long time limit $t \to \infty$ are taken is a subtle operation. For example,

$$q_{EA} = \lim_{t\to\infty} \lim_{N\to\infty} [\langle S_i(0)S_i(t)\rangle]_{av} \tag{9.18}$$

is identified with q(x=1) in Sompolinsky's theory. Note that if the $N \to \infty$ limit is taken before the $t \to \infty$ limit the system remains "confined" to the "valley" in which it was prepared because all the free energy barriers diverge in the thermodynamic limit. But, if the order of these two limits are interchanged, the barriers remain finite; therefore, hoppings over the barriers take place and the "memory" of the initial state is eventually lost. In the latter case

$$q = \lim_{N\to\infty} \lim_{t\to\infty} [\langle S_i(0)S_i(t)\rangle]_{av} = \int q(x)\, dx \tag{9.19}$$

Comparing (9.18) and (9.19) we see that x=0 corresponds to isolated "valleys" whereas larger x implies higher overlap between the "valleys". This observation is consistent with Parisi (1983) and Houghton et al. (1983a,b) interpretation of the parameter x, explained earlier in this chapter. DeDominicis and Young (1983a) argued that (9.19), together with the FDT

$$T \chi_{ij} = 1 - \int q(x)\, dx$$

imply that there is no violation of the FDT according to Parisi's theory. On the other hand, according to Sompolinsky's (1981a) theory

$$q = q(x=0) \tag{9.20}$$

and hence Dasgupta and Sompolinsky (1983) argued that x is a measure of the "distance" between the TAP solutions in the phase space, x = 0 corresponds to solutions which are farthest from each other and x = 1 corresponds to solutions which are highly correlated to each other (for an alternative investigation of the relation between the TAP solutions and Sompolinsky's dynamical solution see Sommers et al.1983). What is the reason for the disagreement between the q computed by DeDominicis and Young (1983a) and by Sompolinsky (1981a) (Houghton et al.1983a,b)? It has been claimed (DeDominicis and Young 1983b, also Houghton et al.1983a,b) that the initial condition plays a crucial role in the SK model. The result (9.19) holds only when the initial distribution of the states is "canonical", i.e.,

$$P_\alpha \sim \exp\{-\beta\mathcal{H}(\alpha)\}$$

But if P_α deviates initially slightly from canonical form, one gets (9.20) instead of (9.19). Since Sompolinsky's theory predicts (9.20) it is believed to describe a nonequlibrium situation. The latter interpretation is consistent with Hertz's (1983a,b) and Sommers' (1983) demonstration that Sompolinsky's solution is only in dynamic equilibrium. For alternative scenario see Horner (1984a).

Does broken ergodicity lead to breakdown of any of the standard thermodynamic relations, e.g., Maxwell's relations ? The susceptibility can be calculated from the specific heat measured in the laboratory using the Maxwell relation

$$(\partial^2 M/\partial T^2)_H = T^{-1}(\partial C/\partial H)_T$$

So far both laboratory experiments (Fogle et al.1981, Gullikson and Schultz 1982, Wenger and Mydosh 1982, Bowman and Levin 1982, Fogle et al.1982, Guy and Park 1983) as well as numerical calculations (Soukoulis et al.1982a,b, 1983) sugget that the Maxwell relations are satisfied quite well by the data for the field-cooled (FC) samples. Note that, as discussed in appendix B, the FC state is not a true equilibrium state.

Very recently, it has been realized that the results derived by the replica method can also be derived without using replicas (Mezard et al.1986). In the latter approach one used the distribution of the local random field for carrying out the configuration averaging. However, the simple-minded distribution

P(h) derived by Klein and Brout (1963) is insufficient because it does not take the hierarchical cluster structure of states of the SK model. Therefore, in the non-replica approach one has to calculate the appropriate distribution of local field. Suppose, the local field on the site 0 in the α-th state

$$h^{\alpha} = \Sigma K_i M_i^{\alpha},$$

is a random variable, and it depends on the state as well as on the sample. Suppose there are R pure states of the system. Therefore, the generalized distribution is formulated as follows: the probability that choosing a sample, the field h^{α} takes the values $h_1, \ldots, h^R$ is given by

$$P(h_1, \ldots, h^R) = \int \Pi [dK_i \surd (N/2\pi) \exp\{-NK_i^2/2\}] \Pi \delta(h^{\alpha} - \Sigma K_i M_i^{\alpha})$$

Introducing the integral representation of the δ function

$$P(h_1, \ldots, h^R) = \int dh \, (2\pi q_0)^{-1/2} \exp\{-h^2/(2q_0)\}$$

$$\Pi[\exp\{-(h^{\alpha}-R)^2/2(q_1-q_0)\}\{2\pi(q_1-q_0)\}^{-1/2}$$

where there is a common factor $h = (1/R)\Sigma h^{\alpha}$ depending only on the sample, whereas the second factor takes the fluctuations h^{α} around this average.

So far we have focussed our attention only on the SK model. We concluded that this model has a large number of metastable states (see Fig.10.5) and the pure states exhibit an ultrametric structure. Do the short-ranged Ising SG also display ultrametricity? The first evidence against ultrametricity in the ±J model comes from the Monte Carlo simulation by Ogielski and Morgenstern (1984b) (see Fig.24.2). The latter would imply that for the short-ranged SG models

$$P(q) = \delta(q - q_{EA})$$

and, absence of replica symmetry breaking (compare with one dimensional SG (Bray and Moore 1985)) and absence of ultrametricity. Direct search for ultrametricity in short-ranged SG by numerical method (Sourlas 1984) has not been quite successful. If the latter results are correct, one would expect replica symmetric solutions for the short-ranged SG for $d < 6$ and replica symmetry breaking would occur only for $d \geq 6$. In that case, all the properties of short- ranged SG would be self-averaging. In short, it seems that Nature refuses "to display the subtleties of the SK model" (Moore and Bray 1985)

CHAPTER 10

p-SPIN INTERACTION AND THE RANDOM ENERGY MODEL

The p-spin interaction defined by (1.21) is a generalization of the SK model, the latter corresponds to p=2. It can be shown that the p-spin interaction model is identical with the random energy model (REM) in the limit $p \to \infty$ (Derrida 1980a,b, 1981). The REM is defined as follows:

(i) the system has 2^N energy levels E_i ($i=1,\dots,2^N$),

(ii) the energy levels E_i are independent random variables,

(iii) these energy levels have a Gaussian distribution

$$P(E) = A \exp(- E^2/NJ^2) \qquad (10.1)$$

where

$$A = (N\pi J^2)^{-1/2}.$$

The latter model can be solved exactly without using replicas. Let n(E) be the number of levels in the interval between E and E + dE. Now, one can easily show that

$$\langle n(E)\rangle = 2^N \exp(- E^2/NJ^2)\, A\, dE \qquad (10.2)$$

Equation (10.2), together with the physical requirement that the energy is well-defined in the thermodynamic limit, and the requirement of smoothness of $\langle n(E)\rangle$, implies that there is an energy $E_0 = NJ(\ln 2)^{1/2}$ such that

for $|E| < E_0$, the average number of levels per unit interval is larger than one, and

for $|E| > E_0$, (almost) no level is available.

Using the relation $S = \langle \ln n(E)\rangle$ and $(dS/dE) = 1/T$ we get

$$\begin{aligned} F/N &= - T \ln 2 - (J^2/4T) \quad \text{if } T > T_1 \\ F/N &= - J (\ln 2)^{1/2} \quad \text{if } T < T_1 \end{aligned} \qquad (10.3)$$

where

$$T_1 = J/\{2\ (\ln 2)^{1/2}\}$$

In other words, the low-temperature phase is frozen, because no level of lower energy is avilable.

In the presence of dominant ferromagnetic interaction J_0/N, in vanishing external field,

$$P(E) \propto \exp[-\ \{E + (M^2 J_0)/(2N)\}^2/NJ^2]$$

Now, four phases appear on the phase diagram:
Paramagnetic (P) phase where

$$M = 0$$

$$F/N = -\ T \ln 2 - (J^2/4T)$$

Ferromagnetic (F) phase where

$$M = \tanh\ (J_0 M/T)$$

$$F/N = -\ T \ln 2 - (J^2/4T) + (J_0 M^2/2) + T \ln\ (1 - M^2)/2$$

Frozen phase I where

$$M = 0$$

$$F/N = -\ J\ (\ln 2)^{1/2}$$

and Frozen phase II where

$$M = M(T_1(J_0))$$

$$F/N = F(T_1(J_0))/N$$

where $T_1(J_0)$ is the temperature corresponding to transition from the ferromagnetic to the frozen phase II.

There are sample-to-sample fluctuations of the SG order parameter in the REM even in the thermodynamic limit (Derrida and Toulouse 1985). The latter phenomenon is analogous to that in the SK model (see chapter 9). Since the energy levels in the REM are independent random variables, each energy level creates its own free energy valley. The thermodynamic weight W_α of each level E_α is given by

$$W_\alpha = \exp(-\beta E_\alpha)/\Sigma\exp(-\beta E_\gamma).$$

The joint probability of finding two states of weights W_1 and W_2 in the same sample is given by

$$P(W_1,W_2) = P_1\, P(W_1)\, \delta(W_1-W_2)$$

$$+\ (1-y)[\{(W_1W_2)^{y-1}(1-W_1-W_2)^{1-2y}\}/\{\Gamma(y)\Gamma(y)\Gamma(2-2y)\}]\theta(1-W_1-W_2) \tag{10.4}$$

where

$$y = \langle\Sigma\ W_\alpha^2\rangle = 1 - (T/T_1)$$

and

$$P(W_1) = [\{W_1^{-1+y}(1-W_1)^{-y}\}/\{\Gamma(y)\Gamma(1-y)\}]$$

is the probability of finding a state of weight W_1. Note that the functional form (10.4) is identical with the corresponding form (9.15) for the SK model. This agreement is not an accidental coincidence; the pure states in the SK model can be decomposed into a sum of independent random variables (Mezard et al.1985, DeDominicis and Hilhorst 1985).

The REM can be generalized to take correlations between the levels into account. Suppose the 2^N configurations are grouped in a hierarchical structure so that at the i-th level of the hierarchy, there are $[2/(\alpha_i,...,\alpha_n)]^N$ groups of $(\alpha_i,\alpha_{i+1},...,\alpha_n)^N$ configurations each. Thus, there are $(2/\alpha_1)^N$ group each containing α_1^N configurations each at the level i=2 and so on. For example, a special case of N = 16, n=3 with $\alpha_1^N = 2$, $\alpha_2^N = 2$ and α_3^N

= 4 is shown in Fig.10.1 where each of the boxes corresponds to a group and the number inside it corresponds to the number of configurations in that group. Notice that this hierarchical construction is "very reminiscent of the ultrametric structure" (Mezard et al.1984a,b). (However, there are several differences between the SK and the GREM models, e.g., the forms of the "AT-line", q(x), etc.(Derrida and Gardner 1986b)). Now, defining $\epsilon_i^{(\nu)}$ to be the contribution of the i-th level of the hierarchy to the energy E_ν of the ν-th configuration, E_ν reduces to the sum of n random numbers, i.e,

$$E_\nu = \epsilon_1^{(\nu)} + \epsilon_2^{(\nu)} + \ldots + \epsilon_n^{(\nu)}$$

where the random numbers $\epsilon_i^{(\nu)}$ are assumed to be Gaussian-distributed

$$P_i(\epsilon_i) = (\pi N J^2 a_i)^{-1/2} \exp\{-\epsilon_i^2/(NJ^2 a_i)\}$$

It should be noted that the width of the latter distribution depends on the level of the hierarchy through a_i. Thus, the model, called generalized REM (GREM), is defined by the two sequences α_i and a_i. Moreover, the condition of fixed number 2^N of the configurations and the condition of the normalization of the distribution impose the following constraints on α's and a's:

$$\Sigma \ln \alpha_i = \ln 2 \qquad (10.5)$$

and

$$\Sigma a_i = 1 \qquad (10.6)$$

For n = 1, (10.5) and (10.6) impose $\alpha_1 = 2$ and $\alpha_1 = 1$ and GREM reduces to REM. For n = 2, Derrida (1985) showed that in the case

$$(a_1/\ln \alpha_1) > (a_2/\ln \alpha_2)$$

the model exhibits two phase transitions at

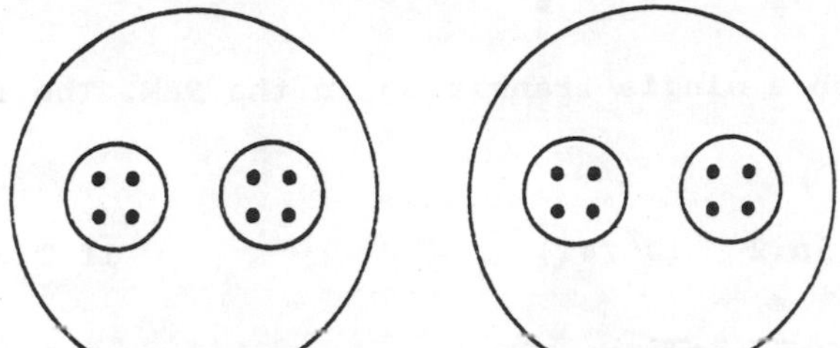

Fig. 10.1. The grouping of the configurations in the GREM (see the text) (after Derrida 1985).

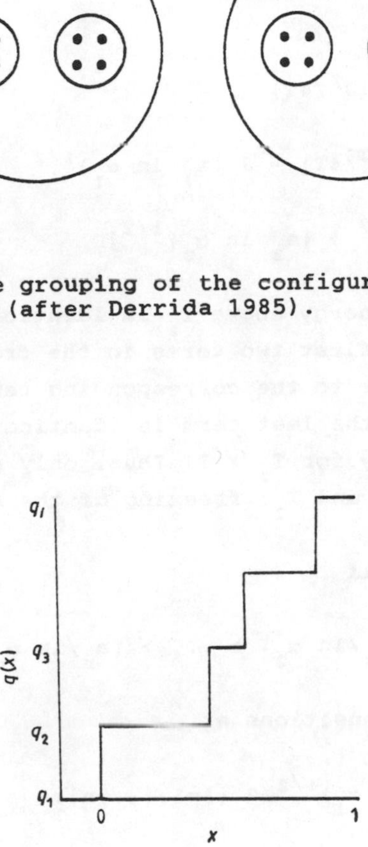

Fig. 10.2. q(x) in the GREM (see the text) (after Derrida and Gardner (1986).

$$T_1 = (J/2)(a_1/\ln \alpha_1)^{1/2}$$

and

$$T_2 = (J/2)(a_2/\ln \alpha_2)^{1/2}$$

in contrast with a single transition in the REM. The free energy is given by

$$F/N = -T \ln 2 - (J^2/4T) \qquad \text{if } T > T_1$$

$$F/N = -T \ln \alpha_2 - (a_2 J^2/4T) - J(a_1 \ln \alpha_1)^{1/2} \qquad \text{if } T_1 > T > T_2$$

$$F/N = -J[(a_1 \ln \alpha_1)^{1/2} + (a_2 \ln \alpha_2)^{1/2}] \qquad \text{if } T_2 > T$$

Notice that the free energy above T_1 is identical with that in the REM above T_1. The first two terms in the free energy at $T_1 > T > T_2$ are similar to the corresponding terms in the free energy for $T > T_1$ and the last term is identical with the first term in the free energy for $T_2 > T$. Thus, only partial freezing takes place between T_1 and T_2; freezing of the system is complete at T_2.

More generally, if

$$(a_1/\ln \alpha_1) > (a_2/\ln \alpha_2) > \ldots > (a_n/\ln \alpha_n)$$

the GREM exhibits n transitions at

$$T_i = J(a_i/\ln \alpha_i)^{1/2}/2 \quad (i=1,\ldots,n).$$

Therefore, one expects that in the limit $n \to \infty$, there will be a densely packed region of transition temperatures. The free energy for the GREM for general n is given by (Derrida and Gardner 1986a)

$$F/N = -T \ln 2 - (J^2/4T) \qquad \text{if } T > T_1$$

$$F/N = \sum_{j=i+1}^{n} [- T \ln\alpha_j - \{(a_j J^2)/(4T)\}] + J\sum_{j=1}^{i}(a_j \ln\alpha_j)^{1/2} \quad \text{if } T_i > T > T_{i+1}$$

and

$$F/N = - J \sum_{j=1}^{n} (a_j \ln\alpha_j)^{1/2} \quad \text{if } T_n > T$$

Thus, for $T_i > T > T_{i+1}$ the system is frozen in a few groups at level i+1 but in each of these groups, it behaves as system in its high temperature phase. As a consequence of this freezing scenario, the order parameter q(x) appears as shown in Fig.10.2.

All the results summarized above have been derived without using the replica methos. However, the same results have also been derived by the replica method (Gross and Mezard 1984, Gardner 1985). Because of this agreement one feels more confident about the applicability of the replica trick. However, we shall not discuss the replica approach to the REM in this book. The effect of an anisotropic interaction on a generalization of the p-spin interaction model will be mentioned in chapter 15. Some dynamical featues of the REM will be summarized in chapter 19.

CHAPTER 11

SEPARABLE SPIN GLASS MODELS

11.1. The Mattis Model:

The simplest among all the separable SG models is the Mattis model where the randomness is incorporated into J_{ij} through the relation (1.9). The classical Mattis model is equivalent to a model for pure system because of the reason explained in chapter 1. However, the properties of the quantum Heisenberg Mattis model is nontrivial because the transformation $S_i \rightarrow \xi_i S_i$ is not necessarily canonical (Sherrington 1979, Johnston and Sherrington 1982, Nishimori 1981). By the latter statement we mean that a transformation

$$\tau_i = \xi_i \; S_i$$

would change the hamiltonian (1.1) into the hamiltonian

$$\mathcal{H} = - \Sigma \; J_{ij} \; \tau_i \; \tau_j$$

corresponding to a nonrandom system but the commutation relation for the τ's would be

$$[\tau_i \; , \; \tau_j] = i \; \hbar \; \xi_i \; \tau_i$$

which is canonical only in the classical limit $\hbar \rightarrow 0$. The dynamical aspects of the Mattis model will be summarized in chapter 18.

11.2. The van Hemmen Model:

The van Hemmen model (1.11) is the simplest possible nontrivial generalization of the classical Mattis model. Let us define (van Hemmen 1982)

$$M = \lim (1/N) \Sigma <S_i>$$

$$q_1 = \lim (1/N) \Sigma \xi_i <S_i>$$

and

$$q_2 = \lim (1/N) \Sigma \eta_i <S_i>$$

(Compare and contrast the definitions of q_1 and q_2 with that of q_{EA} in (3.1)). The free energy is minimum for $q_1 = q_2 = q$ and the self-consistent solutions of

$$M = [\tanh \{K_0 M + Kq (\xi + \eta)\}]_{av}$$

$$q = [\tanh \{K_0 M + Kq (\xi + \eta)\} (\xi + \eta)/2]_{av}$$

(with $K_0=J_0/k_BT$ and $K=J/k_BT$) for apriori assumed distributions for ξ's and η's yield the phase diagram shown in Fig.11.1 (for the detailed properties of this model see van Hemmen 1983 and van Hemmen et al. 1983). The four phases are characterized by

(a) paramagnetic: $M = 0$ and $q = 0$,

(b) SG: $M = 0$ and $q \neq 0$,

(c) ferromagnetic: $M \neq 0$ and $q = 0$,

(d) mixed (II): $M \neq 0$ and $q \neq 0$.

One of the remarkable features of the SK model is that the number of the metastable states increases monotonically with the number of spins, N, as discussed in chapter 6. But the number of solutions of the mean-field equations (analogue of the TAP equations)

$$M_i = \tanh [K_0 M + K (\xi_i q_2 + \eta_i q_1),$$

in the van Hemmen model, is independent of N (Choy and Sherrington 1984). But the latter feature does not necessarily imply any inadequacy of the van Hemmen model to represent a SG (van Hemmen 1986a), because, as we shall see in chapter 24, similar features of the ground states of the short-ranged SG has also been observed. However, the Glauber dynamics of the van

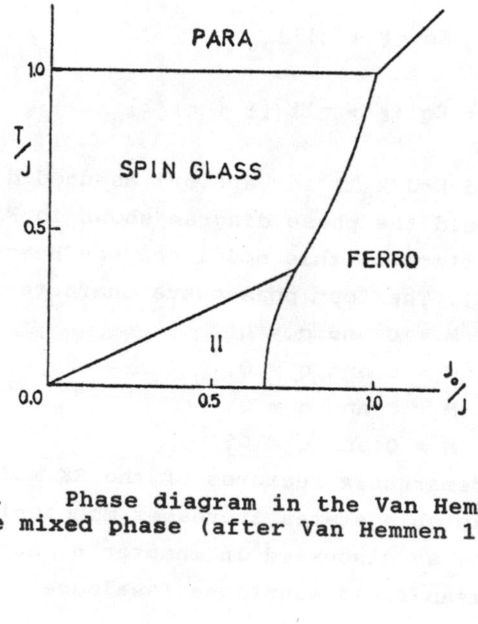

Fig. 11.1. Phase diagram in the Van Hemmen model. II denotes the mixed phase (after Van Hemmen 1982).

Hemmen model yields exponential relaxation at all T except near the transition temperature (Choy and Sherrington 1984). Therefore, the van Hemmen model is inadequate for the description of the dynamics of true SG although, surprisingly, it accounts for several static properties of real SG materials.

Treatment of the short-ranged version of the van Hemmen model in the Kikuchi approximation (Indekeu et al. 1984) is an improvement over the simple mean-field approximation of van Hemmen. Further generalization of the van Hemmen model (van Hemmen and Morgenstern 1984), which incorporates randomness also in the short-ranged part of (1.11), has been solved by the numerical transfer matrix method (the latter method will be explained in chapter 24) and yields interesting phase diagram (see van Hemmen and Morgenstern 1984).

11.3. The Generalized van Hemmen model (Provost-Vallee model):

Provost and Vallee assume the elements J_{ij} of the random exchange matrix to be given by $J_{ij} = \xi_i \, J \, \xi_j$ where J is a p X p symmetric matrix and ξ_i are independent identically distributed p-vectors with zero mean. By varying the dimension p one can interpolate between the van Hemmen and SK models. For Gaussian-distributed components of ξ , and p going to infinity like N, one gets not only the SK model (for judicious choice of J) but a whole family of models without the drawbacks of the Mattis model (Benamire et al. 1985). The salient feature of all such models is the extensivity of the logarithm of the characteristic function of the couplings

$$[\exp\{(1/2)\Sigma \; u_{ij} \; J_{ij}\}]_{av}$$

(where u_{ij} are the elements of a symmetric matrix U). The Provost-Vallee model (1.12) seems to be useful not only as a separable model for SG, but also as simplified models of neural networks, as we shall see later in chapter 27.

CHAPTER 12

THE SPHERICAL MODEL OF SG

The condition

$$\Sigma S_i^2 = N, \qquad (12.1)$$

where N is the number of spins in the system, is called the spherical constraint. Therefore, the Hamiltonian for the spherical model can be written as

$$\mathcal{H} = -(1/2)\, J \sum_i \sum_{j\neq i} S_i S_j + \mu(\Sigma S_i^2 - N) \qquad (12.2)$$

where the spherical constraint has been incorporated through the Lagrange multiplier μ (see Mattis 1985 for an introduction to the spherical model). μ is determined from the requirement that the condition (12.1) be satisfied on the average, i.e., $<\Sigma S_i^2>=N$. Thus, the free energy for the spherical model is evaluated on the surface of an N-sphere. Let us evaluate the free energy without using the replica trick. Suppose, $<\lambda|i>$ is the orthonormal eigenvector of J_{ij} corresponding to the eigenvalue J_λ, i.e.,

$$\Sigma J_{ij} <\lambda|j> = 2\, E_\lambda <\lambda|i> \qquad (12.3)$$

Thus, in the new basis

$$S_i = \sum_\lambda X_\lambda <\lambda|i> \qquad (12.4)$$

so that the spherical constraint (12.1) reduces to

$$<\sum_\lambda X_\lambda^2> = N \qquad (12.5)$$

Let us define the local density of states

$$\rho_i(E) = \Sigma <\lambda|i>^2 \, \delta(E-J_\lambda)$$

and the global density of states

$$\rho(E) = \Sigma \, \rho_i(E) = \Sigma \, \delta(E-J_\alpha)$$

The configuration-averaged free energy is given by (see Mattis 1985)

$$F/N = -(1/2) \, k_B T \ln 2\pi - [\mu]_{av}$$

$$+ (1/2) \, (k_B T/N) \int dJ \, [\rho(E) \ln(2/k_B T)(\mu + E)]_{av} \quad (12.6)$$

Using the well known semicircular law for the eigenvalue spectrum of random matrices we get

$$\rho(E) = \{(2N)/(\pi J)\}[1 - (E/J)^2]^{1/2} \qquad (12.7)$$

in the case of the SK model with vanishing mean of J. Substituting (12.7) into (12.6) we get the free energy for the spherical SG in the MFA. Interestingly, identical results are obtained for the spherical model in the replica-symmetric formalism, therby establishing the absence of replica symmetry breaking in the latter model (Kosterlitz et al.1976). Moreover, the same expression for the free energy is recovered from the m-vector SG in the limit $m \to \infty$ (deAlmeida et al.1978), thereby establishing the fact that $m \to \infty$ corresponds to the spherical model for SG (the equivalence of the $m \to \infty$ vector model with the spherical model for ferromagnet was established long ago). (See Nieuwenhuizen 1985 for a generalization of the spherical model where the long-ranged part of the exchange interaction is identical with that in the van Hemmen model and the short-ranged interaction is nearest-neighbour ferromagnetic exchange interaction).

Consider the model Hamiltonian (Shukla and Singh 1981)

$$\mathcal{H} = -(1/2) \Sigma J_{ij} S_i S_j + \mu\Sigma S_i^2 + \Sigma(P_i^2/2I) \qquad (12.8)$$

where S_i and P_i are conjugate variables satisfying the commutation relations

$$[S_i, P_{i'}] = i\hbar\delta_{ii'},$$

and the second term in (12.8) imposes the sperical constraint (12.1). The most striking difference between the classical spherical SG (12.2) and the quantum spherical SG (12.8) is that the entropy becomes negative at sufficiently low temperatures in the former but not in the latter. Note that the negative entropy is obtained at low temperatures also in the case of nonrandom spherical model. The dynamical aspects of the spherical model will be discussed in chapter 19.

CHAPTER 13

MFT OF VECTOR SG; MIXED PHASE

The Ising model (m = 1) and the sperical model (m = ∞) represent two extreme situations. Intuitively, the Heisenberg model with finite m seems to be the most realistic and therefore deserves special attention. For the economy of words, we shall call the SG models with finite m as "vector SG"

$$\mathcal{H} = - \Sigma J_{ij} \; \Sigma \; S_{i\mu} \; S_{j\mu} - H \; \Sigma \; S_{i1}$$

where μ ($\mu = 1,2,..,m$) denotes spin components and the magnetic field is applied along the direction 1. Moreover, the normalization of the spin magnitude implies

$$\Sigma S^2_{i\mu} = m.$$

Most of our attention will be focussed on the Gaussian model where the exchange interaction J_{ij} are Gaussian-distributed independent random variables with mean J_o and variance J as in the case of the SK model. We have shown in the preceeding chapters that replica symmetry breaking takes place in Ising SG but not in the spherical model. What happens in the case of vector SG? We shall see in this chapter that replica symmetry breking does, indeed, take place in vector SG. Moreover, the latter model exhibits new types of spin ordering (at least for d → ∞) which is impossible in Ising SG.

Let us generalize the definition of the SG order parameters as appropriate for vector SG:

$$q_\mu = [\langle S_\mu \rangle^2]_{av} = q_T + (q_L - q_T)\delta_{\mu,1} \quad (13.1)$$

where q_L is the longitudinal SG order parameter and q_T is the transverse order parameter. A parameter x is defined via

$$x = \{[\langle S_1^2\rangle]_{av} - 1\}/(m - 1) \quad (13.2)$$

First, let us consider the case when $J_0 = 0$, so that no ferromagnetic phase can occur. In the replica-symmetric formalism, two transition lines appear on the H-T phase diagram (Toulouse and Gabay 1981, Gabay and Toulouse 1981). At sufficiently high temperatues, $q_L \neq 0$ because $H \neq 0$ whereas q_T vanishes. At the first transition temperature T_{GT}, the transverse SG order parameter becomes nonzero. The phase remains replica symmetric for $T_{AT} < T < T_{GT}$, where $T_{AT}(H)$ is the so-called AT line. Note that $T_{GT}(0) = T_{AT}(0) = T_g$, the transition temperature in zero field. For small H, we have

$$T_g - T_{GT}(H) \propto [(m^2+4m+2)/\{4(m+2)^2\}]H^2 \quad (13.3)$$

and

$$T_g - T_{AT}(H) \propto [\{(m+1)(m+2)\}/8]^{1/3}\ H^{2/3} \quad (13.4)$$

In the Ising limit (m=1) the GT line must disappear because there are no transverse degrees of freedom, and the AT line reduces to the AT line for the SK model. Assuming that replica symmetry does not break immediately below T_{GT} (this assumption will be shown to be incorrect later in this chapter) one gets (Toulouse and Gabay 1981) the longitudinal susceptibility

$$\chi_L = \beta\ \{1 - q_L + (m - 1)x\}$$

and the corresponding transverse susceptibility

$$\chi_T = \beta\ \{1 - q_T - x\}$$

for $T_{AT} < T < T_{GT}$.

Moore and Bray (1982) showed that the critical behaviour of the m-component SG model in a magnetic field is identical, to all orders in $\epsilon = 6 - d$, with that of an (m-1)-component model in zero field. It is well established that the replica symmetry breaks down in the latter model (Almeida et al.1978). The eigenvalue corresponding to the replicon mode is given by

$$\lambda = -8q^2/(m+2)^2 + \ldots \quad \text{(for T near } T_g\text{)} \qquad (13.5)$$

which is negative for all finite m. (Note that this eigenvalue is nonnegative for $m = \infty$, in agreement with Kosterlitz et al. 1976). Therefore, Moore and Bray suggested that replica symmetry is broken simultaneously with the transverse SG ordering. Cragg et al. (1982) carried out the stability analysis of the GT replica-symmetric solution in the presence of an external field; the latter analysis is similar to that of AT for the infinite-ranged Ising SG. Taking the fluctuations as $x^{\alpha} = x + \epsilon^{\alpha}$ and $q_{\mu}^{(\alpha\beta)} = q_{\mu} + \eta_{\mu}^{(\alpha\beta)}$ and keeping terms upto the second order, as before, we get increment in the free energy to be

$$(1/2)\beta^2 \, [\{\epsilon^{\alpha}\}\{\eta^{(\alpha\beta)}\}] \begin{bmatrix} A & B \\ B^T & C \end{bmatrix} \begin{bmatrix} \{\epsilon^{\alpha}\} \\ \{\eta^{(\alpha\beta)}\} \end{bmatrix}$$

where

$$A^{\alpha\beta} = (1/2)m(m-1)\delta_{\alpha\beta} - (m^2\beta^2/4)\{\lim\langle(S_1^{\alpha})^2(S_1^{\beta})^2\rangle[1+(m-1)x]^2\}$$

$$B_{\mu}^{\alpha(\beta\gamma)} = -(1/2)m\beta^2\{\lim\langle(S_1^{\alpha})^2 S_{\mu}^{\beta} S_{\mu}^{\gamma}\rangle_n - [1+(m-1)x]q_{\mu}\}$$

$$C_{\mu\nu}^{(\alpha\beta)(\gamma\delta)} = \delta_{(\alpha\beta)(\gamma\delta)}\delta_{\mu\nu} - \beta^2[\lim\langle S_{\mu}^{\alpha} S_{\mu}^{\beta} S_{\nu}^{\gamma} S_{\nu}^{\delta}\rangle - q_{\mu}q_{\nu}]$$

The matrix shown above is then diagonalized. For all $T > T_{GT}$, all the eigenvalues are positive. But one eigenvalue vanishes at T_{GT} and becomes negative at $T < T_{GT}$ thereby indicating instability of the replica-symmetric solution for $T < T_{GT}$. This analysis establishes that the perpendicular SG freezing in vector SG and replica symmetry breaking take place simultaneously. The eigenvalue considered by GT for the computation of the AT line is different from the eigenvalue that leads to the replica symmetry breaking at T_{GT} (see Cragg et al. 1982 for the details). Thus, unlike in the case of the Ising spins, the line (13.4) does not correspond to any second order phase transition.

The permutation symmetry among the replicas in the MFT of vector SG can be broken (Elderfield and Sherrington 1982, Gabay et al.1982) following a prescription similar to that of Parisi for the Ising SG. Because of replica symmetry breaking the order parameters $q_\mu \rightarrow q_\mu(r)$ where $0 \leq r \leq 1$. The modification of the form of the expressions of χ_L and χ_T and the corresponding physical interpretations are very similar to those discussed in Parisi's theory for the SK model:

$$T\chi_L = 1 + T^2\{(m-1)x - \int dr\, q_L(r)\} \qquad (13.6)$$

and

$$T\chi_T = 1 - T^2\{x + \int dr\, q_T(r)\} \qquad (13.7)$$

Next let us consider the MFT of vector SG when $J_0 \neq 0$ but $H = 0$. As the temperature is reduced starting from a high value a transtion from the paramagnetic phase to a ferromagnetic phase ($M_L \neq 0$, $q_T = 0$) takes place. On further cooling, a transtion from the ferromagnetic phase to a mixed phase takes place; the mixed phase is characterized by $M_L \neq 0$ and $q_T \neq 0$. In other words, there is co-existence of long-ranged ferromagnetic order in one direction (say, along 1-direction) and SG order in the perpendicular plane. (The effects of the anisotropic interactions on the mean-field phase diagram of the vector SG will be discussed in chapter 15). (See Elderfield and Sherrington 1984 for P(q)).

Now wė shall generalize the phenomenological theory of Ising SG reviewed in chapter 3 to vector SG. The Landau free energy functional introduced by Barnes et al.(1984) is

$$F_{BMB} = F_0 + aM_L^2 + bM_L^4 - cq_L^2 - dq_L^3 - 2cq_T^2$$
$$- 2dq_T^3 + 2fq_L^2q_T^2 + eq_LM_L^2 - H_LM_L \qquad (13.8)$$

where the term $2fq_L^2q_T^2$ introduces coupling between the longitudinal and transverse order parameters. The coupling $eq_LM_L^2$ is the analogue of the coupling between M and q introduced in the case of Ising SG (Suzuki 1977). However, note that no coupling of the type $q_T M_L^2$ has been included in (13.8). Minimizing with respect to q_T we get,

$$q_T = (2c/3d)$$

for small q, and, as in the case of Ising SG, the transition temperature corresponding to transverse SG freezing is given by

$$c(T_{GT}) = 0$$

Suzuki (1985) included the transverse magnetization M_T in the GL free energy functional to write

$$\begin{aligned} F_S = F_L(M_L,q_L) &- M_L H_L + aM_T^2 + bM_T^4 + \ldots - M_T H_T \\ &+ cq_T^2 + dq_T^3 + \ldots + eq_T M_T^2 \\ &+ M_T^2 g_1(M_L,q_L) + q_T^2\, g_2(M_L,q_L) \qquad (13.9) \end{aligned}$$

where $F_L(M_L,q_L)$ is the free energy of the longitudinal spin component and the last two terms in (13.9) denote the coupling between the longitudinal and transverse spin components. In this case the transverse SG transition temperature is given by

$$c(T_{GT}) + g_2(M_L,q_L) = 0 \qquad (13.10)$$

Some interesting predictions of this GL phenomenological theory will be presented in chapter 16.

The experimental situation remains ambiguous. On the one hand, some experimental data were shown to be consistent with the mixed phase. On the other hand, other experimental groups argue that the mixed phase is an artifact of the MFA for the vector SG, real SG exhibit re-entrant SG phase at low temperatures which was predicted originally by Sherrington and Kirkpatrick in the MFT for the Ising model. Van Enter and van Hemmen (1984b) proposed an alternative non-mean-field scenario where the competition between ferromagnetic nearest-neighbour exchange interaction and long-ranged antiferromagnetic exchange interaction leads to spin canting at a temperature lower than the ferromagnetic transition temperature.

All the models considered so far in this chapter are bond-randon models (also see Cieplak and Cieplak 1985). Dunlop and Sherrington (1985) carried out computer simulation of the following site-dilution model

$$H = - \Sigma J_{ij} S_i \cdot S_j c_i c_j \qquad (13.11)$$

where c_i is the site-occupation probability, as before, and

$$J_{ij} = \begin{cases} J_1 > 0 \text{ for i and j nearest-neighbours} \\ J_2 < 0 \text{ for i and j next-nearest-neighbours} \\ 0 \text{ otherwise} \end{cases}$$

The data suggested the possibility of a spin canting at sufficiently low temperature provided the vacancy concentration is low enough. Note that the Dunlop-Sherrington model (13.11) is short-ranged.

I would like to mention that the mixed phase was anticipated before GT. Villain (1979) called it semi-SG, Mookerjee (1980) called it canted random ferromagnet and Medvedev (1979b) called it asperomagnetic phase.

CHAPTER 14

OTHER LONG-RANGED MODELS

We have been exploring the possibility of a finite temperature phase transition in the SG models. We shall see in chapter 24 that the SG transition is possible in short-ranged SG models at $T \neq 0$ only in $d \geq 2$. Therefore, at first sight the study of one-dimensional SG models might seem an irrelevant exercise. However, the long-ranged interaction leads to highly nontrivial behavior of SG models! Kotliar, Anderson and Stein (KAS) (1983) introduced the one-dimensional model (1.17). The range of the interaction is determined by the parameter σ. The distribution of ϵ_{ij} is given by

$$P(\epsilon_{ij}) = (2\pi J^2)^{-1/2} \exp(-\epsilon_{ij}^2/2J^2).$$

The high temperature expansion of the free energy is given by

$$-\beta F = [\ln \mathrm{Tr} \prod_{i<j} \exp\{\beta\epsilon_{ij}S_iS_j/|a(i-j)|^\sigma\}]_{av}$$

$$= \sum_{i<j} \ln\cosh\{\beta\epsilon_{ij}/|a(i-j)|^\sigma\} + N\ln 2$$

$$+ \ln 2^{-N}\,\mathrm{Tr}\prod_{i<j} S_iS_j \tanh\{\beta\epsilon_{ij}/|i-j|^\sigma\}$$

$$\simeq N\,[\ln 2 + \{(\beta J)^2/2\}\Sigma\{1/|na|^{2\sigma}\}] + (N/4)\int\{d(ka)/2\pi\}\ln\{1-g(k)\} \tag{14.1}$$

where $g(k)$ is the Fourier transform of

$$g(j) = [\tanh^2(\epsilon_{ij}\beta J/|aj|^\sigma)]_{av}.$$

The free energy (14.1) converges for $\sigma > 1/2$. According to the definition of the range of exchange interaction mentioned in chapter 1, the KAS model becomes short-ranged for $\sigma > 1$.

Therefore, the most interesting range of the magnitudes of the parameter σ is $1/2 < \sigma < 1$. The Landau-Ginzburg-Wilson effective replica Hamiltonian corresponding to the d-dimensional KAS model is given by

$$\mathcal{H}(q_{\alpha\beta}) = (1/4)\, \Sigma\, [k^{2\sigma-d}\, \mathrm{Tr}\, q_{\alpha\beta}(k)q_{\beta\alpha}(-k)]$$

$$-\{w/(Na^{d})^{1/2}\}\, \Sigma\;\; q_{\alpha\beta}(k_1)q_{\beta\gamma}(k_2)q_{\gamma\alpha}(k_3)\; \delta(k_1+k_2+k_3) \qquad (14.2)$$

RG treatment (see chapter 17 for an introduction) of the Hamiltonian (14.2) leads to Gaussian fixed point for $d/2 < \sigma < 2d/3$. Thus, the one-dimensional model (1.17) exhibits mean-field behavior in the parameter range $1/2 < \sigma < 2/3$. On the other hand, non-mean-field exponents are obtained for the parameter range $2/3 < \sigma < 1$. The phase transition disappears for $\sigma \geq 1$. The latter result should be contrasted with the corresponding result for the one-dimensional long-ranged ferromagnets where the phase transition is destroyed by a higher value of σ (see van Enter and van Hemmen 1983, also see van Enter and Froehlich 1985 for the two-dimensional XY model). Thus, the merit of the KAS model lies in the fact that the inverse of the parameter σ, effectively, plays the role of dimensionality d, and therefore, many properties of the more realistic models are shared also by the one-dimensional KAS model for an appropriate choice of the parameter σ. Keeping d constant and varying σ one can verify (Chang and Sak 1984) that the critical exponents change discontinuously from the MF values to the non-MF values at a certain value of σ. The probability distribution of the overlap of the pure states for the model (1.17), calculated by the ϵ-expansion ($\epsilon = 1-\sigma$), is given by (Moore 1986)

$$P(q) = [\delta(q - q_{EA}) + \delta(q + q_{EA})]/2 \qquad (14.3)$$

and, therefore, the 'geometry' of the ground state(s) of the model near $\sigma = 1$ is similar to that of short-ranged models (i.e., replica symmetric) rather than that of the SK model.

CHAPTER 15

ANISOTROPIC EXCHANGE INTERACTIONS AND SG

Anisotropy of a SG Hamiltonian can arise in two different ways: (a) suppose the exchange Hamiltonian is given by

$$\mathcal{H} = -\Sigma(J_{xx}S_i^xS_j^x + J_{yy}S_i^yS_j^y + J_{zz}S_i^zS_j^z \quad (15.1)$$

where J_{xx}, J_{yy} and J_{zz} are the diagonal elements of the exchange matrix J (the off-diagonal elements are zero). If $J_{xx} = J_{yy} = 0$ and $J_{zz} \neq 0$ the model is called the Ising model, whereas $J_{xx} = J_{yy} \neq 0$ and $J_{zz} = 0$ corresponds to the isotropic XY model, and finally, $J_{xx} = J_{yy} = J_{zz}$ in the isotropic Heisenberg model. The model (15.1) would be anisotropic if the strength of at least one of the exchange constants J_{xx}, J_{yy}, J_{zz} is different from that of the other two. However, we shall be mainly interested in the Hamiltonians of the type (b) where the exchange Haniltonian is given by

$$\mathcal{H} = - J \Sigma \vec{S}_i \vec{S}_j + \mathcal{H}_a \quad (15.2)$$

where $\mathcal{H}_a$ is the anisotropic contribution. In the absence of $\mathcal{H}_a$, the Hamiltonian (15.2) would be isotropic in the spin space.

One of the most common types of exchange anisotropy in nonrandom magnetic systems is the so-called single-ion anisotropy

$$\mathcal{H}_1 = - D \Sigma (S_i^z)^2 \quad (15.3)$$

where D is a measure of the strength of the anisotropy. In amorphous magnetic systems the Hamiltonian (15.3) is generalized as

$$\mathcal{H}_2 = - D \Sigma (\hat{n}_i \cdot \vec{S}_i)^2 \quad (15.4)$$

where $\hat{n}_i$ is the unit vector along the local easy axis at the i-th site. Since the direction of n is assumed to vary randomly from site to site and since (15.4) is invariant under the transformation $n_i \to -n_i$, H_2 is also called the random uniaxial anisotropy (RUA) (see Cochrane et al.1978, Coey 1978, von Molnar et al.1982 for reviews). The randomness in the Hamiltonian (15.4) lies the randomness in the direction of n_i, the magnitude of D is assumed to be nonrandom. (See Moorjani and Coey 1984 for the physical origin of the RUA).

Another anisotropic term used in some models is (Taggert et al.1974)

$$H_3 = \Sigma D_i (S_i^z)^2 \qquad (15.5)$$

where it is the magnitude of D, rather than the direction of n_i, that is assumed to be random.

In canonical SG systems the dominant contribution to the anisotropy comes from the Dzyaloshinskii-Moriya (DM) interaction

$$H_{DM} = \Sigma \vec{D}_{ij} \cdot (\vec{S}_i \times \vec{S}_j) \qquad (15.6)$$

which arises from the spin-orbit interaction (see Levy et al.1982 for a review, also see Goldberg et al.1986, Goldberg and Levy 1986 for more recent results). Note that every term in (15.6) is antisymmetric with respect to the interchange of two spins whereas all the other anisotropies (15.3)-(15.5) are symmetric with respect to such interchange of the spins. DM interaction is unidirectional in contrast to the uniaxiality of the anisotropies (15.3)-(15.5). Also note that the anisotropy (15.3) tends to orient two neighboring spins parallel to each other whereas (15.6) favors mutually perpendicular orientation of the neighboring spins. Hence, a Hamiltonian consisting of (1.1) and (15.6) might lead to spin canting.

The last and almost ubiquitous anisotropy is the so-called dipolar interaction. However, in most of the real SG materials the latter anisotropy is much weaker than the other types of anisotropies. (See Aharony 1978a, 1983 for brief reviews).

So far we have written all the anisotropic Hamiltonians in the discrete form. The Hamiltonian that includes, for example, the RUA, can be written in the continuum form as

$$\mathcal{H} = \int d^d x[(1/2)|\nabla\vec{S}(x)|^2+(a/2)|\vec{S}(x)|^2+(b/4)\{(\vec{S}(x))^2\}^2 - \{\vec{n}(x)\cdot\vec{S}(x)\}^2] \qquad (15.7)$$

where the last term corresponds to the RUA.

In general, there are two different motivations behind the incorporation of the anisotropies in the theoretical studies of random magnetic alloys-

(i) can a random uniaxial anisotropy of the form (15.4) lead to a SG phase in systems with nonrandom nearest-neighbor exchange which would, in the absence of this anisotropy, exhibit long-range order in the low-temperature phase ?

(ii) what is the effect of the various types of anisotropies on the phase diagrams which contain a SG phase even in the absence of such anisotropy ?

What happens if the anisotropy is of the random axis type (15.4) ? The nn ferromagnetic exchange J tends to align the spins parallel to each other. On the other hand, D tries to orient the spins along the corresponding local easy axes which are random functions of position. Thus, J and D compete against each other. The consequence is very clear in the limit $D >> J$; D wins over J and forces the spins to orient along the local easy axes and hence ferromagnetic ordering is impossible even at T=0. In the opposite limiting case $D << J$ the competition between D and J is much more subtle. From a simple-minded consideration of the energetics of a domain, one can show (Grinstein 1985) that the energy cost ΔE_J in creating a domain wall of length L is

$$\Delta E_J \sim J\, L^{d-2}$$

whereas the corresponding gain of anisotropy energy is

$$\Delta E_D \sim - D\, L^{d/2}$$

and hence the ground state is ferromagnetic if $(d-2) > d/2$, i.e., for $d > 4$. This result is consistent with the observation (Derrida and Vannimenus 1980) that there is no SG phase in the infinite-range limit. However, for $d < 4$, arbitrarily weak D is sufficient to destroy the ferromagnetic order. Is the low-temperature phase SG for $d < 4$? There have been very strong indications from the studies of this model by various techniques (Aharony 1975, Harris and Zobin 1977, Chen and Lubensky 1977, Pelcovits et al.1978, Pytte 1978, Pelcovits 1979, Jayaprakash and Kirkpatrick 1979, Pytte 1980, Aharony and Pytte 1980) that the low-temperature phase in $d < 4$ is a SG. If the anisotropy $D \to \infty$, the model is exactly solvable in d=1 (Thomas 1980). Very recently, the possibility of a zero-temperature phase transition in $D \to \infty$ limit of the model in d=2 has been established (Bray and Moore 1985a) by applying a modified form of the large cell renormalization group applied to the study of the Ising model (see chapter 24). See Fischer and Zippelius (1985) for a related model.

If the low temperature phase of the random-axis model is a SG, does any replica symmetry breaking occur at the transition temperature? I am not aware of any general answer to this question. But, the latter problem has been investigated in the special case $m \to \infty$ of the random-axis model by systematic expansion in the parameter $1/m$ (Ginzberg 1981a,b, Goldschmidt 1983, 1984, Khurana et al.1984, Jagannathan et al.1985). We know that the negative replicon eigenvalue in the m-vector random exchange model (Almeida et al.1978) has a $1/m^2$ dependence and, therefore, no replica symmetry breaking takes place in the spherical model $m=\infty$ (Kosterlitz et al.1976) (see chapter 12). It turns out that the replicon eigenvalue signalling the instability in the random-axis model is of the order $1/m$, in contrast to the $1/m^2$ in the random exchange model.

Let us now mention briefly the effects of the anisotropies listed above on the phase diagrams of Ising as well as m-vector

SG models with competing interactions. Following a mean-field approach, similar to that of Sherrington (1975), Ghatak (1976) studied the effect of the anisotropy $\mathcal{H}_1$ on the susceptibility of the EA model. Ghatak and Sherrington (1977) extended the replica-symmetric EA theory by assuming a Hamiltonian of the form

$$\mathcal{H} = - \Sigma J_{ij} S_i^z S_j^z + \mathcal{H}_1$$

where S was assumed to be an integer and it can vary between -S and +S. The phase diagram consists of a second order transition line and another first order transition line, the two lines meeting at a tricritical point. The replica-symmetric solution turned out to be unstable (Lage and Almeida 1982). The nature of this instability is subtle; the Hessian matrix yields even complex eigenvalues! The true nature of the transition is not yet completely understood (Mottishaw and Sherrington 1985) and needs a theory with broken replica symmetry in the same spirit as Parisi's approach to the SK model.

Ghatak and Sherrington's work has been extended to include the effect of uniform external magnetic field (Deo and Mishra 1984) in the replica-symmetric formalism. Assuming Gaussian distribution of D_i in (15.5), Forti et al.(1982) have developed a replica-symmetric MFT that incorporates anisotropy (15.5) instead of (15.3).

The effect of anisotropy (15.3) on the mean-field phase diagram of the classical m-vector SG has also been studied both in the replica-symmetric (Roberts and Bray 1982, Cragg and Sherrington 1982) as well as replica-symmetry breaking formalism (Elderfield and Sherrington 1982, Sherrington et al.1983b). The following phases appear (see Fig.15.1):
paramagnetic (P) phase ($q_L=0=q_T$), longitudinal SG (L) phase ($q_L >$ 0, $q_T = 0$), transverse SG (T) phase ($q_T > 0$, $q_L = 0$) and longitudinal-transverse (LT) phase ($q_L > 0$, $q_T > 0$).
Strongly positive D favors the L phase whereas strongly negative D favors T phase at sufficiently low T.

P-L and P-T transitions have been observed in real SG

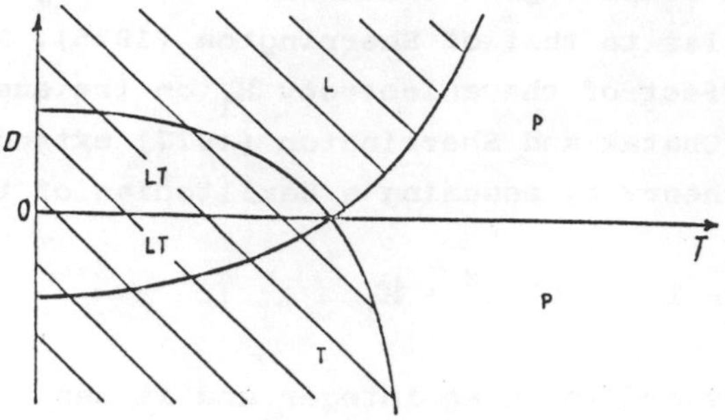

Fig. 15.1. Schematic mean-field phase diagram for a vector SG with the single-ion anisotropy (15.3). The hatched area represents that part of the phase diagram for which replica symmetry is broken (after Elderfield and Sherrington 1982).

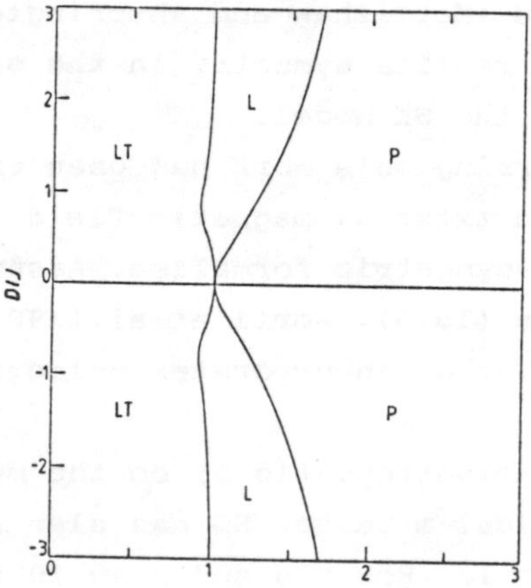

Fig. 15.2. Mean-field phase diagram for a vector SG with the random anisotropy (15.5) with the discrete distribution (15.8) for c = 0.5 (after Viana and Bray 1983)

materials (Albrecht et al.1982, Fert et al.1982, Baberschke et al.1984, see appendix B). But the transition to the LT phase has never been observed so far. Possible interpretation of the latter observation in terms of infinite-ranged interaction has been suggested by Sherrington (1984).

Viana and Bray (1983) studied the mean-field phase diagram (see Fig.15.2) of the m-vector SG in the presence of random anisotropy (15.5) where D_i has a discrete distribution

$$P(D_i) = (1-c)\ \delta(D_i - D_1) + c\ \delta(D_i - D_2). \qquad (15.8)$$

The phase diagram agrees with that derived from the Monte Carlo simulation (Morris and Bray 1984). (See Goldbart 1985 for a model that contains the H_{DM} and an uniaxial part).

The most crucial difference between the phase diagrams of the classical m-vector model with the anisotropy (15.3) and the corresponding quantum model is the destruction of the SG phase for large negative D for S=1,2,3,.. in the latter model (Usadel et al.1983) (see Fig.15.3). However, if the anisotropy is of the form (15.8) with $D_1 = D_2 = D$, $T_g(c)$ is almost independent of D for sufficiently large D/T (Brieskorn and Usadel 1986). The latter behaviour follows from the fact that for large enough D, the sites with local $D < 0$ "condense" into a paramagnet-like phase whereas the other spins have a local anisotropy $D > 0$.

Shirakura and Katsura (1983a,b) obtained the following phases of the anisotropic classical planar model

$$H = - \Sigma J_{ij}\ (\cos\phi_i \cos\phi_j + \eta \sin\phi_i \sin\phi_j), \quad 0 < \eta < 1$$

in the infinite-ranged limit:
paramagnetic phase ($m_x = 0$, $q_x = 0$, $q_y = 0$), ferromagnetic phase ($m_x \neq 0$, $q_x \neq 0$, $q_y = 0$), longitudinal SG phase ($m_x = 0$, $q_x \neq 0$, $q_y = 0$), transverse SG phase ($m_x = 0$, $q_x \neq 0$, $q_y \neq 0$) and mixed phase ($m_x \neq 0$, $q_x \neq 0$, $q_y \neq 0$), where
$m_x = [\langle\cos\phi_i\rangle]_{av}$, $q_x = [\langle\cos\phi_i\rangle^2]_{av}$ and $q_y = [\langle\sin\phi_i\rangle^2]_{av}$.

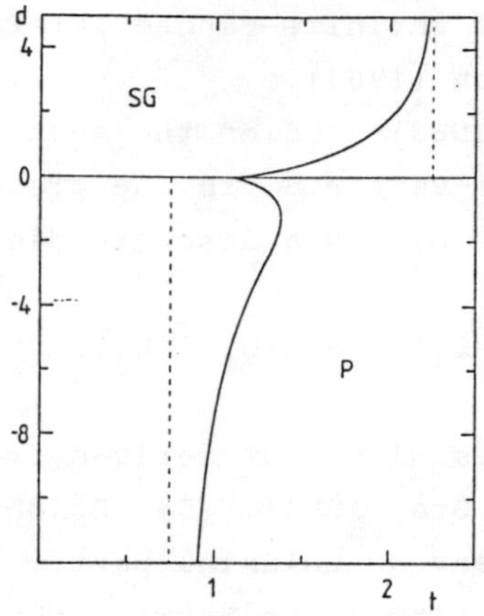

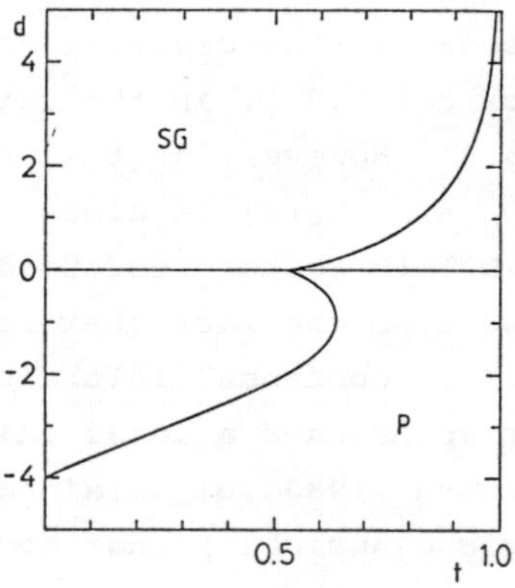

Fig. 15.3. (a) The mean-field phase diagram for a quantum vector SG with S = 3/2 in the presence of the single-ion anisotropy (15.3). (b) Same as (a) but for S = 1 (after Usadel et al. 1983).

The phase diagram for the so-called uniaxial exchange anisotropy model

$$H = - \Sigma (J^T_{ij} \Sigma S_{i\mu} S_{j\mu} + J^L_{ij} S_{i1}) \qquad (15.9)$$

have been studied for the classical spin (see Fig.15.4)(Bray and Viana 1983) as well as for the quantum spins (see Fig.15.5) (Morris 1985).

Dotshenko and Feigelman (1981) studied a two-dimensional model with weak random anisotropy h(x) cos(nθ) where n is the order of anisotropy. For $n \geq 3$ SG order was found neither for $T^* < T < T_c$ nor for $T < T^*$ where $T_c = \pi/2$ is the Kosterlitz-Thouless (see Nelson 1983) transition temperature and $T^* = 4\pi/n^2$. However, SG ordering takes place at T^* for a quasi two-dimensional layered XY model (Dotshenko and Feigelman 1982a,b) where the interlayer coupling is given by $g \cos n(\theta_{i+1}-\theta_i)$, the index i labels the layers and $g \ll 1$.

Mottishaw (1986) has introduced a p-spin interaction model, in the same spirit as in Derrida's p-spin model (see chapter 10), with integer spins $S_i = 0, \pm1, \pm2, \ldots, \pm S$, including an uniaxial anisotropy of the form (15.3). Thus, the p = 2 limit of Mottishaw's model corresponds to the model of Ghatak and Sherrington, discussed earlier in this chapter. The limit S = 1 and $D \to \infty$ is identical with Derrida's model. There is a first order transition from paramagnetic (see Mottishaw 1986 for more details of the nature of this paramagnetic phase) to the SG phase.

Anisotropy plays a crucial role in determining the line $H_c(T)$ line in the H-T phase diagram, as we shall see in the next chapter. The irreversibilities of the short-ranged SG might arise from the anisotropies of the models (Soukoulis et al.1983a,b, Grest et al.1984, Soukoulis and Grest 1984). We shall review the macroscopic effects of anisotropy on the low-temperature spin dynamics in SG in chapter 18.

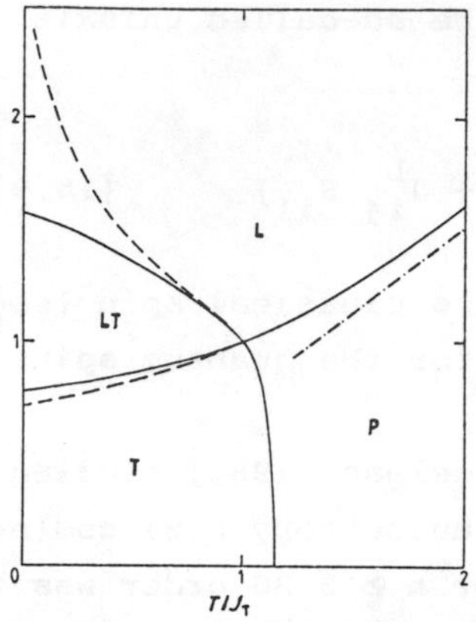

Fig. 15.4. Mean-field phase diagram for the uniaxial exchange model (15.9) with classical spins (after Bray and Viana 1983).

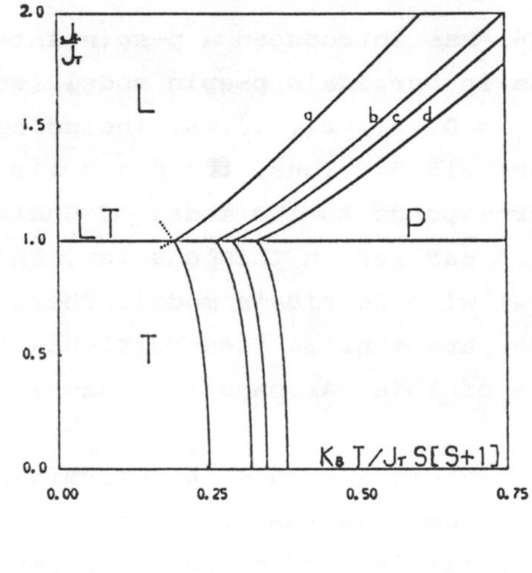

Fig. 15.5. Same as fig. 15.4, but for quantum spins. The curves a,b,c,d represent the P-L and P-T boundaries for S = 1/2,1,3/2, ∞ respectively (after Morris 1985).

CHAPTER 16

NONLINEAR SUSCEPTIBILITIES, AT AND GT LINES AND SCALING THEORIES

As is well known, the ferromagnetic transition in magnetic systems is signalled by the divergence of the linear susceptibility. The field conjugate to the linear susceptibility is the uniform external field. Although the cusp in the linear susceptibility has been used extensively as one of the main criteria for the SG ordering, the corrcet criterion would be the divergence of the EA-order parameter susceptibility χ_{EA} (5.17) because the mean square field $\tilde{h}^2$ is naturally the field conjugate to the EA order parameter. But, it is difficult to "measure" the latter susceptibility in laboratory experiments. However, since near T_g (Chalupa 1977)

$$\chi_{EA} \propto \chi_2 \propto \Sigma\ (\langle S_iS_j\rangle - \langle S_i\rangle\langle S_j\rangle][\langle S_iS_i\rangle - 3\ \langle S_i\rangle\langle S_j\rangle) \qquad (16.1)$$

the crtical behaviour of χ_{EA} can be explored indirectly by studying that of χ_2. (see appendix B for the experimental determinations of χ_2).

It was Katsura (1976) who first showed that χ_2 diverges negatively for a bond-random Ising model in the Bethe approximation. However, for simplicity, we shall sketch Suzuki's (1977) phenomenological arguments. The phenomenological free energy functional is given by (3.13). From (3.15), we have

$$q = -\ eM^2/(2c) + O(h^4) \simeq e\chi_0^2h^2/(2c(T)) \qquad (16.2)$$

Substituting (3.17) and (16.2) into (3.14) we get

$$\chi_2 = e^2\chi_0^4[c(T)]^{-1} - 4b\chi_0^4 \qquad (16.3)$$

Thus, χ_2 diverges at T_g as $[c(T)]^{-1}$. Assuming $c(T) \propto (T - T_g)$ in the spirit of Landau-type phenomenological theories, we get

$$\chi_2 \sim (T - T_g)^{-\gamma_s} \qquad \text{where } \gamma_s = 1$$

Similar negative divergence of χ_2 was also predicted for the infinite range model by Ueno (1980) and Wada and Takayama (1980). Mattis (1976) and Aharony and Imry (1976) showed that for the Mattis model χ_2 diverges for d < 4. The divergences of the nonlinear susceptibilities of an Ising SG, the Mattis model, pure ferromagnet and pure antiferromagnet have been compared by Fujiki and Katsura (1981).

The agreement between the AT line and the experimentally observed crossover lines seems to add to the puzzle. The scaling form $\chi_0 - \chi = H^{2/\delta} f(\tilde{t}/H^{2/\phi})$, together with the mean-field estimate $\phi = 2$, suggests that $\tilde{t} \sim H$ should be the crossover line. On the other hand, the works of Parisi, Toulouse and Vannimenus (explained in chapter 7) indicate $\tilde{t} \sim H^{2/3}$ to be the crossover line, the latter being in apparent agreement with the experimental observations. Does it imply that the prediction of the MFT for the crossover line is wrong? For $T < T_g$, MFT of Ising SG predicts

$$q = -\tilde{t}[1 + (H^2/2\tilde{t}^2)] + O(\tilde{t}^2) \qquad (16.4)$$

to the leading order in H, and in the replica-symmetric theory

$$M/H = (1 - q)/\{T_g(1+\tilde{t})\} \qquad (16.5)$$

Substituting (16.4) into (16.5) we get

$$M/H = T_g^{-1}[1 - (H^2/2\tilde{t}) - (\tilde{t}^2/3) + O(\tilde{t}^3)]$$

$$\sim T_g^{-1} - \tilde{t}^2 f(H^2/\tilde{t}^3) \qquad (16.6)$$

where f is an universal (scaling) function of its arguments. The form (16.6) implies $\tilde{t} \sim H^{2/3}$, which is the AT line, as a crossover line. Note that the term linear in $\tilde{t}$ has been cancelled out of the final expression (16.6) (Malozemoff et al. 1983,

Bouchiat 1983, 1984), thereby explaining why the crossover line in the MFT for Ising SG turns out to be the AT line.

A closer look at the breakdown of scaling from the renormalization group (RG) point of view in terms of the "dangerously irrelevant" variables (Fisher and Sompolinsky 1985) will be taken in the next chapter.

Next we investigate the nonlinear susceptibility in vector SG starting from the phenomenological free energy functional (13.8). Expanding the transverse magnetization M_T as

$$M_T = \chi_0^T H_T + \chi_2^T H_T^3 + \ldots$$

and proceeding as in the case of Ising SG we get

$$\chi_2^T = -\{e q_T H_T^{-2}/(2a^2)\} \propto [c(T) + g_2(M_L q_L)]^{-1}$$

and hence the transverse susceptibility diverges along the GT line.

Now let us write $g_j(M_L,q_L)$ as

$$g_j(M_L,q_L) = g_{j1}\, M_L^2 + g_{j2}\, q_L^2 + \ldots \qquad \text{for } j = 1,2$$

One can argue (Suzuki 1985) that $g_{21}M_L^2 << g_{22}q_L^2$. Then, equation (13.9) reduces to

$$(T - T_{GT}(0)) + g_{22}(T)q_L^2 + .. = 0$$

Then, expanding $(T - T_{GT}(H))$ around $(T - T_{GT}(0))$ we get

$$\{T_{GT}(H) - T_{GT}(0)\} \propto q_L^2$$

On the other hand, minimizing the free energy with respect to M_L, one can easily verify that

$$q_L \propto \chi_L \propto H_L^{2/\delta}$$

Therefore, finally, we get

$$\{T_{GT}(H) - T_{GT}(0)\} \propto H_L^{4/\delta}$$

Using the MFT values $\beta = 1$ and $\gamma = 1$ and the standard scaling relation $\beta\delta = \beta + \gamma$, we get $\delta = 2$ and we recover the GT line.

So far we have not answered the question why the $H_c(T)$ line in the H-T phase diagram is the AT line rather than the GT line. One possibility is that there is anisotropy in all real SG. Assuming that the dominant anisotropy is of the DM type (15.6), with zero mean and variance $D/N^{1/2}$, Kotliar and Sompolinsky (1984) showed (see also Fischer 1985) that
(i) In the strong-anisotropy limit $(D/k_BT) >> (H/k_BT)^{2/3}$, the transition is AT-like, i.e., $\tilde{t} \sim H^{2/3}$ for any fixed D/k_BT,
(ii) the transition line is the GT line in the weak-anisotropy limit $(D/k_BT) << (\mu H/k_BT)^{5/2}$.
These results are in qualitative agreement with the experimental results of de Courtney et al.(1984, 1985) and Yeshurun and Sompolinsky (1986) (see appendix B). In the SG phase one must distinguish between the <u>local</u> transverse response χ^Y and the <u>uniform</u> transverse response Δ^Y (Sompolinsky et al.1984). These two responses are related by (Kotliar and Sompolinsky 1984)

$$\chi^Y \simeq (M/H) - \Delta^Y(1 + HM_r/K)^{-1}$$

where K is the macroscopic anisotropy constant and M_r is the longitudinal (along the z-direction) remanent magnetization. Δ^Y corresponds to a small rotation of the field on a <u>particular</u> spin while keeping the unifprm field, which acts on the rest of the system, unaltered. χ^Y is the response to a uniform rotation of the external field. In the weak-anisotropy regime $\Delta^Z << \Delta^Y$ near $T_{GT}(H)$ and the two become comparable only below a crossover temperature $\tilde{t} \sim H^{2/3}$. It has been argued (Ketelsen and Salamon 1986) that

$$\delta = \Delta^Y(1 + HM_r/K)^{-1} \tag{16.7}$$

should obey a scaling form, provided the system exhibits a thermodynamic phase transition. Such a scaling plot is shown in Fig.16.1.

I would like to stress that the AT line does not necessarily imply a nonzero T_g for the corresponding SG system. Consider the short-ranged Gaussian Ising SG model for which $T_g=0$, as will be shown in chapter 24. From the Monte Carlo simulation of this model, Kinzel and Binder (1983, 1984) computed the line $H_{AT}(t)$ which describes the onset of the irreversibility, on a time scale of observation t. Surprisingly, when scaled appropriately (Fig.16.2), the line $H_{AT}(t)$ appears qualitatively similar to the AT line (5.21) on all time scales. Thus, the AT line observed for the latter model is an artifact of the finiteness of the time of observation and would disapper in the long time limit $t \to \infty$. This observation is also consistent with Young's (1983) observation that points on the H-T plane with the constant relaxation time form AT-like lines (see Fig.16.3).

Suppose γ_ψ is an exponent related to the Morgenstern-Binder order parameter, defined in chapter 3, in the same way γ_s is related to the EA order parameter. The temperature dependence of γ_s observed by Fahnle ans Egami (1982a,b) indicated the possibility that the divergence of χ_2 "has nothing to do with a real critical phenomenon". However, the question of the true phase transition in three dimensional SG is a subtle issue to be investigated in detail in chapter 24. Some ad hoc attempts have been made (Honda and Nakano 1981, Suzuki and Miyashita 1981, Ueno and Oguchi 1980) to interprete Miyako and co-workers' observation that χ_2 diverges logarithmically in a.c. experiments (see appendix B).

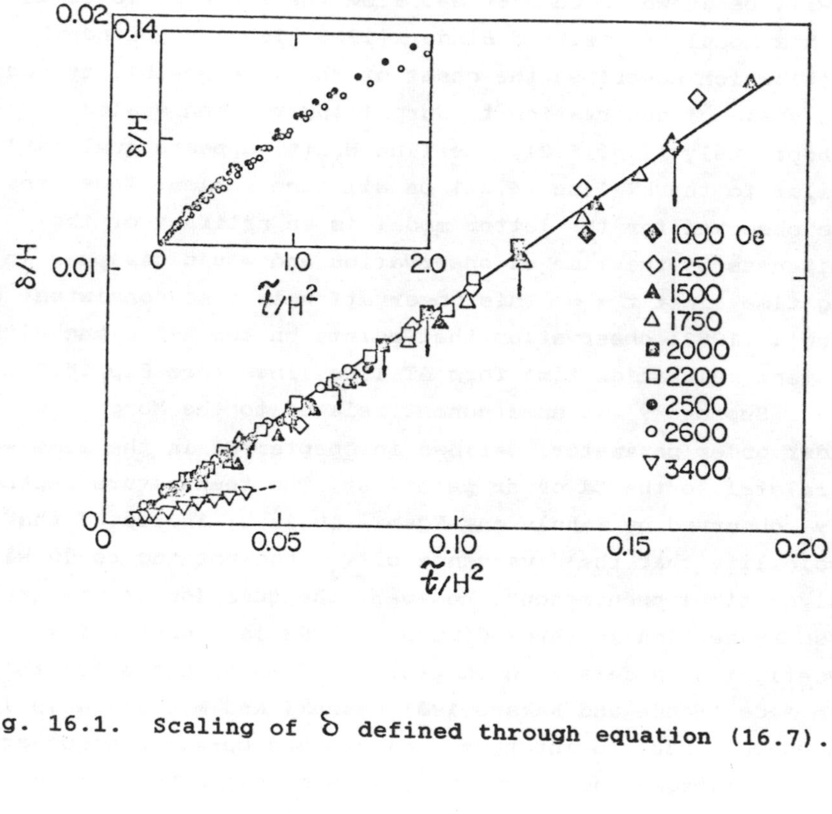

Fig. 16.1. Scaling of δ defined through equation (16.7).

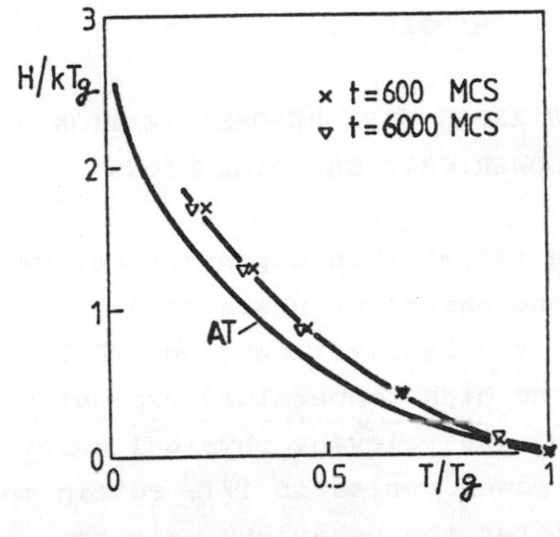

Fig. 16.2. Normalized dynamic critical field plotted against normalized temperature (after Binder and Kinzel 1983a).

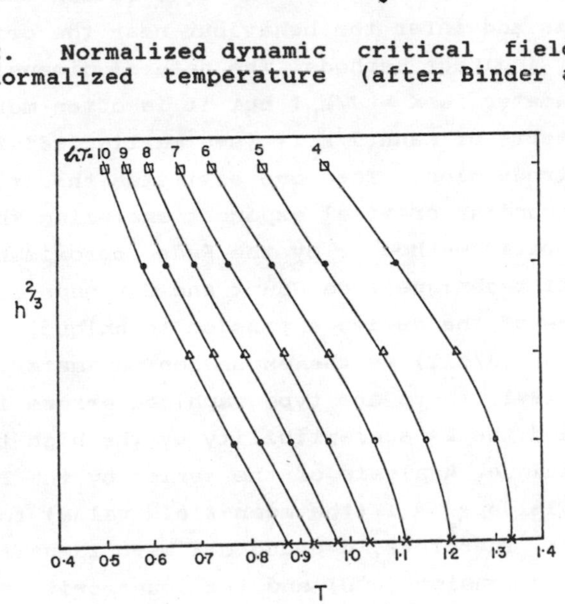

Fig. 16.3. Lines of constant τ plotted in the $h^{2/3}$–T plane showing a series of dynamical AT-like lines for different timescales (after Young 1983a).

CHAPTER 17

HIGH-TEMPERATURE EXPANSION, RENORMALIZATION GROUP; UPPER AND LOWER CRITICAL DIMENSIONS

The technique of high-temperature expansion was quite successful in the study of the phenomenon of phase transition before the powerful technique of renormalization group (RG) was applied. The basic philosophy of the high temperature expansion (1/T) is quite simple. One can expand the relevant physical quantity (say, the susceptibility) in a power series in 1/T, retain manageably large number of terms and infer the behaviour near the critical temperature by indirect methods. The natural dimensionless expamsion parameter is $K = J/k_BT$ but it is often more convenient to expand in terms of $\tanh(J/k_BT)$ (see Mattis 1985 for an elementary introduction). Then one estimates the critical point and the corresponding critical exponent analyzing the series either by the ratio method or by the Pade approximants or by some other efficient technique (see Gaunt and Guttmann 1974 for a detailed review of the series expansion technique).

Using $\tanh^2(J/k_BT)$ as the expansion parameter, Fisch and Harris (1977, 1981, there are typographical errors in the latter errata!) studied the EA susceptibility by the high-temperature expansion technique. Analysis of the series by the Pade approximants yields $\gamma_s = 1$ (the mean-field value) for d=6 and γ_s diverges in d=4. Therefore, d=6 and d=4 were identified as the upper critical dimension (UCD) and the lower critical dimension (LCD), respectively, for SG. The Fisch-Harris (1977) analysis has been extended (a) to the XY and Heisenberg models (Reed 1978), (see also Cherry and Domb 1978, Rapaport 1977)
(b) to all generalized symmetric distributions of J, i.e., for $P(J) = P(-J)$ (Ditzian and Kadanoff 1979)
(c) to asymmetric P(J) (Rajan and Rieseborough 1983),
(d) to models incorporating dilution (Palmer and Bantilan 1985).
All the papers cited in the preceeding paragraph investigated only static thermodynamic quantities. Reger and Zippelius (1985)

have developed a high temperature expansion for the dynamic correlations in the ±J model incorporating the effects of dilution. The series was analyzed by ratio plot as well as by the Pade approximants. These data are consistent with $T_g = 0$ in d=2.

The most recent and the most extensive work on the high temperature expansion of the ±J model is that of Singh and Chakravarty (1986). The order parameter susceptibility is expanded as

$$\chi_{SG}^{-1} = 1 + \Sigma\, L(g_s)\, W(g_s)$$

where the summation is to be carried out over all the topologically distinct star graphs. A graph is said to have a point of articulation (k) provided the graph can be split into disconnected parts by cutting all the bonds incident at k and removing the piece connected to k. A graph without any such point of articulation is called a star graph. $L(g_s)$ is the number of ways per lattice site in which the given graph can be embedded in the lattice and $W(g_s)$ is the weight factor associated with the graph. The series was analyzed upto the nineteenth term in d=2, upto the seventeenth term in d=3 and upto the fifteenth term in d=4. The results agree with the computer simulation data and suggest a nonzer T_g in d=3 but $T_g = 0$ in d=2. We shall compare the above predictions of the LCD with the corresponding prediction of the computer simulations in chapter 24.

Let us now apply the methods of field-theoretic RG (FTRG) (see Amit 1978 for an introduction) to SG. The basic philosophy of the latter technique is to map the effective Hamiltonian, by a repeated application of the transformation R_b, from a calculationally intractable region in the parameter space into a region where conventional perturbation-theoretic techniques (use of Feynaman diagrams simplifies the latter part of the calculation) can be applied. The basic sheme of implementing the FTRG can be summarized by the operation

$$R_b = R_b^s\, R_b^i \qquad (17.1)$$

where b is the so-called scale factor (b > 1), R_b^i denotes an integration over the order parameter field in momentum (k-) space over the range $b^{-1} < k < 1$ and R_b^s denotes the change of scale $k \to bk$.

In the case of vanishing mean of P(J), the effective Hamiltonian for an m-vector SG in the field-theoretic formulation is given by

$$\mathcal{H} = \int d^dx \{(a/4)\mathrm{Tr}Q^2(x) + (1/4)\mathrm{Tr}(\nabla Q(x)\cdot\nabla Q(x)) - w\mathrm{Tr}Q^3(x)$$

$$+ (b/4)(\mathrm{Tr}Q^2)^2 \qquad (17.2)$$

where

$$\mathrm{Tr}Q^2 = \sum_{\alpha,\beta,i,j} q_{ij}^{\alpha\beta} q_{ji}^{\beta\alpha}$$

and

$$\mathrm{Tr}Q^3 = \sum_{\alpha,\beta,\gamma,i,j,k} q_{ij}^{\alpha\beta} q_{jk}^{\beta\gamma} q_{ki}^{\gamma\alpha}$$

where i,j (i, j = 1,..,m) etc. denote the spin-components whereas α,β, (α,β = 1,...,n) etc. label the replicas (see Lubensky 1979 for a detailed derivation of 17.2 using Hubbard-Stratanovich transformation). Using standard ϵ-expansion (ϵ = 6-d) method the critical exponents were computed to the first order in ϵ = 6-d (see Feigelman and Tsvelik 1980 for a diagrammatic approach which does not use the replica Hamiltonian 17.2). Subsequently the critical exponents for the m-vector SG (including m=1) have been computed upto the order ϵ^3 (Chang 1983, Green 1985, also see Elderfield and McKane 1978) In particular, for the Ising SG (m=1),

$$\eta = -\,0.3333\epsilon + 1.2593\epsilon^2 + 2.5367\epsilon^3$$

and

$$\nu^{-1} - 2 + \eta = -\,2\epsilon + 9.2778\epsilon^2 + 4.2336\epsilon^3$$

In the case of Heisenberg SG (m=3)

$$\eta = -0.2\epsilon + 7.7333 \times 10^{-2}\epsilon^2 - 7.8127 \times 10^{-2}\epsilon^3$$

and

$$\nu^{-1} - 2 + \eta = -1.2\epsilon + 1.164\epsilon^2 - 1.4735\epsilon^3.$$

The other exponents can be obtained by using the standard scaling relations $\alpha = 2 - d\nu$, $\beta = (d + \eta - 2)\nu/2$ and $\gamma = (2 - \eta)\nu$. Surprisingly, a FTRG treatment in the case of XY and Heisenberg SG with nonzero mean of P(J) (Chen and Lubensky 1977) yields complex exponents! Moreover, a RG treatment (ϵ = 6-d expansion) of the field-theoretic model in the presence of uniform external magnetic field does not yield any stable fixed point (Bray and Roberts 1980). In fact, the form (5.21) of the AT line breaks down for $6 < d < 8$ where the exponent 3 in (5.21) has to be replaced by $(d/2)-1$; the mean-field value 3 is recovered at $d > 8$ (Green et al.1983). Indeed, b and w in (17.2) turn out to be dangerously irrelevant variables (see Fisher 1983 for an excellent introduction to the concepts of irrelevant and dangerously irrelevant variables) in $d > 6$ leading to the breakdown of scaling mentioned in the preceeding chapter. Also note that UCD = 8 for $m \to \infty$ SG model (Green et al.1982).

Often it is more convenient to apply real space renormalization group (RSRG) to disordered systems (see Burkhardt and van Leeuwen 1982 for an introduction). First let us present a heuristic RG argument based on the rigidity of the arrangement of the spins in the system. Consider a block of the d-dimensional spin system of linear size L. If the exchange interaction in the system is nonrandom and the spins are of Heisenberg type, the energy for rotating the spins on one face of the sample by a small angle, holding those on the opposite face fixed, is proportional to the crosssectional area and inversely proportional to the length L; therefore, $\Delta E \sim L^{d-2}$. The latter implies LCD = 2. Similarly, for nonrandom Ising spin systems, $\Delta E \sim L^{d-1}$, because the interface is sharp. In the latter case LCD = 1. Let us apply similar arguments to SG. Since the exchange is random in SG, we have $\Delta E \sim L^{(d-1)/2-1}$ for Heisenberg spins and hence one would expect that the corresponding LCD = 3. Similarly, for Ising SG one would expect LCD = 1 (Anderson and Pond 1978). These results can be rederived from stronger Migdal-Kadanoff bond-moving arguments.

Next, let us try to formulate more rigorous RSRG arguments for SG. There are several different approaches to the RSRG in spin systems (see Burkhardt and van Leeuwen 1982 for a summary). In disordered magnetic systems one begins with a given distribution of exchange interactions J (i.e., a distribution of the couplings J/k_BT). This distribution gets modified after the application of the RG transformation. In the conventional approaches one keeps track of the distribution (more precisely, the mean K_0 and the variance K of the the distribution) at successive generations of the renormalized spin system. In the general case of the dilute ±J model, the parameter space of the RG consists of the coupling constant and the concentration p of the occupied bonds that are ferromagnetic. We shall not go into the details of the implementation of the RG schemes. The "flow" in the parameter space converges to one of the following fixed points (see, for example, Schlottmann and Bennemann 1982)

(i) $K \to 0$ and $K/K_0 \to 0$ (paramagnetc fixed point)
(ii) $K \to \infty$ and $K_0/K \to 0$ (SG fixed point)
(iii) $K_0 \to \infty$ and $K/K_0 \to 0$ (ferromagnetic fixed point)
(iv) $K_0 \to -\infty$ and $K/K_0 \to 0$ (antiferromagnetic fixed point)

Note that the fixed points corresponding to the various phases are defined by the limiting ratios of the mean and the variance of the distribution of J and not explicitly with respect to the order parameters. The scale factor b chosen for the renormalization can play a crucial role, as we shall see later in this chapter.

In the Migdal-Kadanoff bond-moving method (see Burkhardt 1982 for an introduction), the bonds are moved appropriately and the trace over a subset of the spins is carried out (the so-called decimation) so as to get bigger cells (linear cell size being blown by a scale factor b at each successive step) formed by the left-over spins. For example, on a d-dimensional hypercubic lattice the bonds are moved parallel to one of the axes at a time so that the operation of renormalization can be represented as

$$R_b = [R_{parallel}(b)]^{d-1} R_{series}(b) \qquad (17.3)$$

where the operation

$$R_{parallel}(1+\zeta) = 1 + \zeta L_{parallel}$$

denotes infinitesimal bond-moving operation with the corresponding generator $L_{parallel}$, and

$$R_{series}(1 + \zeta) = 1 + \zeta L_{series}$$

denotes the one-dimensional decimation. Note that (17.3) for RSRG is the analogue of (17.1) for FTRG. Thus, so far as the infinitesimal transformation is concerned, (17.3) reduces to

$$R(1 + \zeta) = 1 + \zeta[(d-1)L_{parallel} + L_{series}]$$

For simplicity, let us consider the Ising SG models with symmetric P(J), i.e., P(K) = P(-K) (Kirkpatrick 1977a). One can easily verify that on a square lattice

$$L_{parallel}[\langle K^2\rangle] = \langle K^2\rangle \qquad (17.4)$$

and

$$L_{series}[\langle K^2\rangle] = -1.5 \langle K^2\rangle \qquad (17.5)$$

Substituting (17.4) and (17.5) into (17.3) we find that $\langle K^2\rangle$ decreases with successive operation (17.3), thereby indicating that T=0 is the corresponding fixed point of the iteration and, therefore, $T_g = 0$ for Ising SG in d=2.

A different RSRG schme also predicted that $T_g = 0$ in d=2 (Young and Stinchcombe 1976) but $T_g > 0$ in d=3 (Southern and

Young 1977). The results of Jayaprakash et al.(1977) which contradicts those of the other RSRG results mentioned above is an artifact of an erroneous choice of the scale factor b. (See Southern et al.1979 for the effects of dilution and compare the results with the corresponding results of Aharony 1978, Giri and Stephen 1978, Aharony and Pfeuty 1979). Later works of Benyousseff and Boccara (1981, 1983, 1984) do not agree with the results of large scale computer simulation to be discussed in chapter 24.

Since there is no natural expansion parameter in RSRG, the errors are often uncontrollable. On the other hand, near the critical temperature the Monte Carlo (MC) simulation technique (to be explained in chapter 24) suffers from finite-size effects. However, an efficient way of avoiding both these difficulties is to use the Monte-Carlo renormalization group (MCRG) (see Swendsen 1982 for an introduction) which combines the techniques of RSRG and MC simulation. Some recent works on large-cell renormalization, which are somewhat similar to MCRG, will be summarized in chapter 24(see also Kinzel and Fischer 1978).

It has been realized for a long time that the implementation of the RG schemes gets simpler on the hierarchical lattices as compared to the conventional Bravais lattices. However, the price one pays is a costly one, the real SG materials need not share any of the essential features exhibited by the hierarchical model. Fortunately, it seems that the renormalization on the hierarchical lattices lead to some physically reasonable results for SG, as we shall see in chapter 26.

We have investigated the critical behaviour of the nearest-neighbor exchange models in this chapter. However, the exchange interaction in some of the most common SG materials, e.g., AuFe, CuMn, etc., are of RKKY type. Therefore, studies of the critical behaviour of the latter model is expected to through light on the nature of the ordering in real SG materials. For generality, let us begin with

$$J_{ij} = J \cos(2k_F r_{ij})/r_{ij}^{(d+\sigma)/2}$$

which reduces to the standard RKKY interaction for d = 3 = σ. The corresponding Landau-Ginzburg Hamiltonian in the replica formalism is given by (Bray et al.1986)

$$-\beta H=-(\beta^2 J^2/4)\sum_{\alpha\beta ab k}\sum(r+sk^2+lk^\sigma)q_{ab}^{\alpha\beta}(\vec{k})q_{ab}^{\alpha\beta}(-\vec{k})$$

$$+(w/6)\sum_{\alpha\beta ab}\sum_{k_1}\sum_{k_2} q_{ab}^{\alpha\beta}(\vec{k}_1)q_{bc}^{\beta\gamma}(\vec{k}_2)$$

$$q_{ca}^{\gamma\alpha}(-\vec{k}_1-\vec{k}_2) + O(q^4)$$

where α,β,γ etc are the replica indices and a,b,c etc. label the spin components. The phase diagram in the d-σ plane is shown in Fig.17.1. Note that the RKKY SG, which correspond to d=σ, are at their lower critical dimension. Moreover, since the lower critical dimension for the short-ranged Ising SG are believed to be smaller than 3, Ising SG with RKKY interaction lie in the same universality class as the short-ranged Ising SG (see also Bray and Moore 1982). On the other hand, the lower critical dimension of the short-ranged vector SG are greater than three. Therefore, the vector SG with RKKY and short-ranged interactions belong to different universality classes.

Let us now go back to the analysis of the short-ranged Ising SG based on the domain wall energies. On long-time and large distance scales, the dominant low-lying excitations in the latter systems are "droplets" which are defined as the clusters of coherently flipped spins (Fisher and Huse 1986). In the same spirit as in the works of Anderson and Pond, Bray and Moore and McMillan, let us assume that the surface energy is given by ΥL^θ with $\theta \le (d-1)$ and Υ is a measure of the interfacial tension. So long as θ is positive, only a fraction of the order of $k_B T/\Upsilon L^\theta$ of the droplets are active, giving rise to the linear specific heat (Fisher and Huse). The nonlinear susceptibility diverges in d=3, and hence the SG transition. However, in contrast to the SK model, there is no AT line in the case of the short-ranged model, when analyzed within the Fisher-Huse droplet picture.

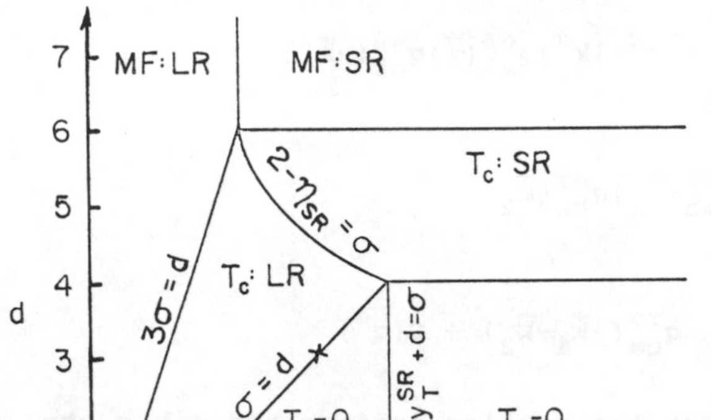

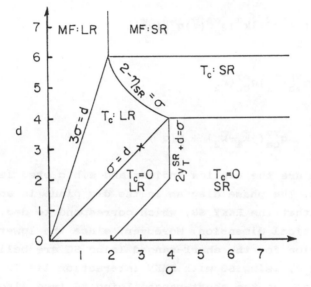

Fig. 17.1. Phases of a vector SG in the d-σ plane. MF denotes a mean-field transition. Tc:LR stands for a finite-temperature phase with exponents in the long-range universality class, Tc=0 LR denotes the region in which a zero-temperature phase transition occurs in the long-range universality class. The crosses mark the point d=σ=3 corresponding to the RKKY interaction.

CHAPTER 18

SPIN DYNAMICS IN VECTOR SG; PROPAGATING MODES

The nonrelativistic version of the Goldstone's theorem (Lange 1965, 1966) states that if the ground state of the system has a lower continuos symmetry than that of the Hamiltonian, there must be an excitation mode whose spectrum in the long-wavelength limit ($k \rightarrow 0$) extends to zero without any gap. However, certain extra conditions must also be satisfied, for example, the range of the interaction must be finite. Such Goldstone modes in ferromagnets with nearest-neighbour (nn) exchange interaction are very well known, these are the spin waves (magnons). We also know that the O(3) symmetry of the state (assuming the spin dimension $m = 3$) is completely broken at the SG transition temperature. Can one expect spin wave excitations in the low temperature phase of the m-vector SG models ? We shall not discuss the theories of spin waves in the Mattis model (see Sherrington 1977, 1978, 1979, van Hemmen 1980, Canisius and van Hemmen 1981). We shall focus our attention to the Halperin-Saslow (HS) (1977) theory and its various extensions. These theories have been developed following closely the hydrodynamic theory of excitations in systems with long-ranged order (LRO) (Halperin and Hohenberg 1969). (Note that there are subtle differences between the hydrodynamic modes and the so-called "collisionless" modes (see Anderson 1984). But for our purposes in this book such distinctions will not be essential.) We shall first explain the regimes of the validity of the hydrodynamic theory.

In the equilibrium states of systems with LRO, the order parameter density and the energy density E are spatially uniform. The hydrodynamic theory describes dynamics associated with small wavelength fluctuations, i.e., fluctuations over length scales k^{-1}, such that $k\xi << 1$, where $\xi(T)$ is the correlation length. The hydrodynamic mode frequencies (ω) must satisfy the following requirement: ω must be much smaller than the internal frequency associated with the "microscopic degrees of freedom" so that in the thermodynamic theory the system can be assumed to be oscillating around a local minimum. The nature of the "microscopic degrees of freedom" depends on the particular system under consideration. If τ_c is the time scale required for such microscopic relaxation, hydrodynamic theory is valid provided $\omega\tau_c << 1$.

In the usual hydrodynamic derivation one assumes that the thermodynamic state of the system is determined by the conserved quantities, $\vec{M}$, E and by any other variable associated with the broken symmetry (see Forster 1975 for a general introduction). The time-derivatives of $\vec{M}$ and E can be expressed as the divergences of the appropriate currents because the former are conserved quntities. In order to solve these hydrodynamic equations one expresses the currents as functions of the hydrodynamic variables, $\vec{M}$, E, etc. through the so-called constitutive equations. Finally, eliminating the currents from the hydrodynamic equations and solving the equations thus obtained one gets the dispersion relation for the long-wavelength excitations (spin waves) (Halperin and Hohenberg 1969).

Let us now focus our attention on the Heisenberg Hamiltonian (1.1) for SG. Let us define the coarse-grained magnetization density, in a region R, of volume v, centered about the point $\vec{r}$ and containing p spins, as

$$\vec{M}(r) = (g\mu_B/v) \sum_{i \in R} \vec{S}_i \qquad (18.1)$$

Let us also define

$$(1/p) \sum_{i \in R} \langle S_i^\alpha \rangle_G \langle S_i^\beta \rangle_G = q\, \delta_{\alpha\beta} + O(p^{-(1/2)}) \qquad (18.2)$$

and

$$(1/p) \sum_{i \in R} \langle S_i^\alpha \rangle_G\, S_i^\beta = t_{\alpha\beta}(\vec{r}) \qquad (18.3)$$

where $\langle S_i^\alpha \rangle_G$ denotes the expectation value of the α-th component of $\vec{S}_i$ in a local equilibrium state. Since q is a measure of the local equilibrium

$$\theta_\gamma(\vec{r}) = (1/2q)\, \epsilon_{\alpha\beta\gamma}\, [t_{\alpha\beta}(\vec{r}) - q\, \delta_{\alpha\beta}] \qquad (18.4)$$

for small θ is a measure of the local rotation angle. In the same spirit as in the case of ferromagnets, let us assume that the state of the system is described by $\vec{M}(\vec{r})$, the energy density E and the angle θ_α. Now, expressing the time derivatives of $\vec{M}$ and E in terms of the divergences of the corresponding currents and then utilizing the constitutive equations for these currents, one gets the dispersion relations for the excitation from the solution of the hydrodynamic equations (see Halperin and Saslow 1977). Instead of the detailed hydrodynamic derivation, we shall present here a heuristic derivation which is suitable for future extensions. Assuming the equilibrium state to be macroscopically isotropic and the spin stiffness ρ_s to be nonzero, one can write the free energy change ΔF, due to the displacement from (local) equlibrium as

$$\Delta F = (1/2)\ [|\vec{M}|^2\ \chi^{-1} + \rho_s\ |\nabla\theta|^2\ ] \qquad (18.5)$$

neglecting higher order terms, where χ is the local susceptibility. Terms proportional to $M_\alpha \nabla\theta_\alpha$ and $\vec{M}\cdot\vec{\theta}$ cannot appear in (18.5) because the former would imply a preferred direction in real space, whereas the latter would violate the rotational symmetry of the Hamiltonian in spin space. The equations of motion for $\vec{M}$ and θ are given by

$$(\partial M_\alpha/\partial t) = -\ \gamma\ (\delta f/\delta\theta_\alpha) \qquad (18.6)$$

and

$$(\partial\theta_\alpha/\partial t) = \gamma\ (\delta f/\delta M_\alpha) \qquad (18.7)$$

where $\gamma = g\mu_B/\hbar$ is the gyromagnetic ratio. The meanings of the equations (18.6) and (18.7) are very clear; (18.6) expresses the fact that "the angular momentum is driven by the torque" whereas (18.7) indicates that the "orientations of the spins is driven to align with the internal field". The corresponding solutions of the equations of motion yield a linear dispersion law. A more detailed hydrodynamic derivation leads to the following equations of motion for M_α and θ_α:

$$(\partial M_\alpha/\partial t) = \gamma\rho_s\nabla^2\theta + K\,\chi^{-1}\nabla^2 M_\alpha \qquad (18.8)$$

and

$$(\partial \theta_\alpha/\partial t) = \gamma\, M_{\alpha\chi}^{-1} + \gamma\,\zeta\,\rho_s\nabla^2\theta_\alpha \qquad (18.9)$$

where K and ζ are two phenomenological coefficients. Solving these hydrodynamic equations of motion, HS concluded that
(i) there are three polarizations of the spin wave, each with
(ii) linear dispersion law, i.e., $\omega = ck$, where $c = \gamma(\rho_s/\chi)^{1/2}$ (see also Dzyaloshinskii and Volovik 1978), and
(iii) the damping is proportional to k^2.
These results can also be derived from symmetry considerations (Andreev 1978). Are such spin waves observable (Villain 1980)? This question has two parts- first whether spin waves exist in real SG and second, if yes, whether such spin waves can be observed in real experiments. It might appear at first sight that because of the absence of long-ranged orientational order spin waves, even if they exist in SG, cannot be observed. However, that is not necessarily true, because, for example, phonons are observable in silicate glasses although there is no positional long-ranged order. Spin waves are expected to propagate so long as the corresponding wavelength is much longer than the length scale of the disorder. In ferromagnets, because of LRO, $\vec{S}_k(t)$, the Fourier transform of $\vec{S}_i(t)$, precesses with an uniform frequency ω_k. In neutron scattering experiments one probes $\vec{S}_k(t)$ directly and hence the spin waves are experimentally observable. Neutrons see $\vec{S}_k(t)$ also in SG whereas the precessing quantities are random combinations of $\vec{S}_k(t)$'s because of the lack of LRO. However, a subtle feature associated with the study of the spin waves in metallic SG through the Heisenberg model is that the effect of the conduction electrons is taken into account only through the RKKY interaction. Consequently, the latter treatment ignores the possibility of spin-relaxation through the s-d interaction. The latter can lead to overdamped spin waves in

metallic SG, thereby making it practically impossible to observe in any laboratory experiments. However, this difficulty does not arise in the case of an insulating SG. In spite of several investigations (Shapiro et al.1980, 1981a,b, Blanckenhagen and Scheerer 1983, Maletta et al.1981, Alloul and Mendels 1985, Alloul et al.1986) the existence of well defined spin waves in SG remains to be established.

So far we have been discussing the hydrodynamic theory of spin excitations in SG. Let us review some of the microscopic approaches to this problem. The most straightforward approach is the application of the standard Holstein-Primakoff transformations:

$$S_i^z = S - a_i^+ a_i$$

$$S_i^+ \simeq (2S)^{1/2}\, a_i$$

$$S_i^- \simeq (2S)^{1/2}\, a_i^+$$

to the Hamiltonian (1.1). Keeping upto the terms bilinear in a^+ and a, and follwing Sherrington's (1977) approach to the study of spin waves in random magnets, one can define the Green function (Takayama 1978) so as to derive the dispersion relation for the spin waves. Although the linear dispersion law thus derived agrees with the HS prediction, the damping turns out to be proportional to k^4 in contrast to the k^2 damping law of HS. (see Barnes 1981a for some criticism of Takayama's work). The effect of uniaxial anisotropy is to yield three branches of the spin wave excitations; the two transverse branches have identical gap. External field splits the two transverse branches further (Barnes 1981). In general the longitudinal branch has a linear dispersion whereas the quadratic dispersion law for the two transverse branches reduce to a linear dispersion law as the anisotropy vanishes.

The validity of the approximations made in the analytical calculations of spin wave excitations can be tested by comparing with the straightforward numerical results obtained in the same model. The classical equation of motion of a spin is given by

$$d\vec{S}_i(t)/dt = \Sigma\, J_{ij}\ (\vec{S}_i \times \vec{S}_j). \qquad (18.5)$$

Now suppose that $\{\hat{n}_i^o\}$ is a configuration of a given spin system such that the associated classical energy $-(1/2)\Sigma J_{ij}\ \hat{n}_i^o \cdot \hat{n}_j^o$ corresponds to a local minimum. Let us choose an orthonormal triad of unit vectors $\hat{a}_i$, $\hat{b}_i$, $\hat{n}_i^o$. Since we are interested only in the small amplitude oscillations about a local minimum we can write

$$\hat{n}_i = \hat{n}_i + \hat{m}_i - (1/2)(\hat{m}_i \cdot \hat{m}_i)\ \hat{n}_i^o + \ldots$$

with the condition $\hat{m}_i \cdot \hat{n}_i^o = 0$. Further, writing

$$\hat{m}_i = \alpha_i\ \hat{a}_i + \beta_i\ \hat{b}_i$$

and substituting into the equation of motion (18.5) we get the required equation for the small amplitude oscillations of the spin system about a local minimum. This approach can also be extended to quantum SG. The density of such small amplitude excitations were computed for

(i) RKKY model (1.7a) with and without (a) the finite mean-free path of the RKKY interaction, (b) dipolar anisotropy, (c) single-ion anisotropy, (d) external magnetic field (Walker and Walstedt 1980),

(ii) EA model for the Heisenberg spins in d=3 (Ching et al.1977, 1981),

(iii) XY model in d=1, 2 and 3 (Huber and Ching 1980, Huber et al.1979),

(iv) CuMn (Ching and Huber 1978),

(iiv) PdMn (Ching et al.1979),

(iiiv) $Eu_xSr_{1-x}S$ (Ching et al.1980),

(ix) $Cd_{1-x}Mn_xTe$ (Ching and Huber 1984, Giebultowicz et al.1985),

(x) Heisenberg-Mattis model in d=1, 2, 3 (Ching and Huber 1979,

Ching et al.1977),
by direct diagonalization of the dynamical matrices and/or by the so-called equation-of-motion method.

Krey (1980, 1981, 1982) computed the excitation spectra of $Eu_xSr_{1-x}S$ system, modelled by the Hamiltonian (1.5), by continued fraction method (see Ehrenreich et al.1980 for an introduction to the continued fraction method). The specific heat, computed using this excitation spectrum, is in agreement with the experimental data (appendix B). The density of states develops a gap, smaller in magnitude than the corresponding Zeeman energy $g\mu_B H$, in the presence of strong external magnetic field H (Krey 1985). The latter result is also in agreement with the experimental observation (appendix B).

An interesting feature of the excitations in disordered systems is the possibility of localization. As for example, disorder in the electronic systems leads to localized electronic states (see Chowdhury 1986). In order to investigate whether the spin waves in SG are localized or extended Walker and Walstedt (1980) studied the localization index

$$L_\mu = \Sigma W_{i\mu}^2 / (\Sigma W_{i\mu})^2$$

where

$$W_{i\mu} = |\alpha_{i\mu}|^2 + |\beta_{i\mu}|^2$$

is the time-averaged value of $(\vec{M}_i \cdot \vec{M}_i)$. L_μ varies between unity (for a completely localized mode) and a value of the order of 1/N (for a mode extended over the entire system). The lowest lying states turn out to be strongly extended for purely isotropic RKKY SG models, whereas the higher excitations are much less extended. This observation is in qualitative agreement with the corresponding result obtained by Bhargava and Kumar (1979) using Green's function technique.

The electron spin resonance (ESR) experiments on magnetic systems are another way of probing the spin wave excitation mode corresponding to k=0 (see van Kranendonk and van Vleck 1958)
Let us begin with the macroscopic theories of low temperature ESR

in SG. Although these theories are extensions of the HS "hydrodynamic" theory, usually these cannot be called hydrodynamic because under most of the circumstances to be discussed below a gap appears in the excitation spectrum because of anisotropic interactions.

First of all, the original HS theory is inadequate to account for the ESR modes observed in real SG systems (appendix B). Two additional ingredients required are remanence and anisotropy. Before we go into the details of the macroscopic theories let us introduce the appropriate order parameters. In the so-called "triad" theory (Henley et al.1982, Saslow 1982a,b, 1983) the order parameter is an SO(3) rotation matrix R which represents a uniform rotation of a spin configuration from a reference configuration. Therefore,

$$Tr(R) = 1 + \cos \theta_T$$

where θ_T is the total rotation angle. For example, denoting the remanent magnetization in the reference state as $\vec{M}_r^{(0)}$, we have the remanent magnetization $\vec{M}_r = R(\vec{M}_r^{(0)})$ in the configuration which is generated by rotating the reference configuration by R. Let us denote the unit vector along the current direction of $\vec{M}_r$ as $\hat{n}$, i.e., $\vec{M}_r = M_r\hat{n}$. We also introduce a unit vector $\hat{N}$ which denotes the direction corresponding to the orientation of $\hat{n}$ for H = 0. Thus, $\vec{M}_r$ returns to the preferred orientation $\hat{N}$ when the external field is switched off. Note that because of the non-collinearity of the ground state configuration in SG the vector $\hat{n}$ by itself is insufficient to describe the spin configuration. Therefore, in the triad theory, one introduces an orthonormal triad of unit vectors $(\hat{n},\hat{p},\hat{q})$, instead of only one vector $\hat{n}$ (Saslow 1982a, 1983). When anisotropy is turned on, the "preferred overall orientation" of the spin triad $(\hat{n},\hat{p},\hat{q})$ is given by the so-called anisotropy triad $(\hat{N},\hat{P},\hat{Q})$. The anisotropy torque tends to align $\hat{n},\hat{p},\hat{q}$ with $\hat{N},\hat{P},\hat{Q}$ respectively and

$$TrR = 1 + 2\cos \theta_T = \hat{n}\cdot\hat{N} + \hat{p}\cdot\hat{P} + \hat{q}\cdot\hat{Q}.$$

On the other hand, in the so-called vector theory (Schultz et al.1980) the vector $\hat{n}$ alone is assumed to describe the ground state so that

$$\cos \theta_V = \hat{n}\cdot\hat{N}.$$

The free energy is written as

$$F = F_0 + F_H + F_a \quad (18.6)$$

where

$$F_0 = (M - M_r)^2/(2\chi) + \rho_s |\nabla\theta_\alpha|^2 \quad (18.7)$$

$$F_H = - \vec{M}\cdot\vec{H} \quad (18.8)$$

and F_a is the contribution from the anisotropy and $\vec{M}_r$ is the remanent magnetization. There are several possible forms of F_a.

In the triad model, the anisotropy is given by

$$F_a^T = - K_1 \cos \theta_T - (1/2) K_2 \cos^2\theta_T \quad (18.9)$$

whereas in the vector model

$$F_a^V = - K_1^V \cos \theta_V - (1/2) K_2^V \cos^2\theta_V \quad (18.10)$$

In a loose terminology, the first terms in (18.9) and (18.10) are called the unidirectional anisotropy and the second terms are called the uniaxial anisotropy. Note that the anisotropy in the triad model is not associated with any particular lattice direction; the anisotropy depends only on the magnitude of the total angle of rotation θ_T no matter whatever the axis of rotation. The latter feature is in agreemnt with the experimentally observed isotropy of the anisotropy energy in real SG (see appendix B). I would like to emphasize that the anisotropy exists in the triad model even in the absence of the remanent magnetization; $\vec{M}_r$ serves only as a "handle" which can be used to rotate the triad by applying external field. Moreover, by observing the $\vec{M}_r$ we can determine the orientation of at least one of the members of the spin triad.

Let us denote the angle between $\vec{M}_r$ and the cooling field $\vec{H}_c$ by θ_r and that between the external field $\vec{H}$ and $\vec{H}_c$ by θ_H. Henley et al.(1982) treated the general case $\theta_H \neq 0$ whereas Saslow (1980) studied only the special case $\theta_H = 0$. The longitudinal mode ω_L gets decoupled from the pair of transverse modes $\omega_\pm$ in the special condition $\theta_H = 0$:

$$\omega_\pm = (1/2)(H-H_r) \pm (1/2)[(H+H_r)^2 + 4H_i^2]^{1/2}$$

and $\qquad$ for $\theta_H=0$ $\qquad$ (18.11)

$$\omega_L = H_i$$

Similarly,

$$\omega_\pm^2 = (1/2)(H^2+H_r^2-H_i^2)$$

$$\pm (1/2)[(H^2+H_r^2-H_i^2)^2 + 4H_rH(H_i^2-H_rH)]^{1/2}$$

and $\qquad$ for $\theta_H=\pi$ $\qquad$ (18.12)

$$\omega_L = 0$$

The mixing of the longitudinal mode with the transverse modes provides an indirect way of detecting the former. This longitudinal mode has been observed experimentally (see appendix B) and in computer simulations (Morgan-Pond 1983). What is the physical origin of the longitudinal mode in SG? It is the small amplitude oscillations about n, which never changes $\vec{M}_r$ that gives rise to the longitudinal mode detected in ESR experiments.
If $K_2 < K_1$, the remanence tries to rotate towards the external field H directly, and hence n lies in the $\vec{H}$-$\vec{H}_c$ plane. This state is called the "planar" solution (Henley et al.1982). On the other hand, if $K_2 > K_1$, (n,p,q) is rotated by 180^0 about an axis midway between $\vec{H}$ and $\vec{H}_c$ in the stable state, called the π-rotation state (Saslow 1982). Which part of the anisotropy energy is more

important- unidirectional or uniaxial? The unidirectional part, which arises from the Dzyaloshinski-Moriya (DM) interaction (Levy and Fert 1981a,b, Levy et al.1982), gives rise to the lateral shift of the hysteresis loops in agreement with experiments (see appendix B). On the other hand, the finite width of the hysteresis loop observed in real experiments strongly indicates the existence of uniaxial anisotropy in real SG. But, ESR experiments strongly suggest that K_2 is negligibly small, so that the anisotropy is almost entirely unidirectional! These two apparently mutually-contradictory observations can be reconciled by explaining the uniaxial anisotropy as a remanence effect (Henley 1983). Therefore, the π-rotation state may not be experimentally observable.

Although the triad model has received partial microscopic justification (Beton and Moore 1984), the microscopic approaches to the calculation of the ESR frequencies (Barnes 1981a,b, Becker 1982a, Beton 1985) remains less satisfactory than the macroscopic theories discussed in detail in this chapter.

As stated earlier, macroscopic theories of the spin wave excitations in SG assume a nonzero spin wave stiffness constant in these systems. Computer simulation of Heisenberg SG with Gaussian-distributed exchange interactions (Reed 1979) and of the RKKY SG model (1.7a) (Walstedt 1981) in d=3 indicated the existence of a nonzero spin wave stiffness. So far as the dependence of the stiffness constant ρ_s on the concentration of the magnetic constituent c in RKKY SG is concerned, computer simulationdata suggest $\rho_s \propto c^{4/3}$ and ρ_s is about 25 times smaller than that of the corresponding value in a comparable ferromagnet (Walstedt 1981), the latter observations are consistent with HS. Consider a Heisenberg SG in d=3. For a twist by an angle $\Delta\theta$ across a sample of length L and cross sectional area A, the change in the free energy ΔF scales as $\Delta F \sim \rho_s (\Delta\theta)^2 A/L$ (Walstedt 1981, Banavar and Cieplak 1982b). The scaling of the exchange stiffness will be discussed in more details in chapter 26. The temperature-dependence of ρ_s is very interesting; it vanishes as $(T_g - T)^{\mu}$ where $\mu = 3$ in the MFA. (Sompolinsky et al.1984).

Moreover, so far as the time-dependence of ρ_s is concerned, the latter has been interpreted as a "remanence effect", if the system could relax by hopping to all the other states ρ_s would vanish.

See Kumar et al.1986, Kumar and Saslow 1986 for the macroscopic theory of spin wave-like excitations in XY SG. Also see Beton and Moore (1985) for the solitons (see Bishop and Schneider 1978, Bullough and Caudrey 1980, Eilenberger 1981 for introduction to solitons) in SG.

CHAPTER 19

SPIN DYNAMICS IN SG: RELAXATIONAL MODES AND CRITICAL DYNAMICS

In the preceeding chapter we have discussed the dynamics of SG at $T << T_g$, mostly using hydrodynamic approach. The latter probe essentially the small amplitude long-wavelength oscillations of the systems about equilibrium. However, such an approach is not applicable near a critical point because the hydrodynamic condition $q\xi << 1$ breaks down in this regime. Besides, the hydrodynamic theory does not describe the decay of a non-conserved order parameter to equilibrium. In the SG terminology, the basic harmonic approximation strictly restricts the spin wave excitation to the small energy excitations within a given (free)energy minimum whereas near the critical temperature sufficient thermal energy is available for flipping of large clusters so that the system can hop from a given minimum to another.

The simplest approach to the study of the relaxational modes is the so-called time-dependent Ginzburg-Landau (TDGL) formalism. The general TDGL equation for an arbitrary conserved order parameter Ψ is given by

$$(\partial\Psi/\partial t) = \Gamma \nabla^2 (\delta G/\delta\Psi) + \eta \qquad (19.1)$$

which totally neglects terms proportional to the stiffness constant ρ_s and hence does not give any propagating solution like spin waves. The statistically-defined noise term $\eta(r,t)$ ensures the relaxation of the system to thermodynamic equilibrium defined by the free energy functional G. Γ is a transport coefficient.

For $k \to 0$, the Fourier components of the order parameter relax by a diffusive process, with a characteristic frequency

$$\omega_k = \Gamma \chi_\psi^{-1}(k,\xi)\, k^2$$

where χ_ψ^{-1} is the static susceptibility associated with the order parameter Ψ. Since χ_ψ diverges at the critical point, one would expect $\omega_k \to 0$ as T approaches the critical temperature, provided Γ does not diverge simultaneously. This phenomenon, first pointed out by Van Hove, is called critical slowing down. Physically, it means that once the system is displaced from (one of its) equilibrium and then allowed to relax, it would take longer to attain equilibrium configuration as it approaches closer to the critical point.

The TDGL equation for a nonconserved order parameter reads as

$$(\partial\Psi/\partial t) = -\Gamma\,(\delta G/\delta\Psi) + \eta \qquad (19.2)$$

If the Ψ^4 term in G could be neglected, equation (19.2) would lead to exponential relaxation of the Fourier components of the order parameter with a rate

$$\omega_k = \Gamma\,\chi_\psi^{-1}(k,\xi)$$

which, in turn, implies critical slowing down.

Thus, the first step towards setting up a TDGL approach for SG would be to construct a free energy functional. As a first approximation one assumes that for Ising SG, the TAP free energy is good enough so that

$$G(q) = J\,[(1/2)\tilde{t}^2Q^2 + (2/3)\tilde{t}Q^3 + (1/4)Q^4] \qquad (19.3)$$

where $\tilde{t} = (T-T_g)/T_g$, and J is the r.m.s. width of the Gaussian distribution of the exchange. Notice that (19.3) contains neither M nor its higher powers. More crucial observation is that (19.3) contains terms proportional to both odd and even powers of q whereas for ferromagnets, from symmetry considerations, only even powers of the corresponding order parameters M appear. Using (19.3) together with (19.2) one gets (Kumar and Barma 1978)

$$q \sim t^{-\nu} \quad \text{with } \nu = 1/2 \qquad (\text{for } T \simeq T_g)$$

and

$$q \sim \exp(-\tilde{t}^2 t/\tau) \qquad \text{(for } T > T_g\text{)}$$

Later Sastry and Shenoy (1978, 1979) generalized the free energy functional (19.3) to include terms involving even powers of M, terms coupling M and q and a term coupling M with the magnetic field :

$$G = G_q + G_m + G_{mq} + G_{mh} \qquad (19.4)$$

where G_q is given by (19.3) and

$$G_m = aM^2/2$$

$$G_{mq} = -bM^2q - cM^2q^2$$

and

$$G_{mh} = -\,MH$$

where $b = b_o\tilde{t}^2$ and $c = c_o\tilde{t}$ to the leading order. The free energy functional (19.4) led to coupled relaxation of the order parameters M and q with different time dependences in the three regimes of time, viz., short, intermediate and long.

Ma and Rudnick (1978) proposed a Ginzburg-Landau-Wilson free energy functional for m-component spin field that incorporates random uniaxial anisotropy as well as magnetic field, viz.,

$$G[\phi] = \int dx\cdot[(1/2)r(x)\left|\vec{\phi}(x)\right|^2 + (1/2)\ |\nabla\vec{\phi}|^2 + (1/8)u((\vec{\phi}(x))^2)^2)$$

$$+ \{\vec{a}(x)\cdot\vec{\phi}(x)\}^2 + \vec{h}(x)\cdot\vec{\phi}(x)] \qquad (19.5)$$

where $\phi(x)$ is the m-component classical spin-density field in a d-dimensional space x; r and u are the usual phenomenological parameters, $\vec{a}(x)$ is a random local anisotropy axis and $\vec{h}(x)$ is a spatially-varying external magnetic field. Starting from (19.5), Ma and Rudnick (1978) claimed to have obtained $t^{-1/2}$ decay of the spin-correlation function in the long t regime. However, the treatment for the static properties of the model (19.5) was erroneous, as pointed out later by Sherrington (1980) and, hence the result for the dynamics of the model might also be questionable. More serious shortcomings of the model (19.5) is that it does not contain frustration in explicit manner.

Hertz and Klemm (1979) considered two models- in the first the dynamics is purely dissipative (spin is assumed to be a non-conserved vector) and described by (19.2) where G is given by the Hertz-Klemm model described in chapter 1. Diagrammatic perturbation theoretic treatment of this model reproduces $t^{-1/2}$ long-time tail of the spin-correlation function. In the second model, the dynamics is described by (19.1) to conserve spin and an extra term is added to the equation of motion in order to describe the precession of a spin in the field of its neighbours, viz.,

$$(d\vec{S}_i/dt)_{pres} = \Sigma K_{ij} \vec{S}_j \times \vec{S}_i$$

In the conventional terminology the latter term is called a "mode-coupling" term for obvious reason. This model predicts that the spin-diffusion coefficient diverges as

$$\Gamma \sim (T - T_g)^{-1/2} \quad \text{for } d > 4$$

and

$$\Gamma \sim (T - T_g)^{2/d-1} \quad \text{for } 2 < d < 4$$

(see also Hertz and Klemm 1978). All the dynamical theories described above break down below T_g because of the same reason as for the SK solution, namely, assumption of a single order parameter to describe the SG phase. We shall discuss the more recent, and successful, TDGL approaches to the SG dynamics later in this chapter. Usually, the relaxation dynamics in SG is explained as a consequence of the hoppings between the different valleys in the phase space over the separating barriers. On the other hand, small amplitude oscillations around local minima that give rise to the spin wave like excitations are assumed to dominate the low-temperature dynamics. However, Bray and Moore (1981c, also see Bray et al.1983) argued that the slow decay of the spin autocorrelation at very low temperature can be explained as a consequence of small amplitude oscillations around free energy minima. They showed that in the case of the XY model

$$\mathcal{H} = -(1/2)\Sigma J_{ij}\cos(\theta_i - \theta_j)$$

with random J_{ij}, the autocorrelation

$$a(t) = N^{-1}\Sigma\langle\cos[\theta_i(t) - \theta_i(0)]\rangle = 1 - T[\rho(0)\ln t + ..]$$

where $\rho(\lambda)$ is the density of states of the Hessian

$$A_{ij} = (\partial^2\mathcal{H}/\partial\theta_i\partial\theta_j)$$

evaluated at a local minimum of the free energy.

Another approach to the relaxational dynamics involves the Master equation using Glauber single-spin-flip dynamics. Notice that since the Ising spins commute with the corresponding hamiltonian $dS_i/dt = 0$. In other words, an Ising spin system, by itself, cannot have dynamics. But any realistic system consists of many degrees of freedom other than spin, e.g., phonon, etc.. Therefore, the Ising spins do exhibit dynamics as a consequence of exchange of energy with the other degrees of freedom. We are not intereested in the detailed dynamics of these non-spin degrees of freedom; the only important role these degrees of freedom play is to flip the spins so as to equilibrate the latter system. This purpose is served by assuming that the degrees of freedom other than spin form a "heat bath" which is in contact with the system. (For example, the conduction electrons constitute a part of the heat bath in metallic SG). The Master equation for the probability distribution $P(S_1,..,S_n;t)$ for spin configuration $\{S_1,..,S_n\}$ of an n-(Ising)spin system is written as

$$\begin{aligned}(dP(S_1,..,S_n;t)/dt) = &- \Sigma W(\vec{r}_i,\uparrow|\downarrow)\,P(S_1,.., S_i,..S_n;t)\\ &+ \Sigma W(\vec{r}_i,\downarrow|\uparrow)\,P(S_1,..,-S_i,..S_n;t) \quad (19.6)\end{aligned}$$

where $W(r_i, \uparrow|\downarrow)$ is the probability of a transition from the up state to the down state of the spin at the site $\vec{r}_i$. The thermally-averaged local magnetization m_i is defined as

$$m_i = \Sigma\, S_i\, P(S_1, .., S_i, .., S_n)$$

where the summation is to be carried over all possible spin configurations. The first application of the Glauber model to the Ising SG (Fischer 1977) gave an exponential relaxation of the EA order parameter for both $T > T_g$ and $T < T_g$, except at $T = T_g$ where relaxation was governed by a power law. Similar results have been obtained also for the relaxational dynamics of SG with arbitrary spin value S (Riess et al. 1978) and also for vector SG in external magnetic field (Chowdhury and Mookerjee 1983c). The disagreement of the prediction of these theories with the corresponding experimental results is attributed to the assumption of single order parameter with single relaxation time (also see Bray et al. 1979 where a "hard-to-justify-assumption" led to better agreement with experimental results).

Kinzel and Fischer (1977b), Mody and Rangwala (1981) developed a Master equation approach to the SG dynamics. A distribution of relaxation times could be taken into account by focussing on the evolution of the staggered magnetization m_λ, defined through (3.6)(see Kinzel and Fischer 1977b, also see Shukla and Singh 1981a,b for the quantum spherical model):

$$(1 + \tau_0\, d/dt)\langle m_\lambda m_\lambda(t)\rangle = \beta J_\lambda \langle m_\lambda m_\lambda(t)\rangle$$

This equation leads to exponential decay of the correlation for $T > T_g$. The strong overlap between the various modes below T_g is taken into account through the self-energy $\Sigma(\omega)$ in the Dyson equation (Fischer 1983a,b)

$$\beta\chi_\lambda^{-1}(\omega) = \beta\chi_{\lambda 0}^{-1}(\omega) + \Sigma(\omega)$$

where $\chi_\lambda = dm_\lambda/dh_\lambda$ is the staggered susceptibility. Moreover,

(6.10b) should be taken as the mean-field expression for the local field. There are indications of slow decay of the correlation function for $T < T_g$ (Fischer 1983a,b), in qualitative agreement with the corresponding result of Sompolinsky and Zippelius (1981,1982).

Sompolinsky and Zippelius (SZ) (1981,1982) developed a novel TDGL approach using equation (19.2). This approach is based on DeDominicis' (1978, 1979) functional integral method based on an extension of the Martin-Siggia-Rose formalism (1973)(see also Ioffe 1983). In this approach one begins with the soft-spin Hamiltonian introduced in chapter 1. One defines a generating functional for dynamic correlations and response functions as

$$Z\{J_{ij},l_i,\tilde{l}_i\} = \int DS\, D\sigma \exp[\int dt\; l_i S_i(t)+i\tilde{l}_i\sigma_i(t)+L\{S,\sigma\}]$$

where

$$L\{S,\sigma\} = \int dt\; \Sigma i\sigma_i(t)[-\Gamma_0^{-1}\partial_t\sigma_i(t)-r_0S_i(t)+\beta\Sigma J_{ij}S_j(t)$$

$$- 4uS_i^3(t)-h_i(t)+\Gamma_0^{-1}i\sigma_i(t)] + V\{S\} \qquad (19.7)$$

where

$$V = -\;(1/2) \int dt\; \Sigma\; \{\delta^2(\beta H)/\delta S_i^2\}$$

$$= - \int dt\; \Sigma\; [(1/2)r_0 + 6uS_i^2(t)]$$

and $\int DS\, D\sigma$ denotes $\int \Pi\, [dS_i(t)\; d\sigma_i(t)]$ and $i\sigma$ is an auxiliary field which acts as a response field $\partial/\partial h_i(t)$. Therefore, the response function χ_{ij} is given by

$$\chi_{ij}(t-t') = \langle i\sigma_j(t')S_i(t)\rangle$$

The Dyson equation for the response function is

$$\chi^{-1}(\omega) = G_0^{-1}(\omega) + \Sigma(\omega)$$

where G_0 is the bare propagator and Σ is the self-energy. Unlike the earlier treatments of the dynamics (Hertz and Klemm 1979), SZ calculated the leading singularity of the the self-energy Σ to all orders in u. The most important result of their treatment is that ν is temperature-dependent, viz.,

$$\nu = (1/2) - (1/\pi)(1 - T/T_g) + O((1 - T/T_g)^2)$$

so that $\nu = 1/2$ only at $T = T_g$ (also see Schuster 1981, Hertz and Klemm 1983a,b).

As stated in chapters 9 and 10, the probability for a state α being occupied is given by $P_\alpha \sim \exp(-\beta F_\alpha)$ where F_α is a random variable with exponential distribution. This remarkable feature of the SK model (as well as the REM) leads to the stretched exponential relaxation (De Dominicis et al.1985) in agreement with the experimental observation (see appendix B).

All the works mentioned above in this chapter are mean-field theories (MFT). No well established analytical result is available for the critical dynamics of short-ranged Ising SG in d=3. Extending an earlier work on ferromagnets by Dhar (1983), Randeria et al. (1985) assumed that above T_g the "compact unfrustrated clusters" of Ising spins in SG contribute independently to the correlation function

$$q(t) = (1/N) \Sigma \langle S_i(0) S_i(t) \rangle$$

and would provide a lower bound on q(t) in d-dimension. Unfortunately, their prediction

$$q(t) \geq A \exp[- C (\ln t)^{d/(d-1)}]$$

(A and C are nonuniversal constants) is <u>not</u> supported by the MC simulation of the ±J model in d=3 (Ogielski 1985). On the other hand, there are claims (van Hemmen and Suto 1985) that the relaxation of TRM should follow an enhanced power law

$$M = M_0 \,(t/\tau)^{-A[\ln\,(t/\tau)]^{y-1}} \text{ with } y \geq 1 \text{ at } T < T_g.$$

Binder and Schroeder (1976) observed nonexponential relaxation of M and q in the nearest-neighbour (nn) Gaussian SG model in d=2. The MC data for M could be fitted to logarithmic decay as well as power law decay. When the magnetization data was fitted to a power law of the form $M(t) = M(0)\ t^{-a}$ it was found that the power a depends on temperature (Binder and Schroeder 1976) as well as the field that created the magnetization (Kinzel 1979). Similar results have also been obtained for the two-dimensional ±J model (Jaggi 1980). But we shall see in chapter 24 that there is no SG transition at nonzero temperature in d=2 ! More surprisingly, even the Mattis model and random Ising chains with Gaussian-distributed random exchange, which do not have frustration, exhibit logarithmic relaxation (Fernandez and Medina 1979, Medina et al. 1980) ! Numerical solution of the Glauber equations for finite linear Ising chain with two different probability distributions (Kumar and Stein 1980) throws light on the possible origin of the slow relaxation. In one of the models J was equally likely to be ferromagnetic and antiferromagnetic; this model showed logarithmic relaxation of magnetization after an initial regime of fast relaxation. On the other hand, the second model contained only positive J and the relaxation was exponential for all t investigated. Thus, even if frustration is absent (e.g., in one dimension) presence of both ferromagnetic and antiferromagnetic interactions give rise to slow relaxation. The most interesting case is, however, three dimensional SG because the corresponding transition temperature is believed to be nonzero (see chapter 24). The results obtained from the most extensive MC simulation so far (Ogielski 1985) have been summarized in chapter 24. Simulation of rather small systems of Ising spins randomly occupying f.c.c. lattice sites and interacting via RKKY interaction also indicates logarithmically slow relaxation of the spin auto-correlation (Fernandez and Streit 1982).

As stressed throught this book, SG materials as well as SG

models have a distribution of relaxation times (DRT). Therefore, at all temperatures in the SG phase, the order parameter relaxation is <u>not</u> of the Debye type

$$q(t) = q_0 \exp(-t/\tau) \qquad (19.8)$$

but much slower. The stretched exponential form (also called Kohlrausch law)

$$q(t) = q_0 \exp[-(t/\tau)^{1-n}] \quad (0 < n \le 1) \quad (19.9)$$

seems to be the most popular, although some other forms, e.g., logarithmic decay, are close competitors. It is almost trivial to derive (19.9) assuming an "appropriate" form of DRT, $g(\tau)$, and substituting into the expression

$$q(t) = \int g(\tau) \exp(-t/\tau)\, d\tau \qquad (19.10)$$

But most often it is very difficult to justify the form of $g(\tau)$ from microscopic considerations (Ngai 1979, 1980, Ngai et al. 1984).

In recent years, the differences between "parallel relaxation" and "series relaxation" has been stressed very strongly (Palmer et al. 1984). In parallel relaxation all the degrees of freedom x_i relax simultaneously, and independently, with characteristic time τ_i. On the other hand, in series relaxation relaxation of different subsets of the degrees of freedom take place sequentially and relaxation of one subset constraints that of another in a correlated way. Hierarchical relaxation is an example of series relaxation. The first step towards a description of the hierarchical relaxation is to arrange the relevant deegrees of freedom in a hierarchy, beginning with the fastest and ending with the slowest. Then one assumes that the faster degrees of freedom must relax and "move out of the way of the relaxation of the next slower degrees of freedom". More specifically, in the context of SG, let us assume that Ising spins (or pseudospins) be arranged in an n-level hierarchy so that the n-th level of the hierarchy contains N_n number of spins. Then, the assumption of hierarchical relaxation can be stated as follows: each spin in the (n+1)-th level is free to relax if and only if μ_n spins ($\mu_n \le N_n$) in the n-th level have attained one particular state of their 2^{μ_n} possible ones. Now, (19.10) is replaced by

$$q(t) = \Sigma\, g_n \exp(-t/\tau_n) \qquad (19.11)$$

where $g_n = N_n/N$ as well as τ_n (remember τ_n depends on μ_n) are two sets of parameters which are not derivable from the theory but have to be specified as extra inputs for the theory. Several alternative postulates for μ_n and N_n have been considered by Palmer et al. (1984). The best possible forms of μ_n and N_n, in the context of SG, are

$$\mu_n = \mu_0\, n^{-p} \qquad (n \ge 1)$$

and $$\qquad\qquad (19.12)$$

$$N_n = N_{n-1}/\lambda \qquad (n \ge 0)$$

For p=1, the forms (19.12), together with the assumption $\tau_{n+1} = 2^{\mu_n}\tau_n$ ($n \ge 0$), lead to the Kohlrausch law (19.9) for all $t > \tau_0$ (τ_0 is a microscopic time ~ 10^{-14} or 10^{-13} sec.), in agreement with recent experimental results of Chamberlin et al. (1984). However, several other forms of N_n also lead to the Kohlrausch law or laws indistingushable from the latter experimentally. (For discussions on the time scale where the model predicts Kohlrausch law and for discussions on the approximations made to derive the latter law, see Zwanzig 1985, Palmer et al. 1985). For p slightly greater than 1, relaxation form crosses over to purely exponential form (19.8) for $t > \tau_{max}$, where

$$\tau_{max} = \tau_0 \exp[\text{const.}/(p-1)]$$

which, after linearization of the temperature-dependence of p near T_o (T_o corresponds to p=1) leads to the Vogel-Fulcher law (B.6). Note that in the latter situation it is τ_{max}, rather than the "characteristic time" τ that enters into equation (19.8), that follows the Vogel-Fulcher law.

The nature of the spin dynamics determines the behaviour of several thermodynamic and transport properties. We shall discuss the applications of the ideas developed in this chapter to the study of ultrasonic attenuation, resistivity, magnetoresistance, etc., in chapter 25.

Critical dynamics gives rise to the nontrivial variation of the ESR line width and line shift with temperature near T_g, as summarized in appendix B. The line width and the line shift in an ESR experiment are approximately proportional to the real and the imaginary parts, respectively, of (see, for example, Raghavan and Levy 1985)

$$\Sigma(\omega) = \chi_0^{-1} \int dt \exp(i\omega t) [\dot{S}^x(t), \dot{S}^x(0)]$$

where ω is the resonance frequency, S^x is the total spin in a direction transverse to the applied field. The susceptibility along x is χ_0. The Kubo correlation function [A,B] is defined by

$$[A,B] = \int_0^\beta d\lambda \, Tr\,[\rho A(i\lambda)B^*] - \beta\, Tr\,(\rho A)\, Tr(\rho B^*)$$

where $A(i\lambda) = e^{\lambda H} A e^{-\lambda H}$, ρ is the density matrix, A and B are Heisenberg operators, i.e., $A(t) = e^{iHt/\hbar} A(0) e^{-iHt/\hbar}$. Levy et al.(1983,1984) used the Mori-Kawasaki and Memory function formalism and decoupled the four-spin correlation function in the space of the eigenstates of the exchange matrix J (see Salamon and Herman 1978, Salamon 1979, Becker 1982a,b, Barnes 1981a,b for the earlier works). This approach could not explain the scaling form associated with the line width observed experimentally (see appendix B). On the other hand, Huber (1985) decoupled the four-spin correlation function in the reciprocal space and achieved

good agreement with the experimental data. However, we shall not discuss these works further in this book.

CHAPTER 20

FRUSTRATION. GAUGE INVARIANCE. DEFECTS AND SG

We have defined the concept of frustration in chapter 1 at an elementary level. In this chapter we shall examine its meaning and the deeper implications in a more general context (see also Erzan 1984 for a review).

The geometry of the spin-ordering in a SG has similarity with the "parallel transport" of a tangent vector on a curved surface (e.g., in general relativity). The misfit between the various lines of transport is expressed by the "frustration" and "curvature", respectively, in the two cases. In this sense, frustrated plaquettes are "curved" whereas unfrustrated ones are "flat"; all the plaquettes in the Mattis model are flat. This analogy with the general theory of relativity is not accidental. The analogy, at a deeper level, lies in the gauge symmetries involved in the two theories (Toulouse 1977, 1980b). The essential ingredients of a gauge theory (see Moriyasu 1983 for an elementary introduction to gauge theory) are the following:
(i) a symmetry group defining the underlying gauge symmetry,
(ii) the gauge field defining the "connection", and
(iii) the "matter" field that acts as the sources of the gauge field.

As an example, the gauge group of the electromagnetic field is the group U(1), the matter field is the electron field and the gauge field is the electromagnetic potential. In the lattice formulation of the gauge theories (see Callaway 1985 for an elementary introduction) the "matter" field is associated with each of the lattice sites and the gauge field is associated with each of the bonds. There exist close similarities between the usual statistical mechanics of spin systems and lattice gauge theory (Kogut 1979, 1984, Toulouse 1980b, Creutz 1982).

Let us first consider an Ising spin system. The Hamiltonian (1.1) is invariant under the transformation

$$S_i \rightarrow - S_i \tag{20.1a}$$

and

$$J_{ij} \rightarrow - J_{ij} \text{ for all j adjacent to i} \tag{20.1b}$$

Note that the transformation (20.1) is (a) local and (b) mixed. At first sight it might appear similar to the gauge transformation in field theories, e.g., in electrodynamics. However, there is a crucial difference; the exchange variables which are the analogs of the vector potential are quenched variables. The analogy would be complete if the bonds J_{ij} were annealed variables. In the annealed problem, the thermal average of the frustration function Φ is nonzero whereas the latter vanishes in the quenched problem, i.e.,

$$<\Phi>_a = (\tanh 1/T)^4$$

and

$$<\Phi>_q = 0.$$

It would, of course, be erroneous to treat the exchange bonds in SG as annealed variables. However, Toulouse and Vannimenus (1980) argued that the error due to the annealed approximation can be compensated by imposing an extra constraint which would guarantee vanishing of the average frustration. Using a Lagrange multiplier β_p, the so-called plaquette temperature, the "gauge-annealing" free energy is given by

$$F_G/T = - \ln[\Sigma\Sigma \exp\{\beta_p \Sigma JJJJ + \beta_L \Sigma J_{ij} S_j\}] \tag{20.2}$$

and

$$(1/T)(\partial F_G/\partial \beta_p) = <\Phi>_G = 0 \tag{20.3}$$

imposes the quenched constraint. F_G corresponds to the Z_2 gauge theory on a cubic lattice where β_L is the link temperature. The phase transition lines are determined by the free energy (20.2). Equation (20.3) represents a trajectory in the (β_p, β_L) plane. Thus, a SG transition would correspond to this trajectory crossing a phase transition line. Moreover, β_p must be negative to ensure the constraint (20.3). MC simulation (Bhanot and Creutz 1980) of Z_2 lattice gauge theory in d=4 showed that the trajectory given by (20.3) touches a phase boundary at $\beta_p = -\infty$ and $\beta_L = \infty$ therby indicating the possibility that the LCD $d_l = 4$ for Ising SG. However, in a more recent simulation (Aeppli and Bhanot 1981) in d=6 no intersection of any of the transition lines with the trajectory (20.3) was observed at any finite value of either of the two couplings. If we believe that $d_u = 6$ for Ising SG, the latter result might imply inapplicability of the lattice gauge theoretic approach to this problem.

Since the Hamiltonian (1.1) is invariant under the transformation (20.1), the partition function is a functional of only the gauge-invariant quantities. The most general gauge-invariant quantity is frustration. Therefore, it is more sensible to work with the frustration content of the given lattice rather than the individual concentrations of the ferromagnetic and the antiferromagnetic bonds. For example, in the case of square lattice, the fraction of the frustrated plaquettes, c_p, is given by (Vannimenus and Toulouse 1977, Kirkpatrick 1977b)

$$c_p = 4[x^3(1-x) + x(1-x)^3]$$

where x is the concentration of, say, the ferromagnetic bonds. It is also possible to define gauge-invariant correlation functions $\langle S_i(\Pi J)S_j\rangle$ where C(i,j) is the path connecting the sites i and j (Fradkin et al.1978). Moreover, the difference between the correlation function evaluated along two different paths connecting the same end points i and j is given by $(-1)^{f_P}$ where f_P is the number of elementary frustrated plaquettee within the closed loop formed by the two paths.

The high ground state degeneracy of SG is a consequnce of the existence of frustration. Now we shall a geometrical interpretation of this fact. Let us draw a line perpendicular to all the unsatisfied bonds in the frustrated plaquettes connecting the centres of the two adjacent plaquettes on the two sides of the bond. The string thus created by such lines will be called frustration string. As stated before, the free energy of a ±J SG, for example, depends only on its frustration content and not on the detailed distribution of the individual bonds. Therefore, he minimum energy configuration corresponds to the minimum length of the frustration string. In general, there are N_s number of ways of achieving the same minimal length, leading to N_s-fold degeneracy of the ground state.

The Ising SG model remains invariant under the gauge transformation (20.1). Let us investigate the corresponding gauge symmetry of the XY model. One way of introducing the disorder into the standard XY model is to write the Hamiltonian as

$$\mathcal{H} = - J \Sigma \cos (\theta_i - \theta_j - \psi_{ij}) \qquad (20.4)$$

where the randomness arises from the random "difference angle" ψ_{ij} rather than the exchange bonds. Defining a link-gauge degree of freedom

$$U_{ij} = \exp(i\psi_{ij})$$

Hamiltonian (20.4) can be rewritten as

$$H = (J/2) (S_i U^*_{ij} S^* j + \text{hermitian conjugate})$$

with $S_i = \exp(i\theta_i)$. The latter Hamiltonian is invariant under the local transformation

$$S_i \rightarrow V_i S_i \quad \text{and} \quad U_{ij} \rightarrow V_i U_{ij} V^*_j \qquad (20.5)$$

where $V_i = \exp(i\Theta_i)$. In other words, under this local transformation applied at the i-th site the corresponding spin gets rotated by an angle Θ_i and each of the connecting link degrees of freedom by an angle $\Theta_i - \Theta_j$ such that the Hamiltonians remains invariant under such transformation (Fradkin et al.1978). Note that the transformation (20.5) for the XY model is the analog of the transformation (20.1) for the Ising model. Now, one defines the frustration angle by the relation

$$\exp(i\ 2\pi\Phi_{ijkl}) = U_{ij}U_{jk}U_{kl}U_{li} \qquad (20.6)$$

Thus, a plaquette is frustrated if the corresponding Φ is a non-integer. By a duality transformation (see Savit 1980 for an introduction to duality) frustration of a plaquette in the XY model can be expressed as a monopole on the dual lattice (Izuyama 1980). For the detailed results on the energetics and the correlations by duality transformations in the frustrated models in d=2 and d=3 see Fradkin et al.(1978).

Our discussion on frustration in SG so far has been based on spin models on discrete lattices. Let us now discuss the continuum counterpart (i.e., two-component spin field on contonuum) of Ising SG on a discrete lattice. The starting point is the Hamiltonian (1.15) (Hertz 1978). First of all, one must distinguish "trivial disorder" (e.g., that in the Mattis model) from non-trivial disorder. If the Fourier components were completely longitudinal, i.e., $\vec{Q}(k) \propto \vec{k}$, then a gauge transformation

$$\vec{Q}(x) \rightarrow \vec{Q}(x) - \nabla\alpha(x)$$

and

$$\phi(x) \rightarrow \exp\{i\alpha(x)\}\ \phi(x)$$

where $\alpha(x)$ is a scalar function, can remove $\vec{Q}$ completely from the effective Hamiltonian; the problem would reduce to that of a nonrandom ferromagnet. In other words, the disorder in such cases is "trivial" and can be "gauged away". This would be the continuum version of the Mattis model. The structure of the free energy functional for the Hertz model (1.15) is very similar to the action in Eucledian boson electrodynamics. Condensed matter physicists would recognize the Hertz free energy functional as the analogue of the Landau-Ginzberg free energy functional for superconductors. However, $\vec{Q}(x)$, although analogous to the vector potential, is not a true gauge variable because of the quenched nature of the disorder. Moreover, there is no "mass generation" by the Higgs mechanism (the latter mechanism leads to the Meissner effect in superconductors). The distribution of the random variable Q, given by (Hertz 1978)

$$P(Q) \propto \exp[-\ (1/2f)\ \int d^dx\ \Sigma F^2_{\mu\nu}(x)]$$

involves gauge-invariant quantities

$$F_{\mu\nu}(x) = \partial_\mu Q_\nu(x) - \partial_\nu Q_\mu(x)$$

In this model f, the mean-square vorticity, is a measure of the frustration. Thus, the Mattis model corresponds to f=0. One can generalize these concepts to vector spin-fields and to situations where ferromagnetic interactions dominate over the antiferromagnetic ones (Hertz 1978).

A macroscopic theory of frustration in m-vector (m>2) spin models (Dzyaloshinskii and Volovik 1978, Dzyaloshinskii 1979) has been developed by extending the concepts of Yang-Mills gauge theories (see Jackiew 1980, Moriyasu 1983 for introduction to the Yang-Mills fields) to SG. The latter theory is a generalization of the theory of electromagnetic fields. To ensure local gauge invariance, one introduces the so-called Yang-Mills fields $\vec{b}_i(x)$, in addition to the vector spin field $\vec{S}(x)$. In contrast with the theory of the electrodynamic fields, the generators of the rotation now form a non-abelian group. Under infinitesimal coordinate-dependent rotation $\theta(x)$ the spin and the gauge fields transform as

$$\delta\vec{S}_i = (\vec{\theta} \times \vec{S})$$

and

$$\delta\vec{b}_i = (\vec{\theta} \times \vec{b}_i) + (\partial\vec{\theta}/\partial x_i),$$

respectively, which are the analogues of (20.1). If in addition to the spatial local invariance temporal local invariance is also required, one introduces another Yang-Mills field $\vec{a}(x)$ which transforms as

$$\delta\vec{a} = (\vec{\theta} \times \vec{a}) + (\partial\vec{\theta}/\partial t)$$

under time-dependent rotations $\vec{\theta}(t)$. Note that the Yang-Mills fields $\vec{b}_i$ and $\vec{a}$ are the analogues of the vector and the scalar potentials in electrodynamics. Moreover, one can define gauge-invariant quantities

$$\vec{f}_{ik} = (\partial\vec{b}_k/\partial x_i) - (\partial\vec{b}_i/\partial x_k) - (\vec{b}_i \times \vec{b}_k) \tag{20.7}$$

and

$$\vec{g}_i = (\partial\vec{b}_i/\partial t) - (\partial\vec{a}/\partial x_i) - (\vec{a} \times \vec{b}_i) \tag{20.8}$$

which are the analogues of the magnetic and electric fields respectively. In this scenario the frustration density and the furstration current density are defined as

$$\vec{\rho}_i = \epsilon_{ijk} (\partial\vec{b}_i/\partial x_k)$$

and

$$\vec{j}_i = (\partial\vec{b}_i/\partial t) - (\partial\vec{a}/\partial x_i)$$

and, therefore, directly related to the "magnetic" and the "electric" fields (20.7) and (20.8), respectively. Thus, frustration lines are the disclinations (see Harris 1977 for an elementary introduction to the concept of disclination) in the vector fields $\vec{S}$ and $\vec{\theta}$ (Dzyaloshinskii and Volovik 1978). Application of the concepts of defect to SG has proved very useful (Barahona et al.1982). It is Henley (1984a,b) who first emphasized the differences between the "absolute" and "relative" senses in which defect can be defined in SG. One can look at a metastable SG state as a long-ranged ordered state (ferromagnetic) plus topological defects, the latter being responsible for the noncolinear (random) frozen pattern of the actual SG state. This is the absolute use of the defect concept for SG; the defect-free state is not a SG. Thus, the frustration string is a defect in the absolute sense. On the other hand, beginning with a given state, one can generate the possible SG states simply by adding appropriate defects; such defects are relative. The relative defects in m-vector SG have been classified (Toulouse 1979) from topological point of view (see Mermin 1979 for an introduction). There can be three types of such defects for m=3 (Henley 1984a), viz., disclinations, domain walls and Hopf texture. Such defects have been investigated by computer simulation (Henley 1984b).

Villain (1977a) has also used the concept of defect in SG. Consider a two-dimensional classical spin system where $S_i = (\cos \theta_i, \sin \theta_i)$. In the case of the usual XY model the partition function depends on integrals containing $\exp\{-(J/k_BT)\cos\phi\}$, where ϕ is the angle between the two spins. Villain (1977a) argued that the essential physics of this model can be quite effectively simulated by replacing this exponential by

$$\Sigma \exp\{-g/(k_BT)\}[\phi - 2\pi n + \{(1-\epsilon)/2\}\pi]^2, \qquad \epsilon = J/|J| = \pm 1$$

provided n is an integer and the constant g is chosen appropriately. Now instead of the usual XY Hamiltonian, the new model consists of a "fictive Hamiltonian" (Villain 1977a)

$$\mathcal{H} = (1/2)\,\Sigma\, g_{ij}\,[\theta_i - \theta_j - \{(1-\epsilon_{ij})/2\}\pi - 2\pi n_{ij}]^2,$$

where n_{ij} is an integer variable. Carrying out simple transformations this Hamiltonian can be mapped onto an effective Ising Hamiltonian

$$\mathcal{H} = \Sigma\, J_{pp'}\,\tau_p\tau_{p'}$$

where $J_{pp'}$ is the d-dimensional Coulomb interaction between the effective Ising variables

$$\tau_p = 0 \qquad \text{for an unfrustrated plaquette}$$

and

$$(1/2)\tau_p = \pm 1/2 \qquad \text{for frustrated plaquettes}$$

τ is called the "chirality" of the plaquette. It is positive or negative depending on whether the spins rotate clockwise or counterclockwise during a "clockwise trip" around the plaquette (Villain 1977a) (Fig.20.1). The chiralities are reminiscent of the vortices introduced by Kosterlitz and Thouless to describe defect-mediated transitions (see Nelson 1983 for a review). The vortices in the Villain model are defect in the absolute sense; these are present even in the ground state. The Ising nature of the chirality also provides a microscopic justification for the existence of the TLS in SG (also see Kawamura and Tanemura 1985). The treatment has been extended to d=3 and to incorporate the effect of magnetic field (Villain 1978). For later studies of topological phase transitions in the frustrated XY model see Dzyaloshinskii and Obukhov (1982), and for the topological description of glasses see Rivier (1979), Venkataraman and Sahoo (1986).

Most of our attention have been focussed on the frustrated systems with disorder. However, as stated in chapter 1, frustration is possible even without disorder; the simplest example of the latter situation is a triangular lattice with antiferromagnetic nearest-neighbour exchange interactions only. The models in the latter category are usually called periodic frustrated models (Toulouse 1981). The Danielian's model (1964), the "Domino" model (Ander et al.1979, Villain et al.1980), the ANNNI model (Fisher and Selke 1980), metamagnets (Bruinsma and Aeppli 1984) etc. also belong to this category of models.

The fully frustrated models are special cases of periodic frustrated models where each of the elementary plaquettes are frustrated (Wannier 1950, Alexander and Pincus 1980, Derrida et al.1979, 1980, Phani et al.1979, Villain 1977b, Berker and Kadanoff 1980, Diep et al.1985a,b, Lallemand et al.1985). One striking feature of the fully frustrated models is the "overblocking effect". The latter concept can be explained simply

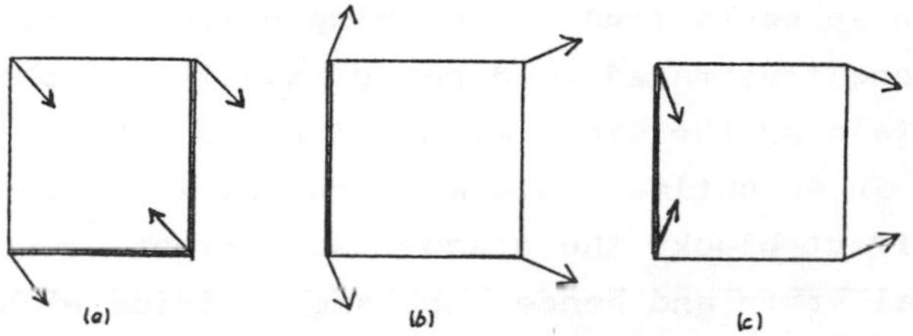

Fig. 20.1. (a) a unfrustrated plaquette, (b) a frustrated plaquette with chirality = 1, (c) a frustrated plaquette chirality = -1 (after Villain 1977a).

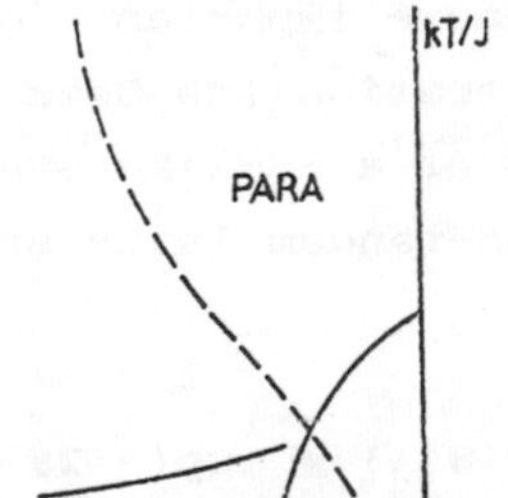

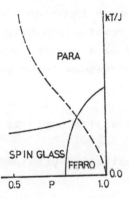

Fig. 20.2. The line exp $(-2\beta J) = (1-p)/p$ for the $\pm J$ model shown as a dotted line (after Nishimori 1981a).

with the illustration of triangular lattice with antiferromagnetic nearest-neighbour interactions. Normally, one would expect at most one frustrated bond per plaquette. But situations may arise where a finite fraction of the plaquettes have all the three bonds frustrated. Simple frustration (that gives rise to one frustrated bond per plaquette) blocks the system from attaining the state with all bonds satisfied and the system settles in an optimal (minimum frustration) state. The overblocking effect blocks the plaquettes further from achieving even the optimal state and hence the name (Derrida et al. 1979). "Superfrustration" is a term (Suto 1981) that describes very stong frustration that destroys phase transition even at $T = 0$ so that the correlation length remains finite at all T (including T=0). The layered frustrated Ising models (Hoever et al.1981, Wolff et al.1981, Hoever and Zittartz 1981, Wolff and Zittartz 1982, 1983, Karder and Berker 1982) have the advantage of being exactly sovable. What is more important for the glassy behaviour-quenched disorcer or frustration (de Seze 1977)? We shall examine this issue in the context of a SG-like system in chapter 27.1.

Let us consider bond-random Ising models for which (Nishimori 1980, 1981)

$$P(-J_{ij})/P(J_{ij}) = \exp(-2a\, J_{ij}) \qquad (20.9)$$

The relation (20.9) is satisfied provided the distribution is of the form

$$P(J_{ij}) = \exp(a\, J_{ij})\, f(J_{ij})$$

with $f(J_{ij}) = f(-J_{ij})$. As an example, the distribution (1.4), with $c_{bf} = p$ and $c_{ba} = 1-p$, can be recast into the form (20.9) with $a = K_p/J$ where K_p is defined through the relation

$$\exp(-2K_p) = (1-p)/p. \qquad (20.10)$$

Using the gauge-invariance of the free energy one can easily check that for $a = \beta = (k_B T)^{-1}$ the internal energy is given by

$$U = - N_B [J_{ij}]_{av} \qquad (20.11)$$

where N_B is the total number of bonds (interacting pairs). Note that for $a = \beta$, the relation (20.10) reduces to the form

$$\exp(- 2 \beta J) = (1 - p)/p,$$

the so-called Nishimori line on the p-βJ phase diagram. Interestingly, this internal energy is nonsingular on the Nishimori line although the latter intersects a phase boundary (see Fig.20.2). The result (20.11) has been investigated by various authors, e.g., in the light of the supersymmetry (Georges et al.1985).

Finally, I would like to end this chapter with a philosophical note. The exchange of the ideas between different branches of physics has often revealed hidden unities in the problems and the methodologies. In this chapter I have tried to provide a glimpse of the concepts common to SG and various other branches of physics.

CHAPTER 21

IS THE SG TRANSITION ANALOGOUS TO THE BLOCKING OF SUPERPARAMAGNETIC CLUSTERS ?

Tholence and Tournier (1974) and Wohlfarth (1977a) propsed that the SG transition is not a true thermodynamic phase transition but is very similar to the phenomenon of blocking of superparamagnetic single domain particles in rock materials (see Morrish 1965 for an introduction). Let us briefly review the latter phenomenon. The magnetic particles in rock materials are small enough to consist of a single domain and large enough to consist of a large number of moment-carrying atoms (or molecules). At high temperatures an assembly of such particles behave paramagnetically where each particle possesses a large magnetic moment and hence the name superpara-magnetism. In nature such particles get an extra contribution to their energy from anisotropy effects - anisotropy induced either by shape anisotropy, or crystalline anisotropy or by external stress. Since no domain wall motion is possible in such single domain particles they acquire magnetization by coherent rotation of magnetic moments against anisotropies present in them. Neel (1947) indicated the possibility of freezing of the moments of these single domain particles. The remanent magnetization relaxes as

$$M_r = M_S \exp(-t/\tau)$$

where M_S is the full magnetization when the field is switched off and the relaxation time τ is given by

$$(1/\tau) = (1/\tau_0) \exp(-KV/k_BT)$$

where K the anisotropy constant and V the volume of the particle, with $\tau_0 \simeq 10^{-9}$ sec. For a particular measurement, characterized by a typical measurement time τ_m, particles with a volume larger than a critical size appear frozen because their relaxation time will be longer than τ_m (compare with the theory of the two-level systems introduced in chapter 2).

Wohlfarth (1980) suggested the following possible sources of anisotropy in SG:
(i) surface anisotropy,
(ii) magneto-crystalline anisotropy,
(iii) shape anisotropy,
(iv) magnetostrictive strain anisotropy,
(v) other magnetostatic dipole-dipole effects,
(vi) anisotropic exchange effects.
This type of phenomenological theories rely heavily on the role of the characteristic time scales of observation in laboratory experiments (as well as in computer experiments). A list of the time scales of some of the experimental techniques is given in appendix B.

The susceptibility is given by (Wohlfarth 1979)

$$\chi(T) = (C/T) \int f(T')\, dT' \qquad (21.1)$$

where C is the Curie constant and f(T) is the distribution of blocking temperatures. Physically, the expression (21.1) stands for the fact that only those clusters contribute to the susceptibility whose blocking temperature $T_B < T$, i.e., those clusters which have not yet 'frozen'. Although this phenomenological theory does not provide any prescription for first-principle calculation of $f(T_B)$, it can be estimated from the experimentally measured values of $\chi(T)$ using the relation (Van Duyneveldt and Mulder 1982)

$$f(T) = \{d(\chi T)/dT\}/\lim_{T\to\infty} (\chi T) \qquad (21.2)$$

The expression 21.1 can be recast in terms of the distribution of the relaxation times $N(\tau)$ as (Murani 1981)

$$\chi(\tau) = (C'/T) \int N(\tau')\, d\tau' \qquad (21.3)$$

A distribution of the relaxation times leads to logarithmically slow relaxation of magnetization (Guy 1977, 1978) and hysteresis (Prejean and Souletie 1980, Souletie 1983).

Note that T_B depends on τ_m; the higher the frequency the higher is the T_B for a given V. This can qualitatively explain the frequency-dependence of the temperature corresponding to the cusp in the $\chi_{a.c}$ (Shtrikman and Wohlfarth 1981). Moreover, since τ_m (neutron scatt.) < τ_m (Mossbauer) < τ_m (a.c. susceptibility), one would expect
T_g(neutron scatt.) > T_g (Mossbauer) > T_g (a.c. susceptibility), if the SG freezing is analogous to the blocking phenomenon. Such a variation of T_g with the time scale of the experimental probes is in qualitative agreement with the corresponding result for most of the real SG systems. (For detailed comparisons of the properties of the SG and superparamagnets see Dormann et al. 1983, Fiorani et al.1986).

So far as the q-dependence of the temperature corresponding to the maximum of $\chi(q)$ is concerned, the phenomenological blocking theory is consistent with the experimental observation. Large q probes smaller clusters which freeze at lower temperatures. Therefore, $T_g(q)$ is a decreasing function of q (Murani 1977, 1978a). Murani (1980a) also criticized the interpretation of $T_g(q)$ by Soukoulis, Levin and Grest (see chapter 3). Murani argued that since in the cluster MFT of Levin et al. a smoothly varying part I_B, which vanishes at a sharply defined temperature, is assumed, the quantity left after subtracting the former will always have a shoulder in $T\chi$ at a temperature independent of q (see Levin et al.1980 for the counter arguments).

The main drawbacks of the cluster blocking models are the following:
(i) the susceptibility peak is rounded, though quite sharp, unlike the sharp cusp observed experimentally;
(ii) while the superparamagnetic particles in rocks are indeed

very dilute, the clusters in SG may not be very far from each other and hence the inter-cluster interaction may not be negligible;

(iii) if the chemical clustering is assumed to be the dominant source of cluster formation it will be inconsistent with the observed scaling laws;

(iv) the origin of the anisotropy is not well established;

(v) since this is a phenomenological theory, $f(T_B)$ cannot be calculated from first principles.

Finally, I would like to emphasize that the existence of the braoad distribution of relaxation times does not necessarily require a distribution of independent superparamagnetic clusters. A braoad distribution of the relaxation times can be incorporated within the MFT of interacting spins, as explained in chapters 8 and 9

CHAPTER 22

IS THE SG TRANSITION ANALOGOUS TO PERCOLATION ?

Smith (1975) propsed a percolation (see Stauffer 1985 for an introduction to the concept of percolation) model of the SG transition. The basic idea behind this theory goes as follows: as the system is cooled from a high temperature a fraction of the spins 'lock' together to form clusters within which spins are very strongly correlated. The clusters keep growing bigger and bigger at the expense of the 'loose' spins as the temperature is lowered more and more. This cluster evolution consists of two processes- more and more 'loose' spins lock together to form clusters and clusters formed at a higher temperature coalesce to form bigger clusters. At a temperature T_g an infinite cluster of the locked spins forms (i.e., percolation takes place) and the corresponding temperature is identified as the SG transition temperature. Abrikosov (1978a,b, 1980), Cyrot (1981) and Mookerjee and Chowdhury (1983b) improved the theory by taking the effect of the finite mean-free path of the RKKY interaction and frustration into account.

Around every spin let us draw an imaginary circle of radius $R = (A/k_BT)^{1/3}$, where A/R^3 is the RKKY interaction between two spins a distance R apart. If the spheres around two spins overlap with each other the two spins are said to be locked to each other. A cluster of n spins is formed if the spheres around n spins overlap with each other. Therefore, the SG transition temperature is determined by the threshold for the percolation of overlapping spheres (this percolation should not be confused with the ferromagnet (or antiferromaget) to SG phase transition with the percolation of the ferromagnetic (or antiferromagnetic) bonds in the ±J model). Therefore, the susceptibility is given by

$$\chi = \chi_p \; [1 - P(T)\; q(T)]$$

where P(T) is the fraction of spins included in the infinite cluster. In the percolation model $f(T) = -\,d(Pq)/dT$ is identified (Chowdhury and Mookerjee 1983a) as the distribution of the blocking temperatures in the same spirit as that in Wohlfarth's model described in the preceeding chapter. Unfortunately, $f(T)$ could be estimated only by fitting the theoretical prediction with the experimentally measured susceptibility (Chowdhury and Mookerjee 1984c). $f(T)$ was found to be sharply peaked near T_g.

In a dynamical experiment a cluster appears to be frozen provided the corresponding relaxation time is longer than the characteristic time of the measurement. The relaxation time τ was argued to follow the Vogel-Fulcher law. Moreover, the magnetization at time t is given by

$$M(t) = (1/N)\ \Sigma\ M(0) \exp\{-t/\tau(n)\}\ n\ P(n) + M_\infty$$

where $\tau(n)$ is the relaxation time of a cluster of size n, $P(n)$ is the cluster size distribution and M_∞ is the contribution to the magnetization from the infinite cluster. Using the form $P(n)$ known for the percolation one can show that (Chowdhury and Mookerjee 1984b) $M(t)$ relaxes logarithmically with time t. The latter also leads to 1/f noise spectrum (Chowdhury and Mookerjee 1983b) in agreement with the experimental observation (see appendix B).

The clusters in the percolation theory of SG have been argued (Chowdhury and Bhattacharjee 1984) to have a fractal geometry (Mandelbrot 1982). More precisely, the cluster size n_ξ is related to the characteristic correlation length ξ (note that $\xi \sim |T-T_g|^{\nu}$) through the expression $n_\xi \sim \xi^{d_f}$ where d_f is the fractal dimension (see Mandelbrot 1982 for an introduction to the fractal geometry). Similar ideas have also been developed independently by several authors. According to the scaling theory the cluster size distribution $P(n) \sim n^{-\tau} f(n/n_\xi)$, where $f(n/n_\xi) \sim \exp\{-(n/n_\xi)^{v}\}$ for sufficiently large clusters. The noise spectrum has been argued (Continentino and Malozemoff 1986a) to be of the form $S(\omega) \sim \omega^{y}$ where $y = (\nu z + 2\phi v)/(\nu z+\phi v)$ where $\phi = \nu d_f$ and z is the standard dynamic critical exponent

(Hohenberg and Halperin 1977). (See Continentino and Malozemoff 1986b for the dynamical susceptibility).

CHAPTER 23

IS THE SG TRANSITION ANALOGOUS TO THE LOCALIZATION-DELOCALIZATION TRANSITION?

Let us first briefly review the eigenstate-space technique of studying the phase transition in nonrandom systems. In such systems the eigenstate of the exchange matrix are plane waves $\exp(i\vec{k}\cdot\vec{r})$ characterized by the corresponding k values. The susceptibility is given by $\chi(\vec{k}) = 1/\{T-J(\vec{k})\}$. As the temperature is lowered starting from a high value, a magnetic ordering takes place at $T_c = J_{max}(\vec{k})$, $J_{max}(\vec{k})$ being the largest eigenvalue of the matrix J. If $J(\vec{k})$ is maximum for $k = 0$, the ordered phase is ferromagnetic, for $k \neq 0$ it is antiferromagnetic. Below T_c the corresponding eigenstate is macroscopically magnetized. This type of macroscopic condensation into one particular mode is reminiscent of the Bose condensation.

The situation changes drastically in SG systems. First of all, plane waves are no longer the eigenstates of J and hence one must use the general "eigenstate-space" technique instead of the k-space technique. Secondly, the random matrix J in SG may have both extended and localized eigenstates (see Fig.23.1). If the eigenstate corresponding to J_{max} is localized, as is usually the case, one gets a hypothetical transition temperature corresponding to to the localized eigenfunction. It is these eigenstates which so convincingly mimic the "clusters" of the Neel intèrpretation of the SG phenomenology above T_g (Anderson 1979). But, since no true phase transition can take place in finite systems the hypothetical transition mentioned above is meaningless. The true SG transition may take place at a lower temperature corresponding to the mobility edge. However, if all the eigenstates are localized no SG transition would be possible (Anderson 1970). Hertz et al.(1979) argued that the SG transition transition is associated with the mobility edge where macroscopic staggered magnetization appears in the first extended eigenstate of the exchange matrix. On the other hand, Sompolinsky (1981b)

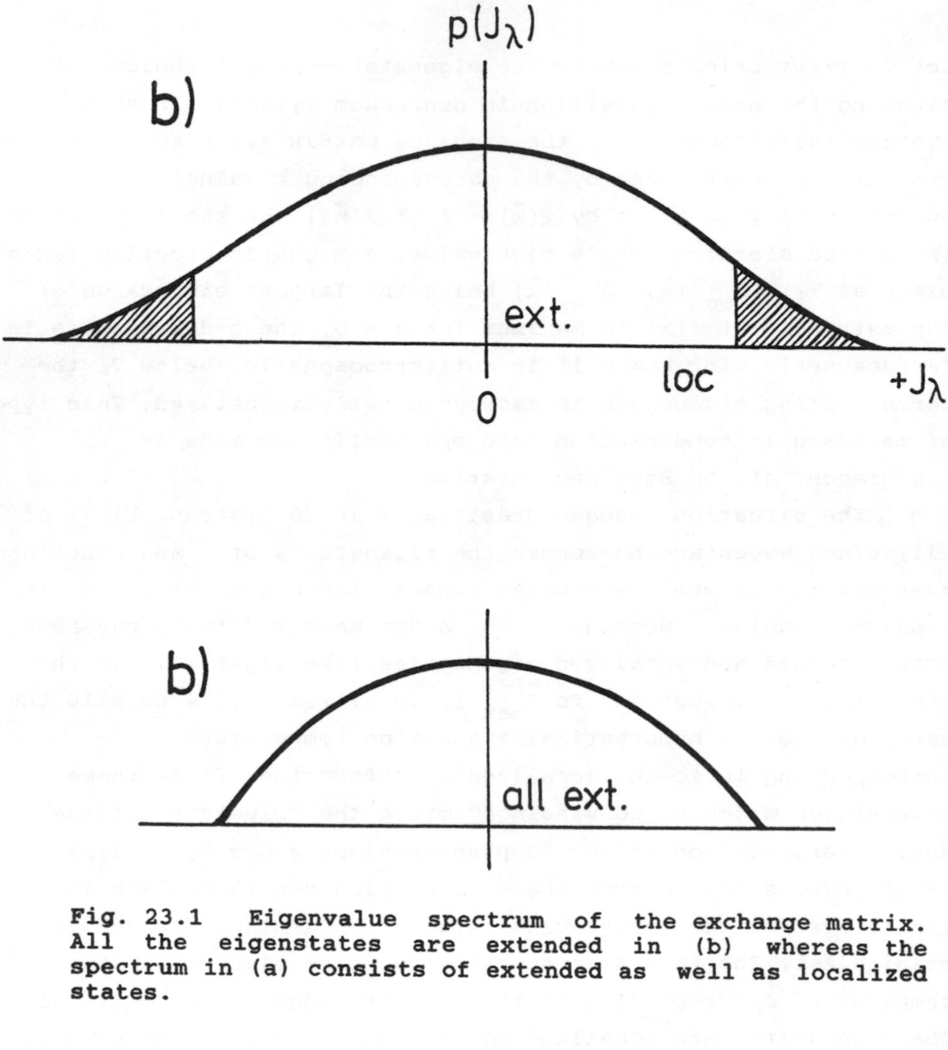

Fig. 23.1 Eigenvalue spectrum of the exchange matrix. All the eigenstates are extended in (b) whereas the spectrum in (a) consists of extended as well as localized states.

and Ueno (1983a,b) stressed that the SG order is characterized by a distribution of magnetizations throughout the whole spectrum, although the distribution is sharply peaked at J_{max} for all T < T_g (DasGupta and Sompolinsky 1983).

Bray and Moore (1982a) recast the same problem in terms of the eigenstates of the susceptibility matrix χ rather than those of the exchange matrix J. In this case the critical temperature is that at which an eigenvalue of χ^{-1} first vanishes. The "speckle picture" of SG proposed by Provost and Vallee (1983b) has been argued (Chowdhury 1984b) to be closely related to the localization- delocalization picture developed by Anderson (1970) and the other authors mentioned above.

Now we shall describe the formal structure of the theories developed along these lines. The free energy functional can be written in the basis of the eigenstates of J as (Hertz et al. 1979, Ueno 1983a,b, Chowdhury and Mookerjee 1984d)

$$F = (1/2) \Sigma\, r_\alpha S_\alpha^2 + (1/4) \Sigma\, u_{\alpha\beta\gamma\delta} S_\alpha S_\beta S_\gamma S_\delta$$

where $r_\alpha = (\beta J_\alpha - 1)$

and

$$u_{\alpha\beta\gamma\delta} \propto (k_B T)^{-3} \Sigma \langle\alpha|i\rangle\langle\beta|i\rangle\langle i|\gamma\rangle\langle i|\delta\rangle\, J_\beta\, J_\gamma\, J_\delta$$

Note that the terms involving $u_{\alpha\alpha\alpha\alpha}$ are the largest among all possible u's. The next largest ones are the $u_{\alpha\alpha\beta\beta}$. Indeed, if the modes α and β overlap strongly, $u_{\alpha\alpha\beta\beta}$ is of the same order as $u_{\alpha\alpha\alpha\alpha}$. On the other hand, $u_{\alpha\alpha\beta\beta}$ are negligibly small compared to $u_{\alpha\alpha\alpha\alpha}$ if the modes $\alpha$and β are localzed in regions far from each other. The couplings of the type $u_{\alpha\beta\beta\beta}$ are smaller than $u_{\alpha\alpha\beta\beta}$ because of the fluctuating sign entering through the odd powers of the wave functions. Roughly speaking, $u_{\alpha\beta\beta\beta}$ are smaller than $u_{\alpha\alpha\beta\beta}$ by a factor of the order of the square root of the overlap between the modes α and β. The three state elements of the form $u_{\alpha\beta\gamma\gamma}$ and four state elements of the form $u_{\alpha\beta\gamma\delta}$ are negligibly small unless all the corresponding modes are in the same region. The approximation in which all the couplings except $u_{\alpha\alpha\alpha\alpha}$ are neglected is called the independent-mode approximation (IMA) (Hertz 1983c). The effect of overlaps have been taken into account independently by Chowdhury and Mookerjee (1984d), Chowdhury (1985) and Hertz (1985). The most crucial effect of the overlaps is that the freezing of the modes becomes a cooperative phenomenon; the α-th mode, for example, does no longer "freeze" at $T_\alpha = J_\alpha/k_B$ but at a temperature T_α' that depends on the overlap of the α-th mode with the others (Chowdhury 1985). An alternative, but equivalent statement can be made about the behavioue of the staggered susceptibility; χ_λ is not given by $\chi = 1/(k_B T - J_\lambda)$ but by the expression

$$\chi_\lambda = [\{\beta\Sigma|<i|\lambda>|^2(1-m_i^2)\}/\{1 - \Sigma<i|\lambda'><\lambda'|i>(1-m_i^2)\beta J_{\lambda\lambda'}$$

where $\beta = (k_B T)^{-1}$. The eigenstate-approach for SG has also been extended to the situations with nonzero external magnetic field (Nambu 1985).

Following a heuristic block-spin renormalization schme where the condensed localized modes correspond to the blocks, Hertz (1985) proposed the recursion relation

$$(J')^2 = b^{d-2}\ J^2$$

which implies a SG transition for $d > 2$. Moreover, the exponent $\nu = 1/(d-2)$; the latter results seem to agree with the most recent Monte Carlo results (see chapter 24).

Hertz (1983c) looked at the problem of mode-freezing from a dynamical point of view. All the condensed modes (including the localized ones) appear frozen to very fast experimental probes. The relaxation time for any given mode λ is given by $\tau_\lambda = \tau_0 \exp(\Delta F_\lambda)$ where ΔF_λ is the corresponding free energy barrier height. In the IMA $\Delta F_\lambda \propto$ size of the mode λ; the size of a mode being determined by the corresponding participation ratio $\Sigma|<\lambda|i>|^4$. (Note that the free energy barrier ΔF_λ is not proportional to the corresponding participation ratio if mode-overlaps are not negligibly small (Chowdhury and Mookerjee

1984d). Therefore, the modes with largest eigenvalue will be the first to relax because of their smallest size. As we wait longer, modes with smaller and smaller eigenvalues relax at successively longer time intervals; finally, after infinitely long time all the localized modes will be found unfrozen. The latter picture of dynamic freezing is very similar to that described in chapter 21. In an independent approach Feigelman and Ioffe (1985) 'picked up' the critical modes near the transition point. The 'fast variables' are averaged out and the dynamics is dominated by the 'slow modes'. The eigenfunctions of J with eigenvalues J_λ near the boundary of the spectrum are separated out as the critical modes. Moreover, Ising-like behavior is observed even for the vector spin systems provided a weak anisotropy is added to the Hamiltian. This approach to the theoretical understanding of SG is now under further investigation (Kumar 1986).

CHAPTER 24

COMPUTER SIMULATION STUDIES AND 'NUMERICALLY EXACT' TREATMENT OF SG MODELS

Recent advances in the computer technology have provided powerful tools for the physicists, viz., high-speed computers with large core memories (e.g., Cray 2, Cray XMP, Cray 1, Cyber 205, etc.), special-purpose computing machines (e.g., the special purpose machine of Condon and Ogielski at the Bell Labs.) and multi-processor machines (e.g., the distributed array processor (DAP) at the Queen Mary College London, etc.). Computational physics has become the third branch of physics, bridging the gap between theoretical physics and experimental physics (see, for example, Kalos 1985, Binder 1986 for the future prospects of computational physics). The introduction to the computer simulation technique in this chapter is quite long. This is not so because of any personal bias of the author but because of the fact that so far results of computer simulation of SG models have played the most decisive role in estimating the lower critical dimension (LCD) and the physical properties of realistic SG models.

In computer simulation, one begins with a well-defined model and lets the system evolve following some "prescription for (time-) evolution". In general, such prescriptions can be grouped into two different categories: molecular dynamics (MD) and Monte-Carlo (MC) simulations. The former is based on deterministic evolution equations, e.g., Newton's equation, whereas the latter is stochastic in nature . Such evolution of the model in a computer is supposed to simulate the evolution of the corresponding real system in time and hence the name "computer simulation". However, certain subtleties are involved in the identification of the one-to-one correspondences between the discrete-time dynamics of the computer simulation and the continuous-time evolution that is believed to be followed by real systems. We shall come back to this point later.

Why does one do computer simulation? Comparison of the results of the computer simulation of a model with the corresponding results of the laboratory experiments on real materials tells us how realistically (or, unrealistically) the real system is represented by the model studied. On the other hand, comparison between the results of computer simulation and analytical treatments of the same model provides a testing ground for the validity of the various approximations made in the latter. Moreover, often physical quantities defined theoretically cannot be measured directly in laboratory experiments. However, such quantities can be computed directly in computer simulation.

Unfortunately, no man-made tool can be perfect. Computer simulation technique suffers from some limitations. First of all, the finite available size of the computer memory puts upper limits on the size of the systems that can be simulated. Imposing suitable boundary conditions, e.g., periodic boundary condition, leads to mimicing an effectively infinite system consisting of periodic repetitions of the actual piece of the system simulated. The latter, together with the extrapolation to $N \to \infty$, is often satisfactory. However, near the phase transition points, the correlation length ξ becomes longer than the system size and finiteness of the system is felt rather strongly. In spite of this limitation, a great deal of important information regarding the truly infinite system can be extracted from those for the finite systems, using, for example, the powerful technique of finite-size scaling theory (see, for example, Barber 1983 for a detailed account). However, limitations of the system size is not as severe a restriction on the power of computer simulation as the limitations of available CPU time. For example, often it might be possible to accomodate large systems within the available memory of the computer but impossible to equilibrate the system within the available CPU time. Various schemes of extrapolation to the $t \to \infty$ limit may be questionable. However, the use of vector or parallel processors can considerably stretch the 'observation time' in a computer simulation. The best solution to the latter problem seems to be the use of special-purpose computing machines which can be run continuously for months or years at a considerably low cost. Needless to say that special purpose machines become useless after the "special purpose" has been served !

The equation of motion technique used in chapter 18, namely, using equations (18.5) is effectively a MD approach. Therefore, in this chapter we shall discuss only results of MC simulation.

Does the MC simulation really simulate the true dynamical evolution of the system in time? In order to answer this question let us examine the basic principles of the MC simulation in the context of spin systems. For simplicity, let us consider Ising spins only. The configuration of the spin system at a given instant of time t is given by the set of orientations of the Ising spin variables $\{S_i\}$. The thermal average of a given property, say A (e.g., magnetization, susceptibility, specific heat, etc.), of the system is given by

$$\langle A \rangle = \int A\{S_i\} \exp(-\beta H\{S_i\}) \, d\{S_i\} / \int \exp(-\beta H\{S_i\}) \, d\{S_i\}$$

In a naive random sampling procedure one would randomly select a sufficiently large number of the configurations and average A over these configurations, weighting each configuration by the corresponding Boltzmann factor $\exp(-\beta H\{S_i\})$. However, generating such a "sufficiently large" number of states by random sampling is a formidable task even for the fastest computers available. Instead, one uses the "importance sampling" trick introduced by Metropolis et al. (see Binder and Stauffer 1984). In the latter approach spin configurations are selected randomly with a frequency proportional to the Boltzmann factor $\exp(-\beta H)$ and averages are carried out over the configuration so selected with equal weight.

The Metropolis algorithm for Ising spin systems, thus, consists of the following steps:

(i) select a spin configuration randomly,

(ii) compute the energy change ΔE_i in energy if, say, the i-th spin S_i would be flipped; if ΔE_i is negative accept the new configuration with S_i flipped; otherwise select a random fraction uniformly distributed between 0 and 1,and if $\exp(\beta\Delta E_i) < f$ retain the old configuration, otherwise accept the new configuration. The latter step allows "uphill" motion in the generalized phase space.

The MC process is usually interpreted as a realization of the (stochastic) dynamical relaxation (to thermodynamic equilibrium described by Gibbs ensemble) process described by a Markovian Master equation (19.6) (see Binder and Kinzel 1983). However, time is discrete in MC processes and its unit is usually taken to be proportional to one Monte Carlo Step (MCS) per spin. Does this discrete nature of time have any consequence ? Some of the objections raised earlier (Choi and Huberman 1983, Choi and Huberman 1984) against the discrete-time MC processes have not been supported by subsequent works. Is the dynamics of a system observed in a MC simulation sensitive to the scheme used for updating the spin states (Gawlinski et al.1985, Ceccatto 1985)? It is generally believed that the random updating of the spins represents the true time-evolution of the system more realistically than the sequential updating.

The simulation of a random magnetic system is much more complicated because the latter requires averaging over configurations and consequently needs longer CPU time. For such systems, even if the number of quenched configurations simulated is very large, equality of the average value and the most probable value of a thermodynamic quantity is not guaranteed (Derrida and Hilhorst 1981, van Hemmen and Morgenstern 1982).

We have already quoted the results of MC simulations and compared them with the corresponding results of analytical treatments in various different contexts earlier in this book. In this chapter we shall focus our attention on the direct as well as indirect evidences in favor of (or against) the existence of the SG transition in a given space dimension d for a given model.

Can we conclude anything about the nature of the ordered state of the SG by computing the time-dependent EA order parameter $q_{EA}(t)$ directly by MC silmulation? There is a fundamental difficulty involved in the case of the SG that makes it a harder problem compared to an Ising ferromagnet. If a system is truly ergodic the order parameter would vanish when averaged over infinitely long time. On the other hand, if the barriers separating the different ordered states are infinite the time-averaged order parameter is nonzero and differs from the ensemble average. Now consider a finite ferromagnetic Ising system. If the time t_{flip} for which the system remains trapped in one of the two ordered states is much longer than the time of observation τ_m, then the time-averaged magnetization is nonzero, but would vanish provided $\tau_m > t_{flip}$. Such a clear-cut separation of the time scales is not possible in SG and hence the difficulty in exploring the nature of the ordered state of a SG by direct computation of $q_{EA}(t)$ in any MC simulation (see Larsen 1983). Historically, Binder and co-workers interpreted the MC data for nearest-neighbour (nn) Gaussian exchange models of Ising spins in d=2 (Binder and Schroeder 1976) and d=3 (Binder and Stauffer 1976) in the light of a static phase transition of EA type. Bray and Moore (1977) repeated Binder and Schroeder's (1976) simulation for various observation times and showed that the temperature T_g corresponding to the cusp in χ varies with the observation time (MCS/spin) τ_m - the higher τ_m, the lower is T_g. Therefore, Bray and Moore suggested that the apparent phase transition at non-zero temperature observed by Binder and Schroeder is only an artifact of the finiteness of the observation time; if one could observe the system for very long time (i.e., $\tau_m \to \infty$), the EA order parameter would vanish and the susceptibility would follow the Curie law. This claim was supported further from the consideration of the 'defect energy' (Reed et al. 1978). Now consider a simpler situation where the exchange couplings on a two-dimensional lattice are distributed periodically with periodicity m in both the x and y directions. The exchange couplings within a cell of size m X m are chosen randomly according to the Gaussian distribution and repeated periodically in both directions. A realistic SG model corresponds to m = ∞. The transition temperature of such a simple model can be computed exactly for a finite-size system using a computer. Following this method (not MC simulation), Reed (1979) showed that $T_g \to 0$ as $m \to \infty$ and, therefore, suggests that a SG transition would be possible in d=2 only at T=0. All these observations led to more extensive MC simulations of Ising SG for all space dimensions d=2 to d=5 (d=6 is the UCD) (Stauffer and Binder 1978,1979). No qualitative difference between the results of d=3 and d=5 was observed and the EA order parameter was observed to relax to a non-zero value after the observation time $\tau_m = 10^5$ MCS/spin. We shall not discuss the results of these early simulations in any further detail in this book (see the reviews by Binder and co-workers). It is Young's MC simulation of the two-dimensional ±J Ising model (1983a,b) (the sample sizes were 68 x 68 and 128 x 128) which convincingly confirmed the earlier claims based on MC simulation as well as based on numerical transfer matrix (TM) calculations (Morgenstern and Binder 1979, to be discussed later in this chapter) that $T_g = 0$ for short-ranged Ising SG in d=2. This also refuted some of the earlier claims (e.g., Fernández 1982) of finite T_g in d=2. More specifically, the EA-susceptibility data for temperatures $T \geq 1.0$ (note that J = 1) was found to be consistent with $\chi_{EA} \propto T^{-4}$ in agreement with Morgenstern and Binder. Moreover, the average relaxation time τ was found to obey the Arrhenius law, viz., $\tau = \tau_0 \exp(\Delta E/T)$. These observations have been confirmed by further extensive simulations (McMillan 1983). Good fit to the data was obtained for $\nu = 2.42 \pm 0.10$ which is much smaller compared to the value $\nu = 4.2 \pm 0.5$ obtained by the numerical TM method (Huse and Morgenstern 1985). The following questions arise naturally: if $T_g = 0$ in d=2, why do the MC simulations of the nn models yield (a) a cusp at non-zero T_g in χ which, as mentioned in chapter 1, has been used most often by the experimentalists as one of the main experimental signals for the SG-transition, and (b) the AT-line along which the SG transition takes place in the SK model (see chapter 5-9)?

From exact computation for small spin-clusters (Kinzel 1982a) and MC simulation of the Gaussian model on a square lattice of

size 60 X 60 (Kinzel 1982b) it was concluded that
(i) T_g, signalled by the cusp in χ, decreases with increasing observation time τ_m (MCS/spin) and would ultimately vanish in the limit $\tau_m \to \infty$ (this would require astronomically large CPU time for reasonably large system sizes!),
(ii) for a given finite τ_m (i.e., for a given non-zero apparent T_g) spins freeze in small clusters just below T_g, and the freezing is complete only at T=0,
(iii) the reversible susceptibility χ is given by $\chi = (1-P)/T$ where P is the fraction of spin "frozen" over the observation time.
Thus, in short, the question (a) above is answered by saying that the apparent finite-temperature transition in the simulation of nn SG models is a consequence of "dynamical freezing" (see also Morgenstern 1983b). The latter also answers the question (b) as explained in chapter 16.

The issue of a finite-temperature SG transition in d=3 has been more controversial than in d=2 not only bacause the former required larger computer memory and longer computer time but also because of the inconclusiveness of the numerical TM calculations in d=3. The analysis of the MC data is a very subtle process because often the same set of data can be fitted to quite different analytical expressions. The numerical TM results in d=2 strongly indicated, a-priori, what should be the best fit for the MC data and these fits did, indeed, work! No such convincing guidelines were available for d=3 and the early results even for systems as large as 64 X 64 X 64 remained inconclusive (Young 1984). Inspired by the apparent success of the Arrhenius law in d=2, Binder and Young (1984) proposed similar behavior also for d=3, viz.,

$$\xi_{EA} \sim T^{-\nu}$$

and

$$\ln(\tau/\tau_0) \sim \Delta F(\xi_{EA})/T \sim \tilde{J}\, T^{-1}\, \xi_{EA}^{z-1/\nu} \sim T^{-z\nu}$$

which is based on the assumption that $T_g = 0$. However, we shall see soon in this chapter that now there are quite convincing evidences from large-scale MC simulations that $T_g \neq 0$ in d=3 and that conventional scaling , viz., $\tau \sim \xi_{EA}^{z}$ holds in the same spatial dimension.

Bhatt and Young (1985) carried out MC simulation of the ±J model for Ising spins in d=3 using lattice sizes between 3^3 and 20^3 on the distributed array processor (DAP) at the Queen Mary College, London. In principle, one can estimate T_g , ν and η by fitting the MC data for the order parameter susceptibility χ_{EA} with the finite-size scaling form

$$\chi_{SG} = L^{2-\eta}\ \bar{\chi}\ (L^{1/\nu}\ (T - T_g))$$

However, a much better approach is to use a quantity that involves fewer parameters to be fitted. Suppose, we simulate two sets of spins $\{S_i^{(1)}\}$ and $\{S_i^{(2)}\}$ which have the same set of bonds but which are uncoupled from each other. Then one can compute the overlap

$$Q = L^{-d}\ \Sigma\ S_i^{(1)}\ (t_o - t)\ S_i^{(2)}\ (t_o + t)$$

and hence the distribution

$$P_L\ (q) = (1/t_o)\ \sum_{t=1}^{t_o} [\ \delta\ (q - Q(t))]$$

The finite-size scaling form for the quantity

$$g = (3 - \langle q^4\rangle\ /\ \langle q^2\rangle^2)/2$$

is given by

$$\bar{g} = g(L^{1/\nu}\ (T - T_g))$$

which, clearly, does not involve η. From the analysis of g, Bhatt and Young (1985) concluded that $T \simeq 1.2$ and $\nu \simeq 1.3$. Then, from the analysis of the data for $P_L(q)$ at T=1.2, they obtained $\eta \simeq -0.3$. However, although the data for $T \geq 1.2$ were consistent with the above values of T_g and the experiment, the data for $T < 1.2$

were somewhat unusual and quite different from the corresponding behaviour for the SK model. Therefore, Bhatt and Young (1985) conjectured that the system is close to criticality for all $T < T_g$, which would imply that the LCD is very close to d=3.

The size of the systems simulated by Bhatt and Young are quite small and hence the results may not seem very convincing. Ogielski and Morgenstern (1985a,b) simulated much larger systems (upto 64^3) on a special purpose machine designed and built at the AT&T Bell laboratories (Condon and Ogielski 1985). The physical quantities computed directly are

(1) the time-dependent local magnetization

$$m_i(t) = (1/t) \int_0^t dt' \langle S_i(t')\rangle$$

(2) the average correlation function

$$G(r) = (1/N) \Sigma \langle S_i S_{x+r}(t)\rangle^2 \qquad \text{and}$$

(3) the dynamic correlation function

$$q(t) = (1/N) \Sigma\, S_i(0)\, S_i(t)\rangle$$

for the standard ±J model.

At $T > 1.175$ the probability distribution, $P(m_i)$, of the local magnetization m_i is Gaussian-like for short times but shrinks with increasing time and finally reduces to a delta (function-) peak at $m_i = 0$ after sufficiently long time. On the other hand, $P(m_i)$ remains almost flat over much longer time interval of observation at $T < 1.2$. Such difference in the behaviour of $P(m_i)$ above and below T=1.175 strongly indicates the possibility of a transition from the paramagnetic to the SG phase at the latter temperature. Further support for the existence of the phase transition at T=1.175 comes from the observation (Fig.24.1) that G(r) approaches a plateau at a non-zero value of r for $T < 1.175$ whereas it decays to zero for all $T > 1.175$. Moreover, good fit to the scaling forms were achieved for $T = 1.175 \pm 0.025$, ν =

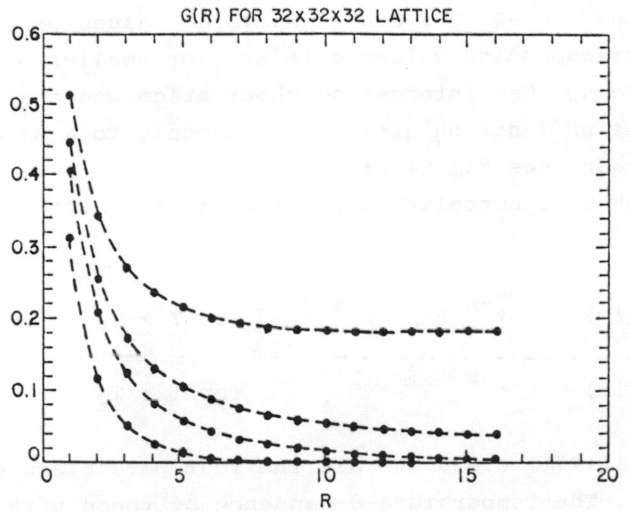

Fig. 24.1. Correlation function G(r) vesus distance r for temperatures T = 1.65 (lowermost curve), 1.35, 1.25, 1.10 (topmost curve) (after Ogielski and Morgenstern 198b).

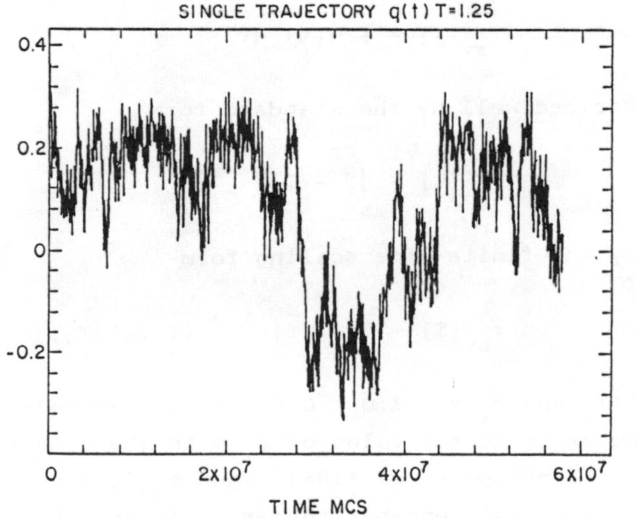

Fig. 24.2. Autocorrelation function q(t) for a single trajectory at T = 1.25 (after Ogielski and Morgenstern 1985b).

1.3 ± 0.1 and η = -0.22 ± 0.05, all these values being consistent with the corresponding values obtained for smaller systems by Bhatt and Young. One interesting observation was that the dynamic autocorrelation function q(t) flips suddenly to a negative value and back again (see Fig.24.2).

The dynamic correlation function q(t) decays as (Ogielski 1985a,b)

$$q(t) = c\, t^{-x} \exp(-\omega t^{\beta}) \qquad \text{for } T > T_g$$

and

$$q(t) = c\, t^{-x} \qquad \text{for } t < T_g$$

where T_g = 1.175 ± 0.025 and all the four parameters c, ω, x and β depend on T. The temperature-dependence of these parameters might arise from the corrections to scaling (Ogielski 1985a,b).

The average thermal relaxation time $\tau_{av}(T)$, computed from (Ogielski 1985a)

$$\tau_{av}(T) = \int q(t)\, dt$$

could be described well by the standard form

$$\tau_{av}(T) \sim |T - T_g|^{-z_{av}\nu}$$

as well as by the finite-size scaling form

$$\tau_{av}(T) \sim L^{z_{av}} f(L^{1/\nu}\, (T-T_g)/T_g)$$

with T_g = 1.175 and $z_{av}\nu$ = 7.0 ± 0.8. The latter value agrees well with the experimental value of $z_{av}\nu$ in the case of insulating SG (Bontemps et al.1984). Thus z_{av}=5.4 ± 0.2, a rather large value ! One must remember that τ_{av} is not the correlation time τ that appears in conventional dynamic scaling laws (see Ogielski 1985). The latter τ leads to z = 6.1 ± 0.3. The latter values are also consistent with the scaling laws relating the dynamic and the static exponents.

Next let us consider the XY model and the Heisenberg models. In the case of the ±J distribution of the exchange interactions, $T_g = 0$ for the two-dimensional XY model (Jain and Young 1986). However, there is a crucial difference between the Ising and the XY models in d = 2. As stated earlier, $\ln \tau_{av}$ scales with T in the case of the Ising model whereas in the case of the XY model $\tau^{av} \propto T_{av}^{-z\nu}$. Note that Jain and Young's MC simulation does not incorporate the intrinsic dynamics due to the vector nature of the spins. The data suggest $z_{av}\nu = 5.16 \pm 0.18$, $\nu = 1.08 \pm 0.27$, and $\eta = 0.3 \pm 0.3$ for the XY model in d = 2. However, the data for d = 3 remains inconclusive.

The "defect energy method" has been applied at T=0 to the m-vector SG for m=2 and 3 in d=2 and 3 (also $m \to$ in $1 \le d \le 4$) assuming Gaussian distribution of J_{ij} (Morris et al.1986). It turns out that the correlation length ξ diverges as $\xi \sim T^{-\nu}$ for $T \to 0$, indicating the absence of any finite temperature SG transition. Some typical values of $\nu(m,d)$ are $\nu(2,2) \simeq 1.1$, $\nu(2,3) \sim 2.2$.

The MC simulation of the RKKY SG for Heisenberg spins revealed the crucial role of the anisotropy (Walstedt and Walker 1981, Walker and Walstedt 1983). There is no cusp in χ at any finite temperature in the absence of the anisotropy and the system remains trapped near a single energy minimum for a long time. But even a small amount of anisotropy gives rise to the cusp in χ, in qualitative agreement with experiments. Moreover, in the presence of the anisotropy the nonlinear susceptibility exhibits a maximum at the temperature corresponding to the cusp in the linear susceptibility. These observations suggest a nonzero SG transition temperature in the RKKY model for Heisenberg spins only in the prssence of anisotropy. This observation has been supported by more extensive work of Chakrabarti and Dasgupta (1986). Finite-size scaling arguments, explained above, lead to $\nu = 0.87 \pm 0.08$ corresponding to the T = 0 critical point.

So far as powerful a 'hammer' as Transfer Matrix (TM) method (see Mattis 1985 for an elementary introduction) has failed to 'crack' the SG problem analytically. However, Morgenstern and Binder (1979) pioneered the so-called "numerically exact" solution of the nn SG models by applying the basic principle of the TM technique to manageably large, but finite, lattices on a computer. The basic procedure consists of a recursive scheme for computing the exact partition function for a given finite lattice with a quenched random-bond configuration. We shall explain the implementation of this procedure in d=2 although it has also been applied to d=3.

The partition function for a nn Ising SG can be written as (Morgenstern and Binder 1979, 1980a)

$$
\begin{aligned}
Z_{\{J_{ij}\}} = {} & \sum_{\{S_{11}\}} \sum_{\{S_{12}\}} \cdots \sum_{\{S_{1L}\}} \sum_{\{S_{21}\}} \cdots \sum_{\{S_{2L}\}} \sum_{\{S_{M1}\}} \cdots \sum_{\{S_{ML}\}} \\
& \times \Big[\prod_{j=1}^{L} \exp(K_{1j,1j+1}\, S_{1j}S_{1j+1}) \exp(K_{1j,2j}\, S_{1j}S_{2j}) \Big] \\
& \times \\
& \times \Big[\sum_{j=1}^{L} \exp(K_{2j,2j+1}\, S_{2j}S_{2j+1}) \exp(K_{2j,3j}\, S_{2j}S_{3j}) \Big] \times \cdots \\
& \times \Big[\sum_{j=1}^{L} \exp(K_{Mj,Mj+1}\, S_{Mj}S_{Mj+1}) \Big]
\end{aligned}
\qquad (24.1)
$$

where S_{ij} is the spin in the i-th row and the j-th column, $K = J/kT$, and the system dimensions are L in the horizontal direction (i.e., j=1,l) and M in the vertical direction (i.e., i=1,M). Periodic boundary condition is applied in the horizontal direction and free boundary condition is applied in the vertical direction. The basic scheme, as evident from (24.1), consists of the following steps (see Morgenstern 1983 for the details and for a computer program):

(i) one starts with the first row and generates all the 2^L configurations of this row using a computer; then one computes the "horizontal factor"

$$\prod_{j=1}^{L} \exp(K_{1j,1j+1}\, S_{1j}\, S_{1j+1})$$

for all the 2^L states of the $\{S_{1j}\}$.
(ii) Next one computes the first "vertical factor" $\exp(K_{11,21}S_{11} S_{21})$ for the two possible states of S_{21} and then performs the trace over S_{11}; then the second vertical factor $\exp(K_{12,22}S_{12}S_{22})$ is computed for the two states of S_{22} and the trace over S_{12} is performed. Thus, when S_{2L} is reached, the trace over all the spins $\{S_{1j}\}$ is completed.
(iii) The program proceeds from second row to the third row and so on, following the steps (i) and (ii) above until the trace over all the spins is carried out. The requirement for the storage of 2^L states, mentioned in step(i) above, puts an upper limit to the system size (more precisely, an upper limit for L) that can be handled with a given computer.

We have explained the numerical TM technique in quite great detail because the results obtained from this technique were the first convincing evidences against the existence of a SG transition in d=2. Using L X L systems with $6 \leq L \leq 18$, Morgenstern and Binder (1979, 1980a) computed the exact partition function $Z\{J_{ij}\}$ and averaged over several configurations $\{J_{ij}\}$. They observed, for Ising spins in d=2, that
(i) the Binder order parameter (see chapter 3) decreases monotonically with the increasing system size thereby indicating the absence of SG transition at any non-zero temperature in d=2,
(ii) the correaltion function $[\langle S_0 S_R\rangle^2]$ decays strongly (exponentially) with increasing R for all T > 0 for both ±J model and the Gaussian model whereas at T=0 it becomes a constant in the Gaussian model and decays algebraically in the ±J model thereby indicating a SG transition at T=0 in d=2,
(iii) the zero-temperature entropy $S(0) \simeq 0.075$; the latter suggesting high degeneracy of the ground state at T=0, a property essential for SG state.

Cheung and McMillan (1983a,b) applied numerical TM technique to the two-dimensional nn Ising models with $3 \leq L \leq 11$ and M upto

10^5, and used finite-size scaling arguments to extrapolate to L = ∞. The best fit to the data was obtained with

$$\xi(T) = a[J/(T - T_g)]^{\nu} + \xi_0$$

with $T_g = 0$, $\nu = 2.59 \pm 0.13$, $a = 4.14$ and $\xi = 0.822$ in the case of the ±J model (Cheung and McMillan 1983a) and $T_g = 0$, $\nu = 2.96 \pm 0.22$, $a = 1.68$ and $\xi = 2.97$ in the case of the model with a square distribution (Cheung and McMillan 1983b). Since both the scaling function $g(\ln(L/\xi(T)))$ and the exponent ν of the two models differ significantly from each other Cheung and McMillan suggested that the two models belong to two different universality classes. Would the latter suggestion, if true, mean that SG models are unusual because universality classes are believed to be determined only by the space dimensionality d, symmetry of the order parameter and the range of the interaction? Not necessarily; it would be unusual if the corresponding T_g were nonzero. At this point I would like to mention that there are strong evidences (van Hemmen and Suto 1984) that, at least for the infinite-ranged model, the configuration-averaged free energy does not depend on the details of the probability distribution of the quenched random variables.

Morgenstern and Binder's (1980b) "numerically exact" solution for d=3 was inconclusive because of the size-limitation of the systems studied.

One must remember that the crucial differences between computer simulations and the numerical TM approach. The latter are exact calculations for the given size of the system whereas MC simulation is a dynamic approach based on statistical random (importance) sampling that might sample one or a few "valleys" in the phase space and hence might overestimate the order. This explains why the earlier MC simulations for relatively small number of MCS/spin observed SG ordering at nonzero temperatures in d=2 whereas numerical TM approach ruled out such possibility.

Now we shall summarize the results obtained by using techniques that are combinations of the RG methods (explained in chapter 17) and the numerical techniques described so far in this chapter. McMillan (1984a,b) introduced the so-called "domain-wall RG" method for the Gaussian model with Ising spins. Suppose an n x n system is extended in the x direction by periodic repeatition whereas periodic and antiperiodic boundary conditions are imposed in the y-direction. A domain wall created by imposing the antiperiodic boundary condition leads to the spin reversal over a length scale L. The free energy of a domain wall in zero field is given by

$$W_n(\bar{K},\tilde{K}) = - T \ln(Z_a/Z_p)$$

where Z_a and Z_p are the partition functions corresponding to the antiperiodic and the periodic boundary conditions, respectively., and $\bar{K} = J_0/T$ and $\tilde{K} = J/T$ correspond to the mean and the variances of the distribution of the interactions. Now let us set up a RG transformation in the same spirit as that in chapter 17. Suppose we have two lattices, one n x n lattice with lattice spacing a and the other n' x n' lattice with the lattice spacing a', such that L = an = a'n', where a' = ba (b > 1), b being the rescaling factor. These two lattices should represent the same physical problem with two different microscopic length scales a and a'. This requirement imposes the condition that the distribution of W_n must be invariant under the RG transformation. More specifically,

$$\bar{W}_{n'}(\bar{K}',\tilde{K}') = \bar{W}_n(\bar{K},\tilde{K})$$

and

$$\tilde{W}_{n'}(\bar{K}',\tilde{K}') = \tilde{W}_n(\bar{K},\tilde{K})$$

for the mean and the variance of the distribution of W_n, respectively. The mean and the variance of W_n are computed numarically by generating a large number of configurations. This method yields $T_g = 0$ in d = 2 and $T_g > 0$ in d = 3.

The large cell RG method introduced by Bray and Moore (1984 a,b, 1985) is based on the concept of "defect energy". Consider a d-dimensional Ising spin system of linear length L with a random distribution of the couplings J_{ij}. Suppose periodic boundary condition is imposed on all directions except, say, the x-direction. A random boundary condition is imposed along the x-direction such that the spins at the two ends of the x-direction are randomly up or down. The purpose of the latter boundary condition is to mimic effectively embedding the finite block in an infinitely large system in its ground state. The defect energy E_{def} is defined as the difference between the ground state energy for a given boundary condition and the ground state energy corresponding to the case when the spins on one of the random boundaries are reversed, keeping those on the other boundary fixed. The effective coupling between the finite blocks of an infinite system is defined as $J_{eff} = E_{def}/2$. Note that J_{eff} is a sample-dependent random variable. Therefore, the distribution $P_L(J_{eff})$ is the important relevant quantity, the scaling of which with L determines the nature of the ordered state. Suppose $\bar{J}(L)$ is the average of the modulus of J_{eff} (note that any physically sensible measure of the width of the coupling-distribution, e.g.,r.m.s. of J_{eff}, should scale the same way as does $\bar{J}(L)$). If $\bar{J}$ increases with increasing L, the ordered state corresponds to a SG, otherwise $T_g = 0$. The numerical implementation of this large-cell RG technique gives $T_g = 0$ for d = 2 and $T_g > 0$ for d = 3. Extension of this large-cell RG treatment to nonzero external magnetic field H (Bray and Moore 1984b) indicated that possibly there is no AT line for the short-ranged models in d = 3. Finally, I would like to remind the reader that the "opinions of the majority" of the SG-experts has oscillated over the last decade. Therefore, don't get prejudiced by the recent successes!

CHAPTER 25

TRANSPORT PROPERTIES OF SG AND SOUND ATTENUATION IN SG

So far in this book we have not explicitly treated the interaction between the 'localized' d-electrons of the transition metals and the s-electrons in the transition metal-noble metal alloys. In chapter 1 we absorbed the effects of the s-d interaction into the RKKY interaction which is mediated via the s-d exchange interaction. In chapter 19 we have assumed that the s-electrons form a part of the heat bath that causes the spin-flip. In this chapter we shall investigate explicitly the effects of the s-d interaction on the transport properties of the metallic SG. Moreover, so far we have always assumed the spins to be rigidly frozen so that the interaction between the spins and the lattice vibrations were also not taken into account. In this chapter we shall review the theories of the interaction between the spin and the phonon degrees of freedom so as to analyze the velocity and the attenuation of sound waves in SG.

25.1. Transport Properties of SG:

Since our basic understanding of the nature of the ground state of SG remains controversial, any study of the transport properties seems to be premature because the latter involve the spectrum of the excited states. However, this situation did not prevent many authors from developing elementary theories of the transport properties of SG based on the simple EA picture. Taking only the contributions from the potential fluctuations V and from the s-d interaction J_{sd} into account, the effective Hamiltonian can be written as (Fischer 1979)

$$H = H_p + H_{imp}, \qquad (25.1)$$

where

$$H_p = \sum_{ks} \epsilon_k \, c^+_{ks} \, c_{ks},$$

$$H_{imp} = (1/\mathcal{N}) \sum_{kk'} \sum_{ss'} c_i \times \exp\{i(\vec{k}'-\vec{k})\cdot\vec{r}_i\}$$

$$c^+_{k's'} c_{ks} \{V\delta_{ss'} - (J_{sd}/2) S_i \sigma_{ss'}\}.$$

c^+_{ks} and c_{ks} are the creation and the annihilation operators, respectively, for the conduction electrons in the Bloch states $\vec{k}$,s of a single band with energy ϵ_k, c_i is unity for an impurity at the site $\vec{r}_i$ and zero otherwise. $\mathcal{N}$ is the total number of lattice sites. Thus, the first term in (25.1) represents the energy of the conduction electrons in the pure host lattice, whereas the second term represents the contribution to the Hamiltonian from the impurities due to the alloy formation. The probability of transition from a state $\vec{k}$s to another state $\vec{k}$'s' is given by

$$\Gamma_{ks\to k's'} = f_{ks}(1-f_{k's'}) \Sigma c_i c_j \exp\{i(\vec{k}-\vec{k}')(\vec{r}_i-\vec{r}_j)\}$$

$$\int d\omega \; \delta(\omega-\epsilon_{ks}+\epsilon_{k's'}) \mathcal{N}^{-2} \int dt$$

$$\times \; [(J^2/4)\langle S_i(t) S_j\rangle + V^2 + sJV\langle S^z_i\rangle] \exp(i\omega t)$$

within the Born approximation. Using the mean-field results (which were available at that time) for the correlation function, Fischer (1979) showed that $\rho(T) = AT^2 - BT^{5/2}$, with positive A and B (see Rivier and Adkins 1975, Larsen 1976 for the earlier works).

Keeping terms upto the order J^3, Fischer (1981a) studied the influence of the Kondo effect on the resistivity of SG. There are two effects: (a) the effective spin becomes temperature dependent, with $S^2_{eff}(T) \to 0$ as $T \to 0$ and $s^2_{eff}(T) \to S(S+1)$ as $T \to \infty$, (b) the Kondo temperature T_K is suppressed by the spin-spin interaction and $T_K \to 0$ for sufficiently strong interaction between the spins. This description of the resistivity qualitatively explains the existence of a maximum of the resistivity at a temperature T_m; T_m depends on the Kondo temperature T_k and on the excitation spectrum of the spin system. Similarly, the experimentally observed behaviour of the thermopower of SG can be explained in this perturbation-theoretic formalism by retaining terms upto the order J^2V, Fischer (1981b).

Next let us consider the magnetoresistance of SG alloys. Suppose, the external magnetic field is applied in the z-direction. The corresponding Hamiltonian is given by (Mookerjee 1980a, Das et al.1982)

$$H = H_p + (V/\mathcal{N}) \sum_{kk'si} \exp\{i(\vec{k}-\vec{k}')\cdot\vec{r}_i\} c^+_{ks} c_{ks}$$

$$- (J_{sd}/\mathcal{N}) \sum_{kk'i} \exp\{i(\vec{k}-\vec{k}')\cdot\vec{r}_i\}$$

$$[(c^+_{k+}c_{k'+} - c^+_{k-}c_{k'-})S^z_j + c^+_{k-}c_{k'-}S^+_j$$

$$+ c^+_{k+}c_{k'-}\ S^-_j] - g\mu_B H \Sigma S^z_j. \qquad (25.2)$$

Mookerjee (1980), Mookerjee and Chowdhury (1983a) computed the magnetoresistance of canonical SG alloys beginning with the Hamiltonian (25.2) but neglecting the complications of spin dynamics. The latter approximation is effectively identical with the assumption that the SG system attains equilibrium instantaneously. Moreover, since the ratio $|J/V|$ is small in canonical SG alloys, Mookerjee and Chowdhury (1983a) neglected terms involving second and higher powers of J/V. Under these approximations the magnetoresistance is given by

$$\Delta\rho = - c\ R_0\ J^2_{sd}\ [\ M(H)\ \tanh(g\mu_B H/2k_B T) + q(H) - q(0)]$$

where

$$R_0 = (3\pi m_e/4\hbar e^2 n E_F),$$

with n being the number of conduction electrons per unit volume. M(H) and q(H) could be computed from the SK-like equations for the RKKY SG (see chapter 4) and hence $\Delta\rho/cR_0J^2_{sd}$. Surprisingly, the $\Delta\rho/cR_0J^2_{sd}$ thus computed from even such a crude theory (see Fig.25.1) agrees qualitatively with the experimental data of Nigam and Majumdar (1983) (see Fig.25.2). Finally, the strength of the s-d exchange interaction can be estimated by comparing the experimental value of $(\Delta\rho/cR_0)$ with the theoretically computed value of $(\Delta\rho/cR_0J^2_{sd})$. Although J_{sd} computed this way is in reasonably good agreement with the corresponding estimates from other alternative approaches, there is systematic variation with the impurity concentration. I believe that a large part of the errors arising from the crude approximations mentioned above, probably, cancel out when $\Delta\rho$ is computed by subtracting $\rho(H=0)$ from $\rho(H)$. Recently, it has been argued (Sherlekar et al.1986) that inclusion of the higher order terms in J/V yields better results. Chowdhury and Mookerjee (1984e) extended Fischer's theory to compute the magneto- resistance of canonical SG alloys within MFA. An alternative, but equivalent, approach based on the double-time Green function

$$G_{k+k'+}(t) = - i\theta(t)\langle[c_{k+}(t)\ c^{+}_{k'+}(0)]\rangle,$$

(θ is the step function) and the appropriate decoupling schemes for the higher order Green functions, has been developed by Das et al. (1982).

A phenomenological theory for the resistivity of SG has been proposed by Campbell (1981)(see also Campbell et al.1982a). The inelastic contribution to the resistivity is given by

$$\rho_{in}(T) \propto \Sigma\ [\{P(E)(E/k_BT)\}/\{\exp(E/k_BT)-1\}]$$

where P(E) is the density of the elementary excitations. Using P(E) computed numerically by Walker and Walstedt (see chapter 18), Campbell observed good quantitative agreement with the experimental data. Moreover, the specific heat has been

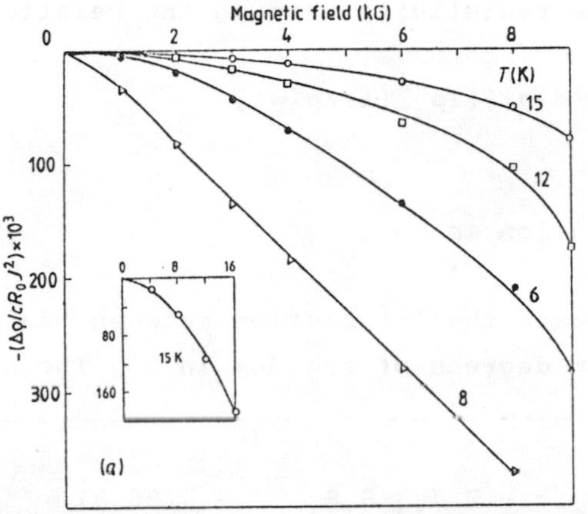

Fig. 25.1. Theoretical value of $\Delta\rho/cR_0J^2_{sd}$ for CuMn, shown as a function of magnetic field for various temperatures (after Mookerjee and Chowdhury 1983b).

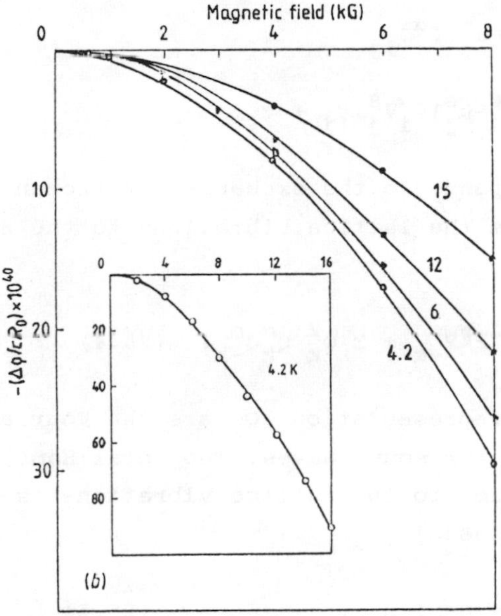

Fig. 25.2. Experimental curves for $\Delta\rho/cR_0$ for CuMn (after Nigam and Majumdar 1983).

correlated with the resistivity through the relation

$$C(T) = R(d/dt)[T\rho_{in}(T)/\rho(\infty)]$$

25.2. Sound Attenuation in SG:

Let us now investigate the interaction between the spin and lattice translation degrees of freedom in SG. The spin system is described by

$$H_s = -\Sigma J_{ij} S_i S_i \qquad (25.3)$$

where

$J_{ij} = J(\vec{r}_i - \vec{r}_j)$ depends on the vector $\vec{r}_i - \vec{r}_j$. Let us write $\vec{r} = \vec{r}^0_i + \vec{u}_i$ where $\vec{r}^0_i$ is the equilibrium position of the i-th spin and $\vec{u}_i$ is the small displacement of the i-th spin. Then,

$$J_{ij} = J^0_{ij} + \Sigma(u^\alpha_i - u^\alpha_j)\nabla^\alpha_i J_{ij}$$

$$+ (1/2)(u^\alpha_i - u^\alpha_j)(u^\beta_i - u^\beta_j)\nabla^\alpha_i \nabla^\beta_j J_{ij} + \ldots$$

where J^0_{ij} corresponds to the exchange for frozen equilibrium. The contribution from the lattice vibrations to the Hamiltonian is given by

$$H_{ph} = (1/2)\ \Sigma(M^{-1}P^\alpha_k P^\alpha_{-k} + M\ \omega^2_k\ Q^\alpha_k Q^\alpha_{-k}) \quad (25.4)$$

in the momentum representation, Q_k are the Fourier transforms of u_i and $\omega^2_k = v^2_0 k^2$ for sound waves. The total Hamiltonian of the spin system coupled to the lattice vibrations is given by (Fischer 1981c, 1983c)

$$H = H_s + H_{ph}. \qquad (25.5)$$

An equivalent, but phenomenological, TDGL approach has been developed by Hertz et al.(1981). Suppose S(x) is the local spin density field and $\vec{\psi}$ is the phonon field. Then the Landau-Ginzburg Hamiltonian can be written as (Hertz et al.1981)

$$H = \int[d^d x/(2\pi)^d](1/2)\{[r_0+\phi(x)]S^2(x,t) + |\nabla S(x,t)|^2$$

$$+ (u/4)[S(x,t)^2]^2 + c_0^2 \Sigma [\partial_\mu \psi^\mu(x,t)]^2 + |\vec{\Pi}(x,t)|^2$$

$$+ g_0 S^2(x,t) \nabla\cdot\vec{\psi}(x,t)\}$$

$$+ \int d^d x [h(x,t)S(x,t) + \vec{f}(x,t)\cdot\vec{\psi}(x,t)] \quad (25.6)$$

where $\vec{\Pi}(x) = (\partial\vec{\psi}/\partial t)$ is the phonon momentum density. $\phi(x)$ is a time-independent Gaussian random variable, and g_0 represents the strength of the coupling between the spin and the phonon fields. The equations of motion for the spin and the phonon fields are given by

$$(\partial S(x,t)/\partial t) = - \Gamma_0\{\delta H/\delta S(x,t)\} + \zeta(x,t)$$

$$(\partial\vec{\Pi}(x,t)/\partial t) = - \{\delta H/\delta\vec{\psi}(x)\} + \gamma_0 \nabla^2\{\delta H/\delta\Pi(x,t)\} + \xi(x,t)$$

with

$$\langle\zeta_i(x,t)\zeta_j(x',t')\rangle = 2\Gamma_0 \delta(x-x') \delta(t-t') \delta_{ij}$$

$$\langle\xi_\mu(x,t)\xi_\nu(x',t')\rangle = - 2\gamma_0 \nabla^2 \delta(x-x') \delta(t-t') \delta_{\mu\nu}.$$

The change of the sound velocity $\delta v(\omega,T)$ and the damping of sound waves $\gamma(\omega,T)$ in SG have been calculated for the model (25.6). δv decreases linearly with $(T - T_g)$ and γ diverges as $(T - T_g)^{-1}$ as T_g is approached from above. At T_g,

$$\delta v(\omega,T) \sim \omega^{1/2} \text{ and } \gamma(\omega,T_g) \sim \omega^{-1/2}$$

These results have been derived by Fischer (1981c, 1983c) from the microscopic model (25.5).

Khurana (1982) has calculated $\delta v(\omega,T)$ and $\gamma(\omega,T)$ for two models: (a) the random uniaxial anisotropy model where the spin part of the Hamiltonian is given by (15.7), and (b) the random bond model. In both these models the results for $\delta v(\omega,T)$ and $\gamma(\omega,T)$ are identical with those quoted above for the model (25.6). (See Beton and Moore 1983 for sound attenuation at temperatures well below T_g).

CHAPTER 26

MISCELLANEOUS ASPECTS OF SG

26.1 Local Field Distribution in SG:

We have mentioned the features of the local field distribution P(h) in SG in various different contexts in the earlier chapters in this book. In this chapter we shall summarize the works of different authors on P(h) and compare with one another. Almost all the analytical works have been carried out within the MFA. As mentioned in chapter 2, Marshall (1960) and Klein and Brout (1963) derived P(h) for the Ising SG with RKKY exchange interaction and argued that taking proper care of the lower cut-off on the inter-spin separation imposed by the lattice P(h) turns out to be Gaussian. The Gausiian form of P(h) was a consequence of the simplifying assumptions in this theory. Later Gaussian P(h) has also been derived by Mookerjee (1978). However, all these theories do not take replica symmetry breaking into account and hence the P(h) does not reflect the hierarchical organization of the states. The nature of the P(h) near $h = 0$ has attracted lot of attention in the past decade. Most of the treatments indicated that $P(h) = 0$ at $h = 0$ (Palmer and Pond 1979, Bantilan and Palmer 1981, Roberts 1981) for m-vector SG (including $m = 1$). This is in agreement with the P(h) suggested by TAP (1977) (see chapter 6). This feature is also shared by the P(h) for the RKKY SG (Walker and Walstedt 1980). Moreover, it has been argued (Palmer and Pond 1979, Bray and Moore 1981d) that $P(h) = 0$ for $-\delta \le h \le \delta$ ($\delta > 0$) for the m-vector SG ($m \ge 2$) in the MFA. From an analysis of the stability of the ground state of the SK model with respect to the flipping of small cluster of spins, Palmer and Pond (1979) (also see Anderson 1979) showed that $P(h) \propto h^{\alpha}$ with $\alpha \ge 1$ (see also Schowalter and Klein 1979).The existence of a "hole" in the P(h) has also been proved by Kaplan and d'Ambrumenil (1982) for general form of J_{ij} which are bounded but otherwise arbitrary. They derived the lower bounds on δ. The "hole" in P(h) persists upto high external magnetic fields and temperatures of the order of $T_g/2$ (Grest and Soukoulis (1983). (See Hudak 1983 for the effect of the random uniaxial anisotropy on P(h) and Gulacsi and Gulacsi 1986 for the effect of the dipolar interaction). The effect of replica symmetry breaking in the SK model can be taken into account by arranging the pure states in a hierarchical manner as described in chapter 9 (see Mezard et al.1986, also see de Alemeida and Lage 1983).

Computer simulation has provided insight into the nature of P(h) in various SG models, e.g., for the three models shown in Figs.26.1, 26.2, 26.3.

26.2 Yang-Lee Method for SG and Other Frustrated models

The basic principle behind the studies of the statistical mechanical models by the Yang-Lee method is the following (see Mattis 1985 for an elementary introduction): defining the free energy as a function of a complex variable (either temperature or the field) one studies the distribution of the zeroes of the partition function in the complex plane. As long as the system remains finite the zeroes of the partition function lie at finite distances from the real axis. However, if the zeroes do not lie on the real axis even in the thermodynamic limit, the free energy will remain analytic function of the corresponding variable and no phase transition will take place. Moreover, the free energy and the other thermodynamic quantities can be calculated provided the distribution of the zeroes of the partition function is known. Unfortunately, this method is very tedious for almost all models.

As an example, consider the long-ranged Hamiltonian (Caliri et al.1985)

$$\mathcal{H} = - (J/2N^{1/2}) \Sigma \Sigma \epsilon_{ij} S_i S_j \qquad (26.1)$$

where ϵ_{ij} is a random variable. The corresponding free energy is

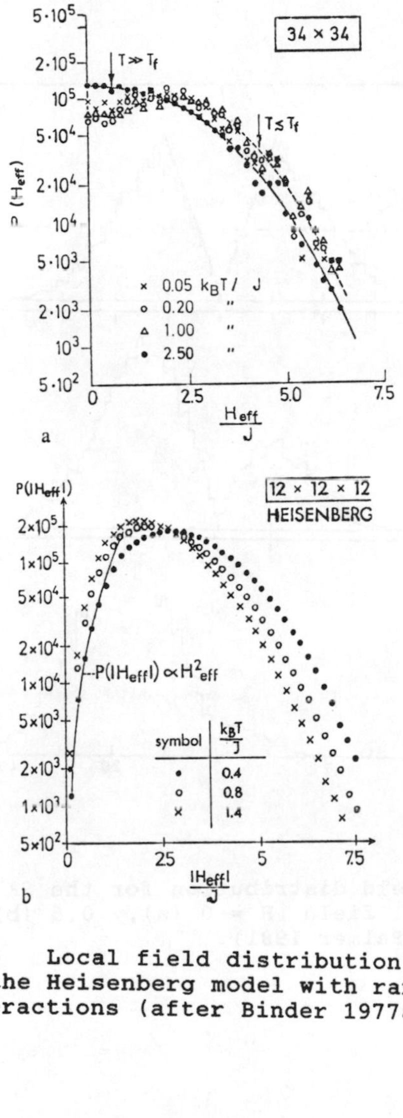

Fig. 26.1. Local field distribution for the Ising model (a) and the Heisenberg model with random nn Gaussian exchange interactions (after Binder 1977a).

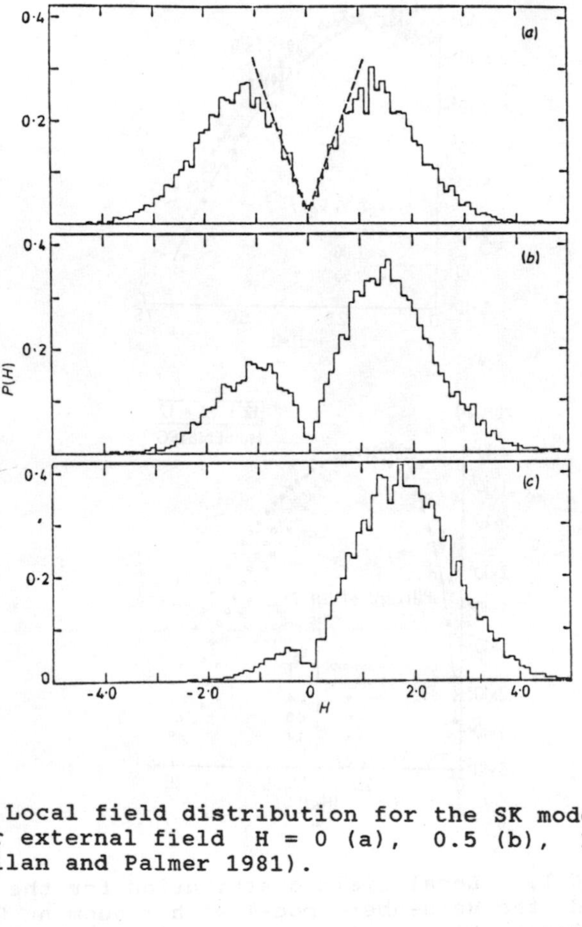

Fig. 26.2. Local field distribution for the SK model at T = 0 and for external field H = 0 (a), 0.5 (b), 1.5 (c) (after Bantilan and Palmer 1981).

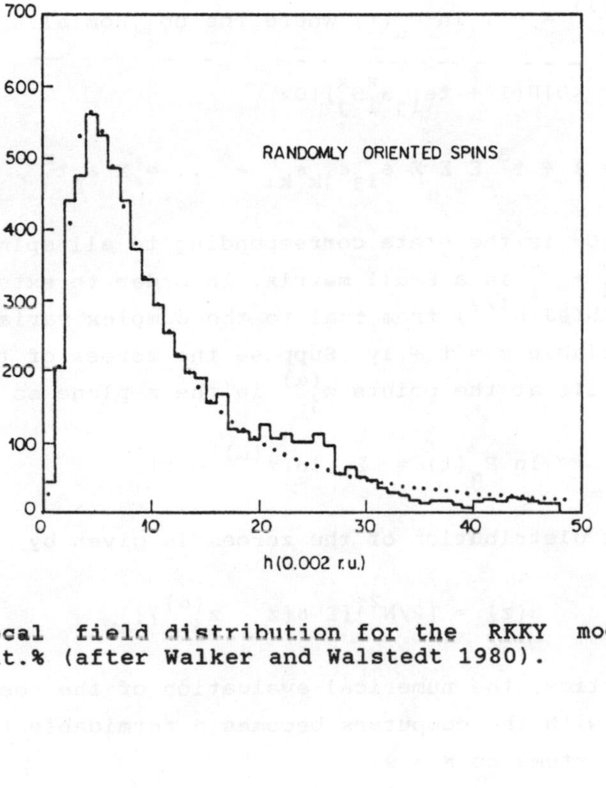

Fig. 26.3 Local field distribution for the RKKY model with c = 0.3 at.% (after Walker and Walstedt 1980).

given by

$$F = [F_\alpha]_{av} \qquad (26.2)$$

where F_α, the free energy for the α-th configuration, is given by

$$F\{J\} = -\,T\,\ln[\mathrm{Tr}\{\exp(-\beta\mathcal{H}_\alpha)\}]$$

$$= -\,NT\ln 2 - N(N-1)(T/2)\ln[\cosh(\beta J/N^{1/2} + F_\alpha^{(1)}] \qquad (26.3)$$

with $F_\alpha^{(1)} = -\,T\ln P_\alpha(t)$ where the polynomial

$$P_\alpha(t) = \langle 0|\Pi(1 + t\epsilon_{ij}S_i^x S_j^x)|0\rangle$$

$$= 1 + t^3 \Sigma\,\Sigma\,\Sigma\,\epsilon_{ij}\epsilon_{jk}\epsilon_{ki} + \ldots = \Sigma\, a_n t^n \qquad (26.4)$$

where $|0\rangle$ is the state corresponding to all spins "up" and $S_i^x = S_i^+ + S_i^-$ is a Pauli matrix. In order to extend $t = \tanh(\beta J/N^{1/2})$ from real to the complex variable, we consider the variable $z = t + iy$. Suppose the zeroes of the polynomial (26.4) lie at the points $z_j^{(\alpha)}$ in the z-plane so that

$$\ln P_\alpha(t) = \Sigma\ \ln(z_j^{(\alpha)} - t)$$

and the distribution of the zeroes is given by

$$g(z) = (2/N^2)[\Sigma\ \delta(z - z_j^{(\alpha)})]_{av}$$

In practice, the numerical evaluation of the coefficients a_n in (26.4) with the computers becomes a formidable task even for as small systems as $N = 9$.

The absence of phase transition even at $T = 0$ in some superfrustrated Ising models has been established by the Yang-Lee method by Suto (1984).

26.3. SG on Hierarchical Lattices and Fractals

We have summarized the RSRG approaches to the SG models in

chapter 17. But, usually, the RG treatment is never exact on Bravais lattices except for one dimension. However, exact implementation of the RG scheme is possible on the hierarchical lattices (McKay et al.1982a,b, McKay and Berker 1984). A hierarchical lattice is constructed by repeatedly replacing each bond by a basic unit (an example is shown in Fig.26.4). Therefore, the coordination number in hierarchical lattices is site-dependent. The randomness in the signs of the bonds leads to frustration.

Chaotic RG trajectories observed in the case of Ising SG models on hierarchical lattices has been interpreted as a signal for the SG phase characterized by random sequence of strong and weak correlations at successive length scales.

The Ising model on the hierarchical lattice shown in Fig. 26.5 has been studied by the Migdal-Kadanoff RG (Collet et al.1984, Collet and Eckmann 1984) and exactly in the limit $p \to \infty$, where p is the number of rungs (Gardner 1984). The model does not exhibit replica symmetry breaking. Note that in this hierarchical model, a spin is frustrated if only one of the two bonds connecting it are satisfied relative to the spins at higher levels in the hierarchy.

Statistical mechanics on fractal lattices (see Mandelbrot 1982 for an elementary introduction to fractals) has attracted lot of attention during the last few years. Banavar and Cieplak (1984) analyzed frustrated Ising models on Sierpinski gaskets (a fractal lattice).

26.4. Transverse Ising SG Model

The transverse Ising model is defined by the Hamiltonian

$$\mathcal{H} = - \Sigma\, J_{ij}\, \sigma_{iz}\, \sigma_{jz} + (\Delta/2)\, \Sigma\, \sigma_x^i \qquad (26.5)$$

where Δ determines the strength of the transverse field and σ are the Pauli matrices. In order to model a SG one usually assumes (Chakrabarti 1981, dos Santos et al.1985, Ishii and Yamamoto

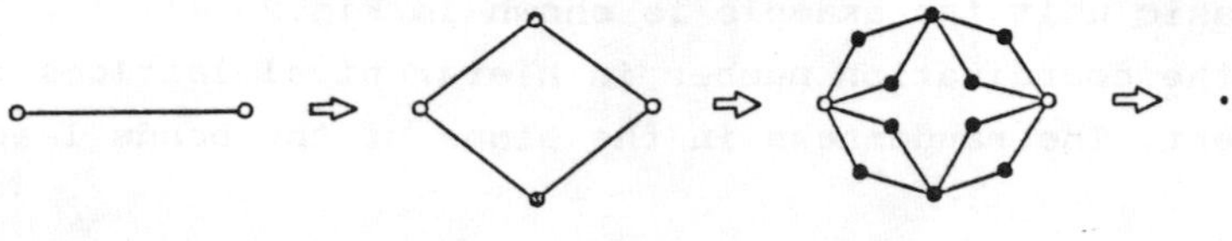

Fig. 26.4. **Construction of a simple hierarchical model.**

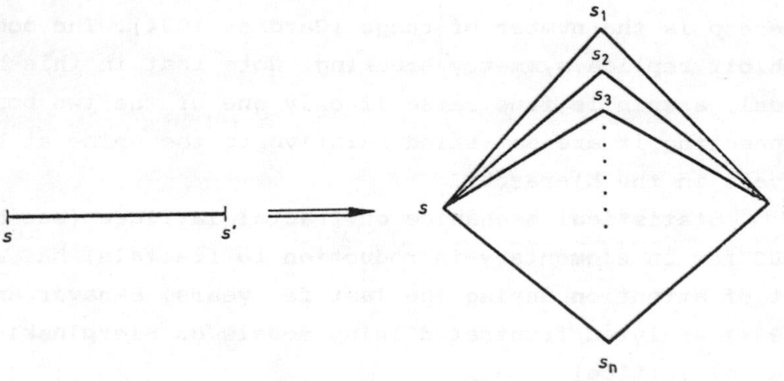

Fig. 26.5.

1985, Usadel 1986) the J_{ij} to independent quenched random variables with distribution (1.3) or (1.4). Because of the transverse field the quantum fluctuations are nonnegligible in this model. Two characteristic features have attracted the main attention:
(a) the variation of the transition temperature with Δ in the long-ranged model, and
(b) the correspondence between the critical behaviour of the d-dimensional transverse Ising SG and the d-dimensional and the (d+1)-dimensional EA-type SG.

26.5. SG in The Clock Model

The s-state clock model is defined by

$$\mathcal{H} = - \Sigma J_{ij} \cos(\eta_i - \eta_j) \qquad (26.6)$$

where $\eta_i = (2\pi/s)n_i$ ($n_i = 0,1,...,s-1$). In order to extend this model to SG, Nobre and Sherrington 1986 assumed the J_{ij} to be quenched independent random variables with Gaussian distribution (1.3). The free energy can be calculated by the replica trick. The critical temperature for the SG transition is

$$T_g = \tilde{J}/(2 - \delta_{2,s})$$

The replica symmetry breaking can be treated appropriately extending Parisi's scheme for the Ising SG to the model (26.6). The case s=3 corresponds to the Potts glass (see section 26.6) and s ≥ 5 correspond to the XY SG. Some common features of the Clock SG and Potts glasses will be discussed in section 26.6.

26.6. Potts Glass

The m-state Potts model (see Fu 1982, 1984 for an introduction to the Potts model) is defined by the Hamiltonian

$$\mathcal{H} = -(m/2)\,\Sigma\, J_{ij}\,\delta_{\sigma(i),\sigma(j)} \qquad (26.7)$$

where $\sigma(i) = 0,1,..,m-1$ are the Potts variables. Frustration is introduced into this system through the assumption that J_{ij} are quenched independent Gaussian-distributed random variables (Elderfield and Sherrington 1983ab, Erzan and Lage 1983ab, Gross et al.1985). The corresponding infinite-ranged model was puzzling in the beginning because of the subtleties involved with the application of Parisi's scheme for replica symmetry breaking to the infinite-ranged Potts glass model. The latter model does not have the reflection symmetry possed by the Ising model. Consequently, terms of the form $\beta_3 \Sigma q^{\alpha\beta}(q^{\beta\gamma})^2 q^{\gamma\alpha}$, which vanish in the case of the Ising SG, appear in the free energy (Goldbart and Elderfield 1985). Therefore, application of Parisi's replica-symmetry breaking scheme to the Potts model needs special care.

The analogue of the EA order parameter for the Potts glass is given by (Gross et al.1985)

$$Q_{rs} = [(m<\delta_{\sigma(i),r}> -)(m<\delta_{\sigma(i),s}> - 1)]_{av} \qquad (26.8)$$

whereas the analogue of the spontaneous magnetization is

$$M_r = m[<\delta_{\sigma(i),r}]_{av} - 1 \qquad (26.9)$$

Below the transition temperature T_g the PG phase is described by the order parameter function $q(x)$ which is a step function. Such a form of $q(x)$ implies that below T_g the phase space splits into an infinite number of nonoverlapping pure states. This PG phase becomes unstable at a lower temperature T_2, where each of the original pure states split into a hierarchical manifold similar to that for the SK model. The mean-field transition at T_g is second order for $m = 3$ and $m = 4$ but turns first order for $m > 4$. This should be contrasted with the infinite-ranged nonrandom Potts model for which the transition becomes first order for $m > 2$. The Potts glasses in the limit $m \to \infty$ are identical with the REM (see chapter 10) in that both have the free energy

$- J(m \ln m)^{1/2}$ for all $T < (1/2)J(m/\ln m)^{1/2}$.

Nishimori and Stephen (1983) introduced a slightly different version of the Potts model, often called the random chiral model, defined by

$$\mathcal{H}_{ij} = - J\, \delta_{\lambda_i,\lambda_j}, \qquad (26.10)$$

where the λ_i are m-component spins. Using the representation $\lambda_i = \omega^{k_i}$ where $\omega = \exp(2\pi i/m)$ with $k_i = 0, 1, \ldots, -1$, the Hamiltonian (26.10) can be expressed as

$$\mathcal{H}_{ij} = (-1/m) \sum_{r=1}^{m-1} J_{ij}^{(r)} \lambda_i^r \lambda_j^{m-r}, \qquad (26.11)$$

where $J_{ij}^{(r)}$ can be assumed to be random variables in order to incorporate randomness in the model. The ±J model and the Gaussian model, defined earlier for the m-vector spin models can also be generalized for the m-state Potts models. Suppose,

$$J_{ij}^{(r)} = J\, \xi_{ij}^{r} \qquad (26.12)$$

where ξ_{ij} is 1 with probability p and one of the other powers of ω (i.e., $\omega, \ldots, \xi^{m-1}$) with probability $(1-p)/(m-1)$. Similarly, the generalization of the Gaussian model for the m-vector spin models is expressed by

$$P\{J_{ij}^{(r)}\} = (2\pi J^2)^{-(m-1)/2}$$

$$\times \exp[- \sum_{r=1}^{m-1} (J_{ij}^{(r)}-J_0)(J_{ij}^{(m-r)} - J_0)/2J^2] \qquad (26.13)$$

with $J_{ij}^{(m-r)} = J_{ij}^{(r)*}$ ensuring the realness of the Hamiltonian although the individual $J_{ij}^{(r)}$ can be complex. The magnetization is defined by

$$M^2 = \lim_{|i-j|\to\infty} [\langle\lambda_i^r \lambda_j^{m-r}\rangle]_{av}$$

whereas the SG order parameter is defined by

$$q^2 = \lim_{|i-j|\to\infty} [\langle\lambda_i^r \lambda_j^{m-r}\rangle\langle\lambda_i^{m-r}\lambda_j^r\rangle]_{av}$$

(Also see Chang 1983 for a field-theoretical formulation of the Potts glass). The model (26.10) with (26.12) can be solved exactly on the Bethe lattice (Nishimori and Stephen 1983).

26.7. SG On The Bethe Lattice

The infinite-ranged model is exactly solvable and provides a starting point for the studies of more realistic SG models. However, because of this special feature of the interaction, there is no boundary condition and hence there is no way of infering the property of the infinite system by "enlarging" a finite system.

Let us focus our attention on a Bethe lattice. Suppose each site is connected to K'+1 (K' > 1) neighbours and let us define a central site 0. Note that the ratio of sites to bonds is 2/(K'+1) for a Bethe lattice (see Thouless 1986a). Then, the effective local field ξ_i satisfies the equation

$$\xi_i = \tanh^{-1}[\tanh K_{1i} \tanh\{h' + \Sigma\xi_j\}] \qquad (26.14)$$

where $K_{1i} = J_{1i}/(k_BT)$ is the coupling of the i-th site to its neighbours further in towards the central site, $h'=H/(k_BT)$ and the summation is to be carried out over the K' outer neighbours of the i-th site. Expanding equation (26.14) for H=0, we get

$$\xi_i = \tanh K_{1i}\, \Sigma\xi_j - (1/3) \tanh K_{1i}\, \mathrm{sech}^2 K_{1i} \{\Sigma\xi_j\}^3 \qquad (26.15)$$

From (26.15), after configuration-averaging, we get

$$[\xi^2]_{av} = t_1[\xi^2]_{av} - (2/3)(t_1-t_2)\{[\xi^4]_{av} + 3(K'-1)[\xi^2]_{av}^2\}$$

and

$$[\xi^4]_{av} = t_2\{[\xi^4]_{av} + 3(K'-1)[\xi^2]_{av}^2\}$$

where

$$t_n = K'[\tanh^{2n}K]_{av}$$

For $(1-t_1) > 0$ the only solution corresponds to all moments zero. On the other hand, for $(1-t_1) < 0$, there is another nontrivial solution

$$[\xi^2]_{av} = [\{(t_1-1)(1-t_2)\}/\{2(K'-1)(t_1-t_2)\}]$$

In order to study the sample-to-sample fluctuations, we proceed as follows: another set of fields η is calculated for the same sample but with different boundary conditions. This leads to

$$[\xi\eta]_{av} = t_1[\xi\eta]_{av} - (2/3)(t_1-t_2)\{[\xi^3\eta]_{av} + 3(K'-1)[\xi^2[\xi\eta]_{av}\}$$

and

$$[\xi^3\eta]_{av} = t_2\{[\xi^3\eta]_{av} + 3(K'-1)[\xi^2]_{av}[\xi\eta]_{av}\}$$

The paramagnetic phase corresponds to $[\xi\eta]_{av} = [\xi^2]_{av}$ whereas the latter vanishes in the SG phase (Thouless 1986). A more careful analysis for $H \neq 0$ gives the paramagnetic solution $[\xi\eta]_{av}=[\xi^2]_{av}$ in the paramagnetic phase. On the other hand, the other solution

$$[\xi\eta]_{av} = -(1/2)[\xi^2]_{av} + [(1/4)[\xi^2]_{av}^2$$
$$+ \{3h'^2(1-t_2)\}/\{K'(t_1-1)(K'+K't_2-2-t_2)\}]^{1/2}$$

is stable for

$$h'^2 < (t_1-1)^3\{K'(K'+2K't_2-2-t_2)\}/\{6(K'-1)^2(1-t_2)\}$$

which is the analogue of the AT line (5.21). However, there are crucial differences between the SG on a Bethe lattice and the SK model; the nonlinear susceptibility is smooth in the former model except for H=0 (see also Bowman and Levin 1982, Chayes et al. 1986).

26.8. Sensitivity of the Free Energy to The Boundary Conditions; Exchange Stiffness versus Scaling Stiffness

The concept of the sensitivity of the free energy to the boundary conditiond has been mentioned on several different contexts in this book. In this section we shall clarify the difference between the concepts of exchange stiffness and scaling stiffness introduced by Banavar and coworkers (Banavar and Cieplak 1982a,b, 1983, Cieplak and Banavar 1983, 198a,b, Caflish and Banavar 1985, Caflish et al.1985).

Consider a block of a magnetic material of length L (free boundary condition is applied along this direction) and of cross-sectional area A (periodic boundary condition is imposed along all the directions of this area). Suppose, the system is in equilibrium. Now, the spins at one end of the block are rotated by a small angle $\Delta\theta$, and then, keeping the spins at both the ends frozen, allowed to relax to the minimum energy configuration consistent with the new boundary condition. The free energy per spin of the system increases by an amount ΔF_e but it remains in the neighborhood of the original equilibrium state provided $\Delta\theta$ is small. ΔF_e is called the exchange stiffness energy. For example,

$\Delta F_e \sim A/L \sim L^{d-2}$ for both Heisenberg ferromagnets and SG.

Next, consider a scaling description where a bulk is created by putting together statistically similar blocks of the material. Now, choose a specific configuration for the spins on the two free boundaries and let the system relax to a minimum energy configuration consistent with the imposed boundary condition. The, rotate the spins on one of the ends of the block by a small angle $\Delta\theta$ and, keeping the spins at the two boundaries unaltered, relax the system again to its new minimum energy configuration. The corresponding energy difference per spin ΔF_s is called the scaling stiffness energy. Note that the basic difference between the exchange and scaling stiffnesses is that while determining the former the block is assumed to be a complete entity in contrast to the latter where the block is visualized as only a part of the bulk. Therefore, the boundary condition imposed in the scaling description mimics the coupling of the given block to the rest of the bulk system.

In a random system the mean and the variances are defined, respectively, by

$$\gamma_m = [\Delta F]_{av}$$

and

$$\gamma_w = \{[(\Delta F - [\Delta F]_{av})^2]_{av}\}^{1/2}$$

In a SG $\gamma_m = 0$, i.e., the average response of the system to the changes in the boundary condition vanishes. Now, it is postulated that

$$\gamma_w = \sigma/A^r L^p$$

Thus, the total free energy change is given by

$$\delta F = N(\Delta F) \sim \gamma_w AL \sim A^{1-r}L^{1-p} \sim L^{d(1-r)+r-p}.$$

Since, δF must be a nondecreasing function of L in the ordered phase, the LCD is given by

$$d_{lcd} = (p-r)/(1-r).$$

For example,

$$\Delta F_s \sim A/L \sim L^{d-2} \text{ for ferromagnets, but}$$

$$(\Delta F)_{\text{short times}} \sim A^{1/2}/L \sim L^{(d-3)/2}$$

and for Heisenberg SG

$$(\Delta F)_{eq} \sim A^{1/2}/L^x \text{ with } x \geq 1$$

Therefore, the LCD can be determined from the numerical estimates of p and r.

26.9. Induced Moment SG

So far we have been discussing the magnetic models which assume the existence of "good moments". We have also mentioned in chapter 1 that we shall not discuss the "Stoner glasses" in this book. In this section we shall discuss the theories for the so-called induced moment SG, e.g., $PrP_{0.9}$.

The crystal field effects are often comparable to or dominant over the exchange effects. We are interested in such systems when the ground state for only the crystal field Hamiltonian is a singlet. In order to get an insight into the effect of the exchange interaction on such a system let us first focus our attention on the so-called induced-moment ferromagnets (Cooper 1969). Let us consider the Hamiltonian

$$\mathcal{H} = \Sigma V_{ci} - (1/2)\Sigma G_{ij} \vec{J}_i \cdot \vec{J}_j - g\mu_B H \Sigma J_{iz}$$

where V_{ci} is the single-ion crystal field potential which leads to a singlet ground state |0> and a triplet first excited state |1> separated from each other by an energy gap Δ. Then one can show that in the absence of the exchange

$$1/\chi = (\Delta/2g^2\mu_B^2\alpha^2)[1/\tanh(\Delta/2k_BT)]$$

where

$$\alpha = |\langle 0|J_z|1\rangle|.$$

The effect of including the exchange is the replacement

$$1/\tanh(\Delta/2k_BT) \rightarrow [1/\tanh(\Delta/2k_BT)] - A$$

where

$$A = 2G(0)\alpha^2/\Delta$$

with

$$G(\vec{k}) = \Sigma\, G_{ij} \exp[i\vec{k}\cdot(\vec{r}_i - \vec{r}_j)]$$

The value $A = A_c$ for which the susceptibility diverges at $T = 0$ gives the threshold value of exchange for the ferromagnetic ordering. Thus, $A_c = 1$ and hence, ferromagnetic ordering is possible for $G(0) \geq (\Delta/2\alpha^2)$. Since $G(0) = zG$ for the model with only nearest neighbour interaction, the condition for the ferromagnetic ordering is given by

$$G \geq (\Delta/2z\alpha^2) \qquad (26.16).$$

In order to extend this theory to the induced moment SG (Sherrington 1979a), let us assume that G_{ij} are Gaussian-distributed independent random variables with mean G_0 and variance G_v. For this model the condition for the ferromagnetic ordering reduces to

$$G_0 \geq (\Delta/2z\alpha^2).$$

On the other hand, SG ordering is possible provided (Sherrington 1979)

$$G_v \geq (\Delta/2z^{1/2}\alpha^2)$$

In other words, SG ordering is possible if the energy splitting Δ is not too large. The latter observation is consistent with the result obtained by Sommers and Usadel (1982) from a TAP-like approach for general quantum spin systems.

26.10 SG Freezing and Superconductivity in SG Materials

It was Davidov et al. (1977) who first experimentally demonstrated that the magnetic ordering in the superconducting materials $Gd_xTh_{1-x}Ru_2$ and $Gd_xCe_{1-x}Ru_2$ is of SG type. In order to model such systems let us consider the Hamiltian (Soukoulis and Grest 1980)

$$\mathcal{H} = \mathcal{H}_{BCS} + \mathcal{H}_H + \mathcal{H}_I$$

where $\mathcal{H}_{BCS}$ is the BCS Hamiltonian without the magnetic impurities, $\mathcal{H}_H$ is the Heisenberg Hamiltonian (1.1) and $\mathcal{H}_I$ is the interaction Hamiltonian representing the s-f interaction between the s-electrons and the localized spins of the rare-earths. The latter interaction is given by

$$\mathcal{H}_I = -(J_{sf}/2) \quad \Sigma \quad \exp\{i\,(\vec{k} - \vec{k}') \cdot \vec{r}_i\}$$

$$[\vec{S}_i \cdot (C^+_{k\sigma}\, \vec{\sigma}_{\sigma\sigma'}\, C_{k'\sigma'})]$$

$\vec{\sigma} \equiv (\sigma^x, \sigma^y, \sigma^z)$ are the three Pauli matrices and J_{sf} is the strength of the s-f exchange interaction. In the Born approximation the superconducting transition temperature T_c is given by

$$\ln(T_c/T_{c0}) = \psi(1/2) - \psi[(1/2) + \{1/(2\pi T_c \tau)\}]$$

where T_{c0} is the transition temperature in the absence of magnetic impurities and ψ is the digamma function. The spin-flip scattering time τ is given by

$$1/\tau = (1/\tau_{AG})\{1/2k_F^2 S(S+1)\} \int \kappa\, d\kappa \int d\omega\; g(\kappa,\omega)\{\beta\omega/(e^{\beta\omega}-1)\}$$

where τ_{AG} is the spin-flip scattering time in the Abrikosov-Gorkov theory that neglects spatio-temporal spin correlations. The function $g(\kappa,\omega)$ is given by

$g(\kappa,\omega) = S(\kappa,\omega) - S_{Bragg}(\kappa,\omega)$ with

$$S(\kappa,\omega) = (1/2\pi N) \int dt\; e^{-i\omega t}\; \Sigma \langle S_i(t) S_j(0)\rangle \exp\{i\vec{k}\cdot\vec{r}_{ij}\}$$

$$S_{Bragg}(\kappa,\omega) = (1/2\pi N) \int dt\; e^{-i\omega t}\; \Sigma \langle S_i\rangle\langle S_j\rangle \text{ e p}\{i\vec{k}\cdot\vec{r}_{ij}\}$$

For SG, the above relations reduce to

$$\tau_{AG}/\tau = [q/\{S(S+1)\}] + [1/\{2k_F^2 S(S+1)\}] \int \kappa \, d\kappa \int d\omega$$

$$\times [\beta^2\omega^2 e^{-\beta\omega}/(e^{-\beta\omega}-1)^2] \, k_B T \, \chi(\kappa) \, F(\kappa,\omega)$$

Assuming that

$$\chi(\kappa) = \begin{cases} S(S+1)/k_B T & \text{for } T \geq T_g \\ S(S+1)/k_B T_g & \text{for } T < T_g \end{cases}$$

and

$$F(\kappa,\omega) = (1/\pi) \, [D\kappa^2/\{(D\kappa^2)^2 + \omega^2\}]$$

with spin doffusion constant D, Soukoulis and Grest (1980) derived the phase diagram for integral as well as nonintegral S values. The SG correlation between the spins are found to enhance the superconductivity in comparison with the Abrikosov-Gorkov theory which ignores spin correlation effects. This result is in contrast to that of Sadovskii and Skryabin (1979) that the SG ordering does not affect the superconducting transition. Soukoulis and Grest's result has been supported by the the experiments of van Dongen et al.(1981) on the $Pd_{1-x}Mn_xH$ system. Moreover, the Soukoulis-Grest theory predicts re-entrant behaviour in agreement with the experimental data of Jones et al.(1978).

The upper critical field $H_{c2}(T)$ has been measured for the system $Th_{1-x}Nd_xRu_2$ has been measured by Huser et al.(1983b). The $H_{c2}(T)$ have been calculated (Crisan 1986) for short-ranged models. The latter model consists of damped RKKY Hamiltonian where the mean-free path δ is the coherence length of the Cooper pairs. The theoretical and the experimental $H_{c2}(T)$ data have been shown in Fig.26.6 (see Gulacsi et al.1983 for the calculation of the specific heat in the latter model).

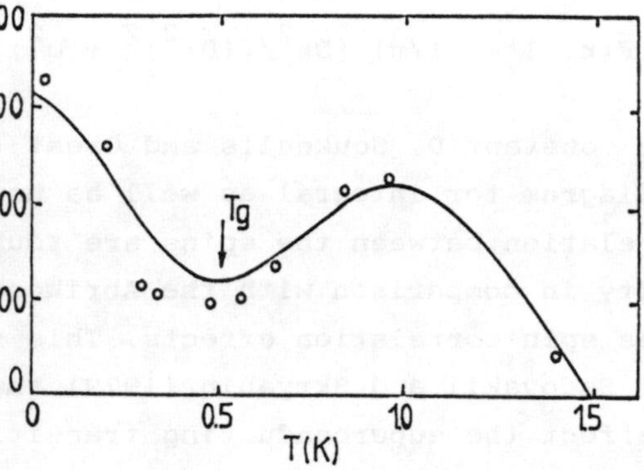

Fig. 26.6. The upper critical field $H_{c2}(T)$ for a spin-glass superconductor. The continuous line is the theoretical prediction and the circles represent the experimental data points (after Crisan 1986).

CHAPTER 27

SG-LIKE SYSTEMS

27.1. Electric Dipolar and Ouadrupolar Glasses:

Several systems with electric moments exhibit glassy behavior over a certain range of composition. Most of the physical properties of such glassy phases are qualitatively similar to the corresponding properties of SG materials. Therefore, in recent years efforts have been made to understand the orientational orderings in these systems in the light of our present knowledge of the ordering of the magnetic moments in SG materials. Examples of materials exhibiting dipolar and quadrupolar glass phases are $(KCN)_x(KBr)_{1-x}$ (see section 27.1.1), solid hydrogen (see section 27.1.2), mixed crystals of $(Rb)_{1-x}(NH_4)_xH_2PO_4$ (see section 27.1.3), $(KCN)_x(NaCN)_{1-x}$ (Luty and Ortiz-Lopez 1983, Loidl et al.1986), $K_{1-x}Li_xTaO_3$ (Hochli 1982), $K_{1-x}Na_xTaO_3$ (van der Klink et al.1983), $KTa_{1-x}Nb_xO_3$ (Samara 1984), etc. We shall discuss only a few of these systems in detail in this book.

27.1.1 $(KCN)_x(KBr)_{1-x}$: A Director Glass ?

The CN molecules in pure KCN have a small electric dipole moment. The electric dipole-dipole interaction is, however, weak compared to the elastic quadrupole interaction. The experimental study of the phase diagram of the $(KCN)_x(KBr)_{1-x}$ system over the whole concentration range $0 \le x \le 1$ is possible because the average size of the rod-shaped CN molecule is almost equal to that of the spherical Br ions. The phase diagram of the latter system has been studied quite extensively by various experimental techniques (Loidl et al. 1980, 1982, 1983, 1984, 1985, Bhattacharya et al. 1982, Birge et al. 1984, DeYoro et al. 1983, Meissner et al.1985, Moi et al. 1984, Mertz and Loidl 1985, Feile et al. 1982, Garland et al. 1982 and references therein).

When cooled from high temperature, pure KCN exhibits two orientational phase transitions :
(a) at the ferroelastic transition temperature T = 168 K the elastic quadrupole interaction aligns the axes of the CN molecules (without introducing dipolar order) as shown in Fig. 27.1b,
(b) at T=83K, the electric dipolar interaction lifts the degeneracy between the two orientations (viz. CN and NC) of the CN molecules and KCN orders antiferroelectrically.

What is the effect of dilution of KCN by KBr ? So long as $x > 0.57$, $(KCN)_x(KBr)_{1-x}$ still exhibits LRO although the phase diagram becomes more complicated (see Sethna 1986 for more details). On the other hand, for $x < 0.1$ the cyanide ions are more or less isolated from each other (see Fig.27.1a) and tunnel between the eight possible equilibrium orientations <111>. We are mostly interested in the intermediate regime $0.1 < x < 0.57$ because $(KCN)_x(KBr)_{1-x}$, in this concentration range, exhibits almost all the "universal" properties of glasses (see Fig.27.1c). For example, at low-temperatures the specific heat is linear in T, the thermal conductivity is proportinal to T^2, etc. The neutron scattering experiments (Rowe et al. 1979) strongly suggest the onset of new type of orientational order, at least on the time-scale of observation. In the latter 'phase', the quadrupoles exhibit SG-like freezing. On the other hand, measurement of the dynamic dielectric constant (Bhattacharya et al. 1982, Birge et al. 1984) indicate that the 180^0 flipping of the dipoles dominate the dielectric loss even below the latter transition temperature.

Michel and Rowe (1980) first stressed the analogy between the glassy states in $(KCN)_x(KBr)_{1-x}$ and SG. The mean-field theory (MFT) for the ferroelastic transition in pure KCN (Sethna et al. 1984) has been extended to describe certain aspects, e.g., the breadth, form and temperature-dependence of the low-frequency dielectric loss peaks, etc., in the glassy state of $(KCN)_x(KBr)_{1-x}$(Sethna and Chow 1985). Suppose the vector $\vec{n}$

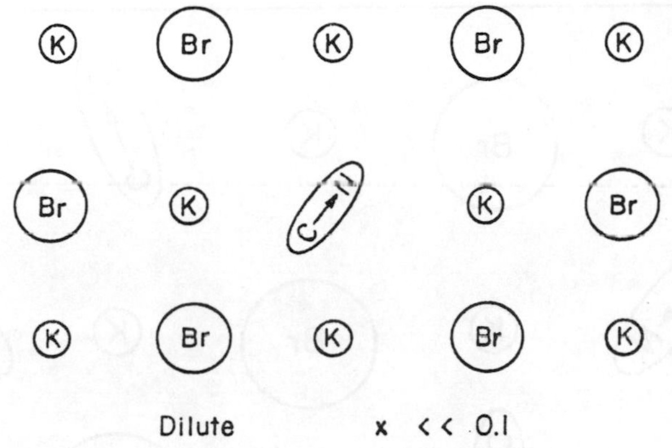

Fig. 27.1(a)

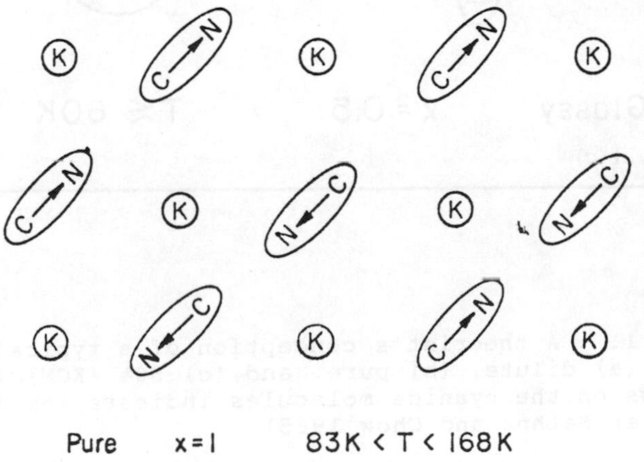

Fig. 27.1(b)

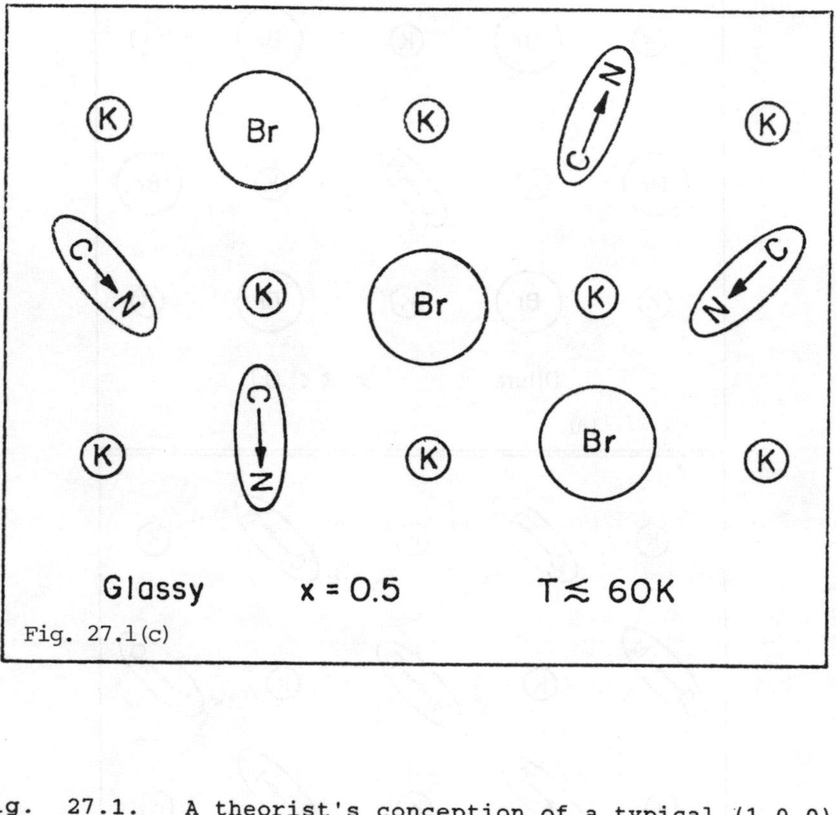

Fig. 27.1(c)

Fig. 27.1. A theorist's conception of a typical (1.0,0) plane in (a) dilute, (b) pure, and (c) 50% $(KCN)_x(KBr)_{1-x}$. The arrows on the cyanide molecules indicate the direction of $\vec{n}$ (after Sethna and Chow 1985).

pointing from the carbon to the nitrogen describes the vectorial orientation of a CN molecule. But, since the cyanides remain free to flip by 180^0, the difference between $\vec{n}$ and $-\vec{n}$ is ignored and the Hamiltonian is an even function of $\vec{n}$ (Sethna and Chow 1985)

$$\mathcal{H}_{SC} = - (1/2) \Sigma Q^i_{\alpha\beta} Q^j_{\alpha\beta} c_i c_j \qquad (27.1)$$

where c_i, c_j are the occupation probabilities of the sites i and j, and the quadrupole $Q^i_{\alpha\beta} = [n^i_\alpha n^i_\beta - (1/3) \delta_{\alpha\beta}]$ is the "director" describing the axial orientation of the CN molecule. However, the latter theory is not completely satisfactory, even at the mean-field level, because, although it takes dilution into account, it ignores the effect of frustration. But, in spite of this shortcoming, Sethna and Chow's mean-field theory yields the distribution of the barrier hights, P(V), hindering the cyanide reorientations in agreement with the experimental data of Birge et al.(1984). Does the latter agreement imply that the role of frustration is irrelevant (or, at least less relevant than the role of dilution) for highly diluted systems? (see Kanter and Sompolinsky 1986 for some criticism of this work).

Kanter and Sompolinsky (1986) proposed another model that ignores site-dilution (it incorporates bond-randomness in the same way as in the EA model) but incorporates frustration. The corresponding Hamiltonian is given by

$$\mathcal{H}_{KS} = \Sigma J^{\alpha\beta\gamma\delta}_{ij} Q^i_{\alpha\beta} Q^j_{\gamma\delta} \qquad (27.2)$$

where $J^{\alpha\beta\gamma\delta}_{ij}$ are independent random variables. Following a replica symmetry breaking scheme (Gross et al.1985), the distribution P(V) was derived within mean-field approximation and found to deviate from that observed experimentally. Several possible reasons for such deviations have been proposed. A more sophisticated theory which incorporates both dilution and frustration, at least under some reasonable and systematically controlled approximations, would throw light on the relative importance of the site-dilution and frustration.

The ferroelastic domain model (Ihm 1985) presents an alternative scenario. In this picture, the CN molecules align, and freeze, with their axes mutually parallel within a ferroelastic domain although the domains remain randomly oriented with respect to one another. Besides, the dipoles keep flipping by 180^0 even at temperatures much lower than that corresponding to the freezing of the axes.

Finally, it is worth mentioning that from a theoretician's point of view, the advantage of studying $(KCN)_x(KBr)_{1-x}$ is that, unlike the SG, this system exhibits properties found to be universal for all glasses.

27.1.2 Solid Hydrogen- A Quadrupolar Glass?

Molecular hydrogen H_2 is diatomic and homonuclear with nuclear spin 1/2 for each proton. Because of the fermionic nature of the protons two species of molecular hydrogen are possible, viz., ortho hydrogen ($O-H_2$) and para hydrogen ($p-H_2$). $O-H_2$ has total nuclear spin unity and odd value of orbital angular momentum J whereas $p-H_2$ has vanishing total nuclear spin and even value of J. In the solid phase only states with lowest J are occupied and therefore

$$J = 0 \text{ for para-hydrogen, and}$$
$$J = 1 \text{ for ortho-hydrogen}$$

in the solid phase. Therefore, ortho-hydrogen molecules are non-spherical in contrast to the spherical shape of the para-hydrogen molecules. Qualitatively speaking, it is the non-sphericity of the ortho-molecules that gives rise to the possibilities of the various types of orientational ordering in solid hydrogen (see Silvera 1980, van Kranendonk 1983). Quantitatively, speaking, the interaction between the spin-1 quantum rotators leads to the orientational ordering.

The orientational orderings in solid hydrogen, with the variation of temperature T and concentration c of the ortho species, have been investigated quite intensively over the last decade, both experimentally and theoretically (Sullivan 1976,

1977, 1983, Sullivan and Devoret 1977, Sullivan et al. 1978, 1979, 1982, 1984, Sullivan 1983, Sullivan and Esteve 1981, Esteve and Sullivan 1982, Devoret et al. 1982, Devoret and Esteve 1983, Ishimoto et al.(1976), Yu et al. 1981, 1983a, 1983b, Washburn et al. 1981, 1982, 1983, Harris et al. 1983, Cochran et al. 1980, Gaines and Sokol 1983, Klenin 1979, 1982, 1983a, 1983b, Klenin and Pate 1981, Kokshenev 1985, Haase and Perrel 1983, Pound et al. 1983, Sokol and Sullivan 1985, and the references therein). For $c > 0.56$, the stable thermodynamic phase just below the melting temperature has h.c.p. crystal structure. At a lower temperature, T_{hc}, a structural phase transition from the h.c.p. crystal structure to the Pa_3 structure (see Fig.27.2) takes place. In the latter structure the centers of the molecules occupy the sites of an f.c.c. lattice where the molecular axes are oriented along the body diagonals. The latter structure can be viewed as four interpenetrating simple cubic sub-lattices such that on each of the sub-lattices all the molecular orbitals are mutually parallel (Silvera 1980, van Kranendonk 1983). Notice that the Pa_3 structure involves not only a new crystal structure but also a new type of long-range orientational ordering of the molecules. Most of the theoretical approaches to the understanding of the nature of the transition at T_{hc} focus attention either only on the h.c.p. → f.c.c. (crystal) structural transition or only on the orientational phase transition. Therefore, our understanding of the double transition (which are so intimately related and probably driven by the same interaction) remains incomplete.

Although the long-range orientational ordering discussed above is interesting in its own right, the ordering which is more relevant in the context of SG-ordering occurs in solid hydrogen for $c < 0.56$ (see Harris and Meyer 1985, Sokol and Sullivan 1985 for reviews). The ortho hydrogen molecules are the analogs of the magnetic moments (spins) in the real SG materials and the dominant electric quadrupole-quadrupole (EQQ) interaction between the ortho molecules plays the role of the exchange interaction in SG. The ortho hydrogen forms the analog of the host lattice (like

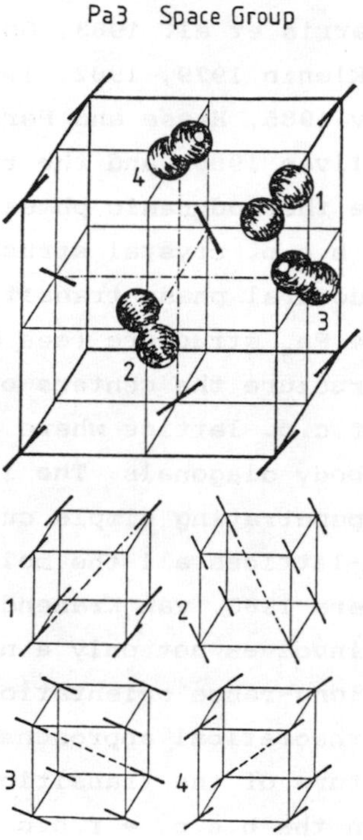

Fig. 27.2. The Pa_3 space group of ordered hydrogen and deuterium. In the upper drawing the spatial distributions of the molecular axes of the four inequivalent molecules is shown. At the remaining sites we show the direction of quantization for the molecules. The division into sublattices is shown in the lower part of the figure (after Silvera 1980).

Au in AuFe SG alloys) and introduces dilution. Besides, solid hydrogen in unique in the sense that, because of the slow spontaneous ortho-to-para conversion, the same sample, can be used at successive time instants for the investigation of samples with different concentrations (c) of the ortho-hydrogen.
In spite of all these analogies between the SG models and solid hydrogen there are some crucial differences. The exchange interaction in (1.1) favors parallel or antiparallel orientation of the two interacting spins depending on the positive or negative sign of J. On the other hand, EQQ interaction favors mutually-perpendicular orientation of the two interacting quadrupoles in order to minimize energy.

Experimental investigation of the orientational ordering is more difficult than that in magnetic SG-materials because the quadrupoles couple with the electric-field gradient and very large electric-field gradient would be required for the direct measurement of the analog of the magnetic susceptibility. Therefore, most of the important conclusions have been drawn so far from nuclear magnetic resonance (NMR) measurements (see Silvera 1980, van Kranendonk 1983, Harris and Meyer 1985 for reviews).

It is easy to see that each of the EQQ bonds on a h.c.p. lattice cannot be satisfied simultaneously giving rise to frustration. Indeed, it is this frustration that leads to the SG-like orientational "freezing" in the dilute system of ortho-hydrogen ($c < 0.56$) at sufficiently low temperature, as verified experimentally (more subtle features that distinguish this freezing from the SG freezing will be discussed later). For convenience, for the time being, let us call the latter frozen state "SG-like phase". Now, the theoretical problem is to describe the latter freezing phenomenon and the corresponding SG-like phase in terms of quantitative concepts. In the absence of interactions which break the time-reversal symmetry, the dipole moments $\langle J_x \rangle$, $\langle J_y \rangle$ and $\langle J_z \rangle$ vanish. Therefore, we need the higher moments for the desired quantitative description. The first nontrivial ones are the components of the second-rank tensor

$\langle Q_{\alpha\beta} \rangle$ (α and β referring to the components in the Cartesian system) which are the expectation values of the operator

$$Q_{\alpha\beta} = (1/2)\,\{J_\alpha J_\beta + J_\beta J_\alpha\} - (1/3)\,J^2$$

Let us assume that the local Cartesian axes are so chosen as to coincide with the principal axes of the tensor $\langle Q_{\alpha\beta} \rangle$ so that $\langle Q_{xy} \rangle = \langle Q_{yz} = \langle Q_{zx} \rangle = 0$. Therefore, there are only two independent nonvanishing intrinsic quadrupolar parameters

$$\sigma_i = \langle 1 - (3/2)J_z^2 \rangle \qquad (27.3)$$

and

$$\eta = (\surd 3/2)\ \langle J_x^2 - J_y^2 \rangle \qquad (27.4)$$

in this coordinate system. These are, respectively, measures of the alignment along z_i and eccentricity. Thus, in this coordinate system, the "ordering" of a single molecule is specified by five independent parameters, viz., the three local principal axes (x,y,z) to specify the orientation of the local symmetry axes, and σ and η to specify the degree of ordering in this reference frame (see Sullivan et al.1984 for an excellent discussion).

In the LRO phase (having Pa_3 structure) the ortho molecules are aligned so that $\eta_i = 0$ for each i and σ_i = constant, independent of i; only the orientation of the local axes differ from one sublattice to another (see Fig.27.3a). On the other hand, in the frozen "SG-like phase" ($c < 0.56$ and at low temperature) not only do the orientations of the local axes vary from site-to-site but also both σ and η become site-dependent (see fig.27.3b).

What is the nature of the transition to the orientational "glass" state ? The latter problem has been the most controversial issue during the last decade (see reviews by Silvera 1980, van Kranendonk 1983, Gaines and Sokol 1983, Sullivan 1983, Klenin 1983b, Harris and Meyer 1984). The reasons for the differences observed in the experiments on the powder samples and on the single crystals need special attention in

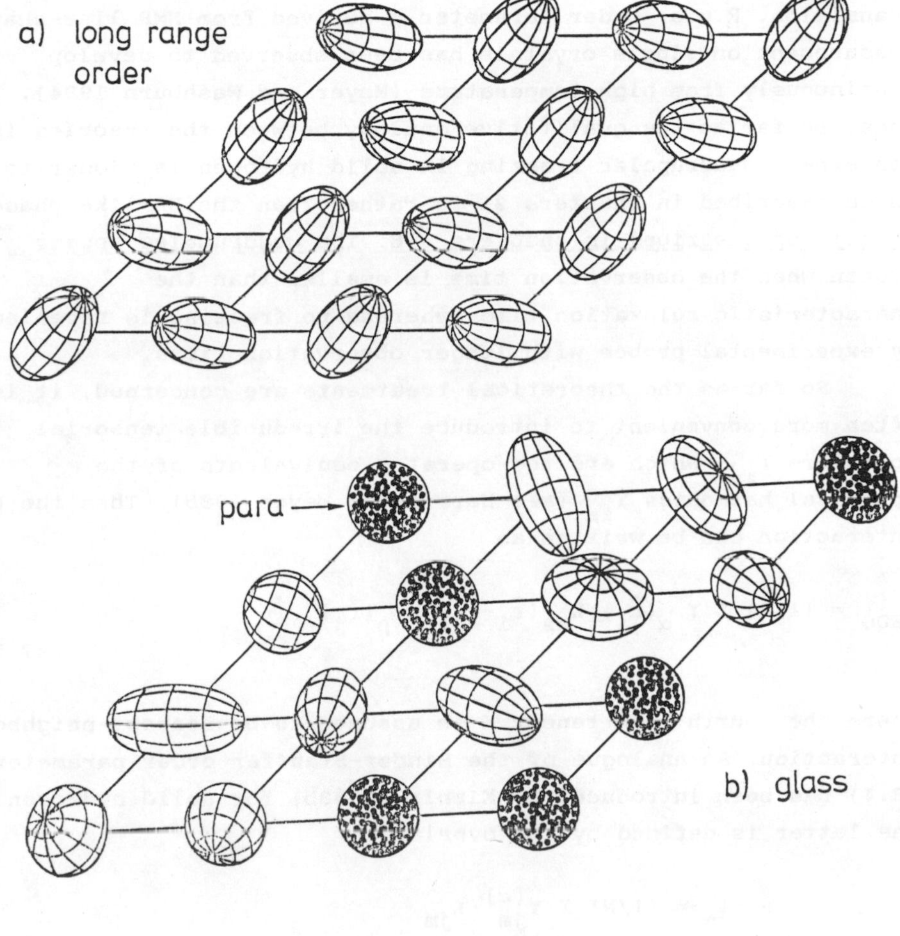

Fig. 27.3. Comparison of the orientational ordering for (a) a plane section of the Pa_3 structure and (b) the quadrupolar glass phase (after Sullivan 1983).

order to clarify the true nature of the transition. At present the view of the majority seems to be in favor of a gradual ordering (a slow relaxation process) rather than a sharp phase transition. R.m.s. order parameter σ derived from NMR line-shape measurement on single crystals has been observed to develop continuously from high temperature (Meyer and Washburn 1984). Thus, so far as the qualitative analogy between the theories is concerned, quadrupolar freezing in solid hydrogen is closer to those described in chapters 21-23 rather than the EA-like phase transition described in chapters 3-8. The quadrupoles appear frozen when the observation time is smaller than the characteristic relaxation time, whereas no freezing is revealed by experimental probes with longer observation times.

So far as the theoretical treatments are concerned, it is often more convenient to introduce the irreducible tensorial operators Υ_{2m}, which are the operator equivalents of the spherical harmonics Y_{2m} (see Harris and Meyer 1985). Then the EQQ interaction can be written as

$$\mathcal{H}_{EQQ} = (1/2)\, \Sigma\, \Upsilon_{\alpha}(\vec{r}_i)\, K_{\alpha\beta}(\vec{r}_i - \vec{r}_j)\, \Upsilon_{\beta}(\vec{r}_j) \tag{27.5}$$

where the fourth rank tensor K is assumed to be nearest-neighbour interaction. An analogue of the Binder-Stauffer order parameter (3.4) has been introduced by Klenin (1983b) for solid hydrogen. The latter is defined by the overlap

$$\psi_{\alpha} = (1/N)\, \Sigma\, \Upsilon^{(\alpha)}_{jm} \cdot \Upsilon_{jm}$$

where the "reference state" is denoted by the superscript α. Such an order parameter is convenient for MC simulation purposes. Several authors (e.g., Devoret and Esteve 1983) proposed an order parameter

$$q_{EQQ} = [x_i^2]_{av} \tag{27.6}$$

where $x_i = \sigma_i + \eta_i$, σ_i and η_i being (proportional to) the intrinsic order parameters σ and η defined by (27.3) and (27.4). However, there is a crucial difference between q_{EQQ} and q_{EA}, the former is always non-zero in contrast to the latter which vanishes in the absence of external field for all $T > T_g$. Therefore, roughly speaking, a ortho-para mixture of solid hydrogen for $c < 0.55$ is similar to SG in an external field (see Harris and Meyer 1984 for more details). In the sophisticated language of modern theoretical physics, the latter phenomenon is expressed by the statement that the order parameter for dilute solid quenched mixture of ortho-para hydrogen "do not display broken symmetry". However, such a gradual transition to the orientational glass state in this system involves strong cooperativity as supported by MC simulations (Devoret and Esteve 1983, Klenin 1984). Kokshenev (1985) has developed a perturbative theory for the quadrupolar glass phase. An AT-like instability line has also been derived (Kokshenev 1986).

Recently, Goldbart and Sherrington (1985) have developed the mean-field theory of the model

$$\mathcal{H} = - \Sigma J_{ij} \Sigma Q^{i}_{\alpha\beta} Q^{j}_{\alpha\beta} \qquad (27.7)$$

(we have dropped the field-term for convenience), where J_{ij} are assumed to be Gaussian-distributed random variables. The latter model differs from (27.5) in the following respects:

(i) Goldbart and Sherrington considered spatial disorder implemented via disorder in J_{ij} whereas the model (27.5) for solid hydrogen corresponds to a definite spatial symmetry; K_{ij}'s are non-random and the system is driven to the glassy state in the latter model by frustration arising from the lattice geometry.

(ii) the coupling K_{ij} depends on the quadrupole components α, β etc., whereas the coupling J_{ij} in (27.7) is independent of the quadrupole components.

More careful experiments are being carried out by the various groups, e.g., at the Duke university and at Florida, to settle the true nature of the transition.

27.1.3. $Rb_{1-x}(NH_4)_xH_2PO_4$ - Hydrogen-bonded Mixed Crystal of Ferroelectric and Antiferroelectric Materials

RbH_2PO_4 (abbreviated RDP) and $NH_4H_2PO_4$ (abbreviated ADP) belong to the so-called KDP (abbreviation for KH_2PO_4) family and have identical lattice structures with nearly the same lattice constants. Therefore, mixed crystals of the type $Rb_{1-x}(NH_4)_xH_2PO_4$ (abbreviated $(RDP)_{1-x}(ADP)_x$) can be prepared with all possible concentrations c, from c=0 to c=1. Pure RDP orders ferroelectrically whereas pure ADP orders antiferroelectrically at low temperature. For an intermediate range of concentration c, $(RDP)_{1-x}(ADP)_x$ exhibits glassy behaviour (Courtens 1982, 1983, 1984, Courtens et al. 1984) which is reminiscent of SG. However, because of some crucial differences and extra complications (as compared to the SG and other dipolar and quadrupolar glasses) we shall not discuss the theories (Prelovsek and Blinc 1982, Matsushita and Matsubara 1984, 1985, Ishibashi and Suzuki 1985, Akhiezer and Spolnik 1983) of such (hydrogen bonded) mixed crystals of ferroelectrics and antiferroelectrics.

27.2. Electron Glass in Doped Compensated Semiconductors ?

Consider a partially compensated n-type semiconductor at a low temperature. Since the compensation is partial all the acceptors have captured an electron and have become negatively charged. On the other hand, a fraction of the donors are ionized and the others remain neutral. Assuming the overlap of the donor states to be very small, one can write the Hamiltonian as (Shklovski and Efros 1980)

$$\mathcal{H}= (e^2/\kappa)[(1/2)\Sigma\{(1-n_i)(1-n_j)/r_{ij}\} - \Sigma\{(1-n_i)/r_{i\nu}\}+(1/2)\Sigma(1/r_{\nu\gamma})]$$

where the first term corresponds to the donor-donor repulsion, the second term describes the donor-acceptor attraction and the last term represents acceptor-acceptor repulsion. The Coulomb repulsion leads to a depletion of the single-particle density of states near the chemical potential; this gap is called the coulomb gap.

For those readers not familiar with semiconductor physics we shall reformulate the problem (Davies et al.1982, 1984, Gruenewald et al.1982, 1983) in the following way: suppose a random energy ϵ_i is associated with each of the sites on a cubic lattice; a charge e/2 is associated with the empty sites and -e/2 is associated with the sites occupied by an electron, so that

$$\mathcal{H} = \Sigma (\epsilon_i - \mu) S_i + \Sigma C_{ij} S_i S_j \qquad (27.8)$$

where μ is the chemical potential, C_{ij} is the long-range coulomb interaction between the i-th and the j-th sites and S_i and S_j are the corresponding occupation numbers.

The Hamiltonian for the short-ranged random-field Ising model (RFIM) (see Imry 1984, Villain 1985, Aharony 1986) is given by

$$\mathcal{H} = - J \Sigma S_i S_j - \Sigma H_i S_i \qquad (27.9)$$

where J is the nonrandom nearest-neighbor ferromagnetic exchange interaction and $H_i = \xi_i H$ is an infinitesimally small random external magnetic field. Note that if the exchange interactions were of the nearest-neighbor type, the model (27.9) could be mapped onto the Mattis model in uniform external field by the transformation $S_i \to \xi_i S_i$. The Hamiltonian (27.8) can be mapped onto the Hamiltonian (27.9) by the following substitutions:

$\epsilon_i \to H_i$, $\mu \to H$, S → Ising spin S, and $C_{ij} \to J_{ij}$.

However, one must keep in mind that C_{ij} is long-range. In all the other chapters in this book we have discussed the SG systems with either bond-disorder or site-disorder. Can SG-like ordering take place in random-field systems ? In other words, is it possible to have nonzero EA order parameter $q = [\langle S_i \rangle^2]_{av}$ in random-field systems in the absence of spontaneous magnetization (Morgenstern et al. 1981, Richards 1984, see Hornreich and Schuster 1982 for the random-field spherical model)?

The mean-field free energy for the model (27.8) is given by (Gruenewald et al.1982)

$$F = \Sigma\, \epsilon_i M_i + \Sigma\, C_{ij}\, M_i M_j + k_B T\, \Sigma\, [\{(1/2) + M_i\} \ln \{(1/2) + M_j\}$$

$$+ \{(1/2) - M_i\} \ln \{(1/2) + M_j\} \qquad (27.10)$$

where $M_i = \langle S_i \rangle$ is the thermal average of S_i. The first two terms in (27.10) are the contributions from the internal energy. The last term in the brackets [ ] is the entropic contribution and is very similar to the corresponding terms in the mean-field equations for SG (i.e., the TAP equations (6.8)). The existence of a large number of solutions of the mean-field equations (27.10) imply that the model (27.8) has a large number of metastable states just like SG (Gruenewald et al.1982). Defining a EA-like order parameter $q \propto \Sigma M_i^2$, these authors studied the low-temperature thermodynamic properties of the model (27.8). On the other hand, Davies et al.(1982, 1984) introduced the order parameter

$$q_{DLR} = [(\langle S_i \rangle - \langle S_i^0 \rangle)^2] \qquad (27.11)$$

where $\langle S_i^0 \rangle$ is the Monte-Carlo average of S_i in the absence of interactions. Although computation of these order parameters has indicated the possibility of the existence of an "electron glass" phase, the studies have not been quite conclusive

27.3. Superconducting Glass Phase in Granular Superconductors and Josephson Junction Arrays

Granular superconductors have attracted lot of attention during the last few years (see Goldman and Wolf 1984, Gubser et al.1980). These materials consist of superconducting grains embedded in a host, the host can be (a) an insulator, or (b) a semiconductor or (c) a metal or (d) a superconductor with a lower superconducting transition temperature. The superconducting grains interact via either Josephson coupling (if the host is an insulator) or proximity effect (if the host is a normal metal). Consider such a granular superconductor system consisting of N grains, the complex energy gap ψ_i of the i-th grain is given by

$$\psi_i = \Delta_i \exp(i\phi_i)$$

The Hamiltonian describing such a system is given by

$$\mathcal{H} = - \Sigma J_{ij} \cos(\phi_i - \phi_j) \qquad (27.12)$$

where J_{ij} is a temperature-dependent coupling that falls as
(a) $J_{ij} \propto (1/r_{ij})$ if the host is an insulator
(b) $J_{ij} \propto \exp(-r_{ij})$ if the host is a normal metal
with the increase of separation r_{ij} between the grains. The Hamiltonian (27.12) is reminiscent of that of the XY model. However, one should remember that J_{ij} is temperature-dependent in contrast to the temperature-independent exchange in the XY model. In the presence of magnetic field, the Hamiltonian (27.12) is modified to

$$\mathcal{H} = - \Sigma J_{ij} \cos(\phi_i - \phi_j - A_{ij}) \qquad (27.13)$$

where

$$A_{ij} = (2\pi/\Phi_0) \int \vec{A}\cdot d\vec{l};$$

$\Phi_0 = hc/2e$ being an elementary flux quantum and $\vec{A}$ is the vector potential.

Disorder can be introduced into the system in various different ways. If J_{ij} is nonrandom and the system is compositionally disordered (i.e., if the superconducting grains occupy the lattice sites randomly in the host lattice) then the system does not exhibit any phase analogous to the SG phase and hence uninteresting from our point of view. The relevant disorder can be introduced in the following two ways:

(i) J_{ij} is nonrandom but the structure of the system is amorphous (Shih et al.1984, Ebner and Stroud 1985); in this case the phase angle A_{ij} far exceeds 2π in the presence of strong magnetic field and therefore, J_{ij} will tend to orient the phase angles at random,

(ii) $J_{ij} = J$ with probability p and $J_{ij} = 0$ with probability 1-p but the superconducting grains occupy the lattice sites on a d-dimensional hypercubic lattice (John and Lubensky 1985).

From the analogy between these models and the SG models one can define the SG-like order parameter

$$\eta = (1/N)\ \Sigma|\langle\exp(i\phi_i)\rangle|^2.$$

Thus, the "superconducting glass" phase is characterized by a random distribution of the frozen-in Josephson currents.

The model (i) above has been studied so far by computer simulation (Shih et al.1984, Ebner and Stroud 1985) as well as by analytical methods (Choi and Stroud 1985, Choi et al.1986). The mean-field (T,H,p) phase diagram (Fig.27.4) of the model (ii) has been investigated by an effective replica Hamiltonian in the continuum approximation (John and Lubensky 1985, John and Lubensky 1986). The (T,p) phase diagram for H > 0 (fig. 27.5a), and the (T,H) phase diagram for $p > p_c$ (p_c is the percolation threshold) (fig.27.5b) consist of three phases, viz., normal (N), superconducting (SC) and the SG phase.

So far we have discussed the existence of "superconducting glass" in granular superconductor arrays that include disorder. The latter models are analogs of disordered XY models. However, it might be possible to have "superconducting glass" phase in Josephson junction arrays without disorder. The latter are the analogs of periodic frustrated models. We mentioned in chapter 20 (see equation 20.5)) that the definition of frustration for the XY model leads to the relation

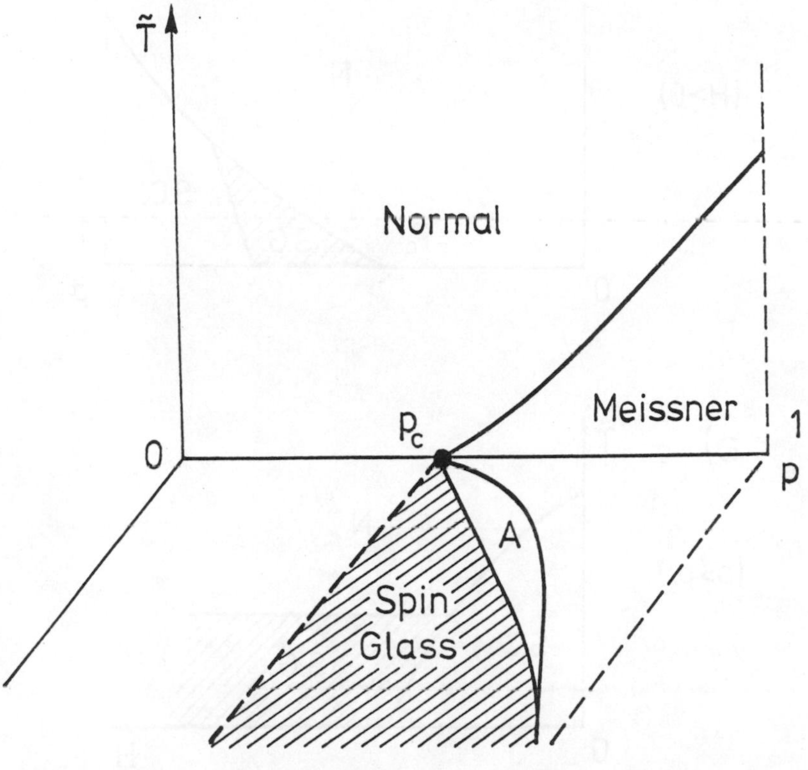

Fig. 27.4. The mean-field phase diagram of randomly diluted granular superconductors as a function of temperature T and applied magnetic field H and the Josephson bond occupation probability p near the percolation threshold p_c exhibiting the normal, Meissner, glassy and Abrikosov (A) phases (after John and Lubensky 1985).

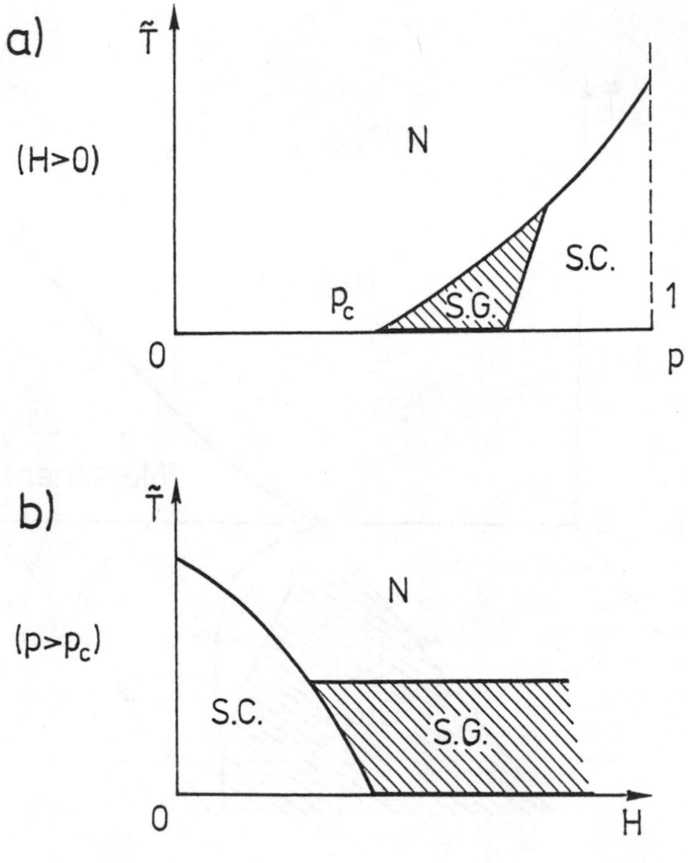

Fig. 27.5. (a) The mean-field T-p phase diagram of randomly diluted granular superconductors for a fixed magnetic field H, showing the SG, superconducting (S.C.) and normal (N) phases. (b) The mean-field T-H phase diagram for fixed concentration of super-conducting grains above the percolation threshold p_c (after John and Lubensky 1985).

$$\Sigma A_{ij} = 2\pi f$$

where f, the index of frustration, is fractional. In Monte-Carlo simulation of a periodic array of Josephson junctions with irrational $f = (3-\sqrt{5})/2$ Halsey (1985) observed a transition to the "superconducting glass" phase at $k_B T \simeq 0.25J$, the latter transition is signalled by the onset of nonzero value of the EA-like order parameter $q = \Sigma\langle\vec{S}_i\rangle^2$ where $\vec{S}_i = (\cos\theta_i, \sin\theta_i)$

27.4 TRAVELLING SALESMAN PROBLEM, COMPUTER DESIGN AND SG

The travelling salesman problem (TSP) can be formulated as follows (see, for example, Kirkpatrick et al. 1983): Given a list of N cities and a means of calculating the distance between any two cities, what is the shortest route that visits each city once and returns finally to the starting point? In other words, if the N cities are at the points $\{\vec{r}_i\}$ and l_{ij} is the distance between two arbitrary cities at $\vec{r}_i$ and $\vec{r}_j$, the problem is to find the permutation P (each permutation P corresponds to a particular sequence in which the cities are to be visited) of cities that minimizes the length $L = \Sigma\, l(\vec{r}_{P(i)}, \vec{r}_{P(i+1)})$.
This is one of the most difficult problems in combinatorial optimization and belongs to the class of the so-called NP-complete problems (nondeterministic polynomial time complete). Let us first explain the meaning of NP-completeness.

The time complexity function for an algorithm is a measure of the time it requires to solve a problem. This definition is not rigorous but pragmatic for our discussion here. An algorithm is called polynomial time algorithm if the magnitude of its time complexity function is less than or equal to some constant times the magnitude of a polynomial function, for a given length of the input information. Algorithms whose time complexity function cannot be bounded in this manner are, in general, called exponential time algorithms. The problems in the latter category are usually called NP-complete problems (see Aho et al.1975,

Garey and Johnson 1979, Papadimitriou and Steiglitz 1982 for details of NP-completeness). Thus, the origin of NP-completeness is clear, the time needed for their solution increases exponentially with the "size" of the problem. In the context of the TSP, the computing time increases exponentially with N, the number of cities.

Readers familiar with the general problems of statistical mechanics must have already noticed the similarity of the TSP problem with those usually encountered in statistical mechanics. In the latter, the problems can be formulated as follws: given an N-particle system and a means of computing the energy of interaction between them, what is the state that corresponds to the minimum of the free energy of the system. We shall see later that the analogy between the TSP and the statistical mechanics of the SG systems is much deeper.

Apart from the purely academic importance, TSP has great technological importance too; the problem of designing a precise and reliable computer can be mapped onto a TSP (see Kirkpatrick et al.1983 for an elementary introduction). Stated in the most general terms, the problem of computer design is as follows: given N components of the system and a method for wiring, what is the configuration of these components that would optimize the "eventual performance" of the computer. The "eventual performance" is quantified by different quantities depending on the purpose of the computer to be designed. For example, length of the wiring should be so small as to minimize the power consumption, the signals should propagate as fast as possible in order to speed up the computation, etc.

One of the convenient ways of getting the best possible solution of such problems is to "divide and conquer".; the task of designing is divided into several stages of manageable complexity and at each stage the "eventual performance" is optimized. Suppose, there are N circuits to be partitioned between two chips placed one above the other. Let K_{ij} be the number of signals that pass between the i-th and the j-th circuits. Next, let us attach a two-valued variable s_i to each circuit i, such that s_i = +1 or -1 depending on whether it belongs to the "upper chip" or the "lower chip". The number of signals crossing a chip boundary is given by $\Sigma(K_{ij}/4)(s_i - s_j)^2$. The difference between the number of circuits on the two chips is given by Σs_i. Therefore, in order to achieve very good "eventual performance", one should minimize the quantity (see Kirkpatrick et al.1983)

$$\Sigma(\lambda - K_{ij}/2)\, s_i\, s_j \qquad (27.14)$$

where λ is a measure of the relative costs of the imbalance in the number of circuits in the two chips and of the boundary crossings. The quantity in (27.14) is very similar to the Hamiltonian for Ising spins where the interacition consists of short-ranged "ferromagnetic-like" interaction K_{ij} and long-ranged "antiferromagnetic-like" interaction λ. The competition between these two types of interactions gives rise to frustration in this problem; and hence the deeper analogy with the SG problem. In analogy with the SG problem, we have an exponentially large number of optimal solutions possible also in computer design. (For NP-completeness of SG see Barahona 1982). Formal similarities in the formulation of the problems of SG and computer design hints at the possibility of similar methods of (approximate) solutions. There are at least two ways of getting the near-optimal solution. In the so-called "iterative improvement technique" one begins with a plausible configuration and tries rearrangement of the circuits. However, only those rearrangements are accepted which yield a better "eventual performance". This is equivalent to rapid quenching of a melt to T=0. (See Canisius and van Hemmen 1986 for an efficient zero temperature algorithm). The disadvantage of this procedure is that most often the system gets trapped in one of the local minima and special ad-hoc tricks have to be applied in order to get out of such traps. Therefore, from a physicist's point of view, a more efficient technique is that of simulated annealing (Kirkpatrick et al.1983, Kirkpatrick 1984) which has proved a very efficient way of getting the nearly degenerate ground states of SG models. In such searching process one has a natural means of not only cooling the system but also of heating it as in the case of Metropolis algorithm for SG (see chapter 24). Usually, such simulated annealing yields better optimal solution of the problem of computer design as compared to the iterative improvement technique (Kirkpatrick and Toulouse 1985). The effect of the cooling rate on the residual entropy of one dimensional SG with random Gaussian exchange has already been investigated by simulated annealing (Ettelaie and Moore 1985). Finite systems of spins were cooled from an initial high temperature T=1 to the final lower temperature T=0. The temperature was reduced in steps of $\Delta T = 0.05$. The cooling rate was varied by varying t, the number of Monte Carlo steps (MCS) per spin at a given temperature. Thus, $r=\Delta T/t$ is a measure of the cooling rate. The data indicated that the slower the cooling(i.e. smaller r) the lower the residual entropy; S_{res}/N seems to decrease logarithmically slowly with r^{-1}. Grest et al.(1986) concluded that for two dimensional ±J and Gaussian models, the ground state energy $E(r) = E_0 + C_1 r^x$ where $x \simeq 0.25$ whereas for the three dimensional ±J and infinite-ranged models, $E(r) = E_0 - C_2 (\ln r)$ 1. This qualitative difference between the two dimensional and three dimnensional results could be a consequence of the fact that the two dimensional models are not NP-complete whereas the other two models are. This speculation of Great et al., if true, would support Barahona's (1982) conclusion.

How does one implement the method of simulated annealing to the iterative improvement in the design of computers? First of all, the following ingredients must be quantified (Kirkpatrick 1984):

(i) a precise definition of the "state" of the system

(ii) a scalar "objective function" that has to be optimized for the best possible "eventual performance",

(iii) a prescription for generatingrearrangements of the system.

In the case of computer design, a state of the system is defined as the list of the assigned circuit locations in the chips. The objective function takes care of minimum wiring, maximum speed, etc. Finally, the simple interchange of the positions of the circuits in the chips according to the Metropolis algorithm is sufficient for the evolution of the system towards the desired (local) optimal state. One can easily recognize the "state", the "objective function" and the "rearrangement prescription" as the analogues of the state of the spin system, the interaction energy and the Metropolis spin-flip prescription, respectively, in the case of simulated annealing of SG. However, one crucial requirement for carrying out the annealing is the introduction of temperature. The effcetive temperature used in implementing the Metropolis algorithm to the computer design is merely a control parameter in the same units as the "objective function"; very high temperature corresponds to a configuration where any of the circuits can be put in any of the locations in the chips. If as a result of cooling the system gets stuck in one of the local mimima, one needs simply to heat up the system so as to take it sufficiently out of the minimum and recool it, the procedure is very similar to that followed for SG. One should keep in mind that in the TSP one wants to get the ground state only and uses the states corresponding to $T > 0$ only as the intermediate steps in the computation. On the other hand, most of the interest in SG lies in the possibility of SG transition at nonzero temperature and the nature of such ordering.

What does such a design of computer have to do with TSP? First of all, it is the prototype of the problems encountered in combinatorial optimization, including those of computer design. Secondly, the definition and computation of the state of the system and the "objective function" in the TSP are much simpler than in many other problems in this class. The state of the system is defined by the set of the coordinates of the cities, $\{\vec{r}_i\}$, and the "objective function" is the total distance L of a tour, the distance being computed in some suitably chosen metric. One additional advantage of dealing with this problem is that the estimation of the bounds on some of the relevant physical quantities is quite straightforward. Let us formulate the TSP as a statistical mechanical problem. One defines the partition function as

$$Z = \Sigma \exp(-\beta L)$$

where L is the length of the tour in a given configuration and the sum is to be carried over all possible configurations.

The strength of the interaction energy J_{ij} in SG is a random variable with an apriori assumed distribution function and the total free energy is to be minimized for the ground state of the system. Similarly, in the TSP the distribution of l_{ij}, where l_{ij} is the distance between the i-th and the j-th cities, has to be specified. In the so-called random link models one assumes apriori the l_{ij} to be independent random variables with an appropriate distribution and minimizes the total tour-length. Kirkpatrick and Toulouse (1985) assumed uniform distribution of l_{ij} over the unit interval (0,1). On the other hand, Vannimenus and Mezard (1985) assumed a Poisson distribution

$$P_r(l) = l^r e^{-l} / r!$$

where r is a parameter. The bounds on the free energy (and hence on the minimal length) can be obtained analytically in the latter model. The two bounds coincide in the limit $r \to \infty$ and hence yield an exact result:

$$\tilde{F}(T)/r \quad \begin{matrix} 1/e \text{ for } \tilde{T} \le 1/e \\ \\ -\tilde{T} \ln \tilde{T} \text{ for } T > 1/e \end{matrix}$$

where $\tilde{F} = F/N^{1-1/(r+1)}$ and $\tilde{T} = T/N^{1/(r+1)}$. This indicates a "phase transition" at 1/e. The physical origin of this transition is very similar to that in the REM (see chapter 10); the average number of available states is exponentially small for $\tilde{T} < 1/e$. Recently, Chakrabarti (1985) has calculated the optimal tour of a travelling salesman excatly in the special case where the salesman is constrained to move on a square lattice such that no move against a particular preferred direction is allowed.

Since the analogy between the TSP and SG is very striking, it is very likely that the ground states of the former have an ultrametric structure like that of the SK model. The meaning of the overlap between any two different ground states in the TSP is very clear: it is the number of the common bonds on the two corresponding (near-) optimal tours. Such quantities are much simpler to compute than those for SG. There is a strong indication that the near-optimal solutions of TSP, at least in some specific models (Kirkpatrick and Toulouse 1985), do have an ultrametric structure.

There is a class of problems, the so-called matching problems, in combinatorial optimization which are usually simpler than TSP. For example, the Bipartite Matching problem (BMP) is defined as follows: given two sets of N points $\{\vec{r}_i\}$ and $\{\vec{\rho}_i\}$ and the distances $l(\vec{r}_i,\vec{\rho}_i)$, what is the permutation P such that $L' = \Sigma l(\vec{r}_i, \vec{\rho}_{p(i)})$ is minimum. Very recently, replica trick has been applied to calculate the "free energy" of the matching problems (Mezard and Parisi 1985, Orland 1985) in the mean-field approximation. Two predictions of such mean-field theory, viz.,

$$\tilde{E}_{r=1}(\tilde{T}=0) = 1.14 \pm 0.01$$

and

$$\tilde{E}_{r=0}(\tilde{T}=0) = 0.825 \pm 0.01$$

where $\tilde{E} = E/N^{1-1/(r+1)}$, have been verified by computer simulation (Mezard and Parisi 1985). Moreover, the order parameter is a function even in the replica-symmetric formulation!

Concepts developed for the SK model have also been applied to the graph partitioning problem. The latter problem is formulated in the follwing way: given a set of vertices $V = \{v_1, v_2, \ldots, v_n\}$ and a set of edges $E = \{(v_i, v_j)\}$ (with N even) how to partition the N vertices into two sets V_1 and V_2 of equal sizes such that the number fo edges joining V_1 and V_2 is minimized. The latter number serves as the cost function (C) of the problem.

This in an NP-complete problem. Fu and Anderson (1986) studied a modified version of this problem where each pair of vertices are connected with probability p independent of whether other pairs are connected. In other words, each vertex has, on the average, a valence $\alpha = \lim pN$. It is straightforward to establish the analogy between the latter model and a SG. One associates an "Ising spin" S_i with each vertex v_i, where $S_i = +1$ or -1 depending on whether v_i belongs to the set V_1 or V_2. Now, one defines a "coupling constant" J_{ij} where $J_{ij} = J$ if the edge $(v_i, v_j) \in E$ (i.e., $J_{ij} = J$ with probability p) and $J_{ij} = 0$ otherwise (i.e., $J_{ij} = 0$ with probability 1-p). Then, the graph partitioning problem reduces to the problem of minimization of the "Hamiltonian" $\mathcal{H} = - \Sigma J_{ij} S_i S_j$.

Different special cases of the above model arise from the different forms of p. In the special case where p is independent of N, the free energy can be calculated using the replica trick (1.25). It turns out (Fu and Anderson 1985) that the expression for the free energy is physically sensible only if $J = J_0 N^{-1/2}$ (compare with the SK model). Indeed, at T = 0 the latter model is equivalent to the SK model. The cost function C(p) is given by

$$C(p) = (N^2/4)\ p - 0.38\ N^{3/2}\ [p(1-p)]^{1/2}$$

This is consistent with the bounds of C proposed earlier and with computer experiment. This model exhibits a phase transition at $T_c = J_0(\{p(1-p)\}^{1/2}$ where replica symmetry breaking takes place.

In the special case where $p = \alpha/N$ (α = constant), the choice $J = J_0 N^{-1/2}$ is ruled out, and hence exact solution of this problem is as difficult as that of the short-ranged SG models.

These applications of the techniques of MFT of SG seems to be the beginning of a new era in the theory of optimization; it would, perhaps, not only provide some analytical means of handling the problems in this class but might clarify the connection between the NP-completeness , replica symmetry breaking and breaking of ergodicity.

27.5. SG MODEL OF CRYSTAL SURFACES

Consider the solid-on-solid (SOS) model of a (100) surface of a perfect cubic lattice at thermal equilibrium with its vapor. If there are screw dislocations these would manifest themselves as sources or sinks of defects on the surface as positive or negative topological point defects. Suppose L(s) is the length of the strings s of defects. The partition function for the imperfect crystal is given by

$$Z_{imperfect} = \sum_{\text{matching } s} e^{-\beta\sigma L(s)} \qquad (27.15)$$

where σ is the energy cost per unit length of the string. The summation in (27.15) is to be carried out over all string configurations that "match" a given quenched distribution of dislocations, i.e., such that an open string connects every positive defect to a single negative defect and vice versa. Note that the parameters $\pm\sigma$ are the analogues of the ferromagnetic and antiferromagnetic exchange interactions in SG. The frustrated plaquettes are the sources and sinks of defects in SG (see chapter 20). From this analogy let us develope a model where N sources and N sinks are distributed independently and uniformly over a volume V in a d-dimensional space. Let D_{ij} be the Euclidean distance between the i-th source and the j-th sink. The corresponding partition function is given by (Bachas 1985)

$$Z = \sum_{\pi \in S_N} e^{-\beta \mathcal{H}(\pi)} \qquad (27.16)$$

where the summation is carried out over all matchings of the sources to the sinks, i.e., all permutations π of N objects and

$$\mathcal{H}(\pi) = \sigma \sum_{i=1}^{N} D_{i\pi(i)}$$

Note that the latter problem is equivalent to the bipartite matching problem defined in the section 27.4. Moreover, one can compute the overlap of any two configurations from the relation

$$Q_{\alpha\beta} = (1/N) \sum_{i=1}^{N} \delta_{\pi\alpha(i),\pi\beta(i)}$$

Thus, the latter model suggests that the crystal-vapor interface, in the presence of screw dislocations, might provide us a real physical system where the overlaps between the various configurations can be measured directly and hence the possibility of probing the geometry of the ground states.

27.6 ULTRADIFFUSION: DYNAMICS ON ULTRAMETRIC SPACE

We have discussed extensively about the various consequences of the ultrametricity in chapter 9. We have also explained the differences between the hierarchical relaxation and conventional relaxation phenomena in chapter 19. In this chapter we shall summarize some of the recent efforts in the understanding of the diffusion on ultrametric spaces which is believed to throw light on the nature of the spin relaxation in SG.

The problem of ultradiffusion was introduced by Huberman and Kerszberg (1985) in the following way:
suppose the energy barriers separating the cells in a one-dimensional array are distributed in a hierarchical manner as shown in figs.27.6 a,b. Let the barriers be labelled by ϵ_i = the probability that the i-th barrier will be crossed in unit time. Defining the ultrametric distance d(a,b) to be the largest barrier between the cells a and b we discover ultrametricity of the system (see fig.27.6 b). Note that in order to travel between two points at the top branches in the hierarchy along the tree, one must go down by a number of levels equal to the ultrametric distance between the corresponding points. The ratio of two successive barriers $\rho=\epsilon_{i+1}/\epsilon_i$ is assumed to be smaller than unity. Suppose $P_a(t)$ be the probability that the diffusing "particle" is at the a-th cell at time t. After a time

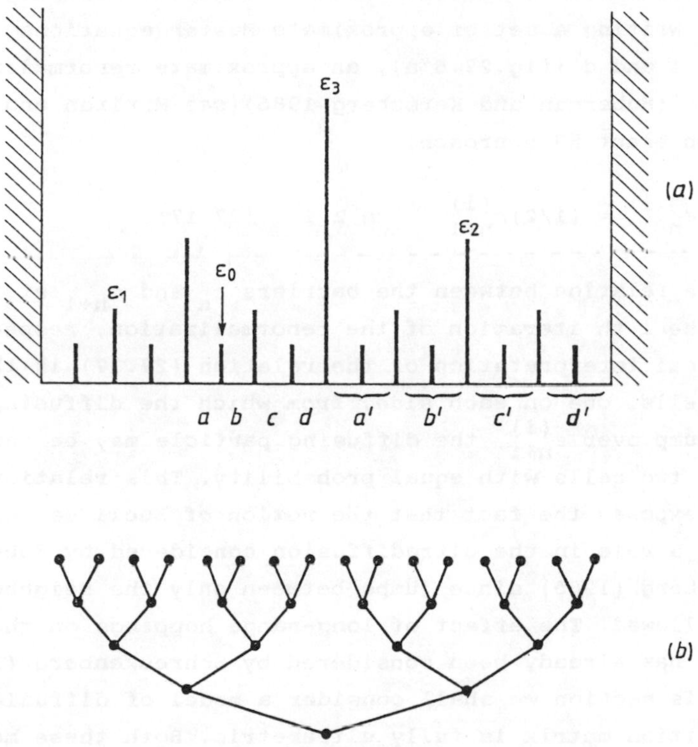

Fig. 27.6. (a) Hierarchical array of barriers over which a particle diffuses. The barriers are labelled by ϵ_i, the probability per unit time that they will be crossed; ϵ_i is small for a tall barrier. (b) Ultrametric structure of the hierarchical arrangement: to travel between two points in the top branches of the tree without leaving the tree, one must go down by a number of levels equal to the ultrametric distance separating the points (after Huberman and Kerszberg 1985).

$\tau_0 \simeq 1/\epsilon_0$ the statistical distribution on both sides of the lowest barriers is roughly equalized, leading to the renormalization of the net flux through the next higher barrier. Then, writing a set of approximate Master equations for the cells a, b, c and d (fig.27.6 a), an approximate renormalization scheme yields (Huberman and Kerszberg 1985)(see Maritan and Stella 1986 for an exact RG approach)

$$\epsilon_n^{(i+1)} = (1/2)\epsilon_{n+1}^{(i)}, \qquad n \geq 1 \qquad (27.17)$$

as the relation between the barriers ϵ_n and ϵ_{n+1} at the (i+1)-th and the i-th iteration of the renormalization, respectively. The physical interpretation of the relation (27.17) is that there are two cells, one on each side, from which the diffusing particle can jump over $\epsilon_{n+1}^{(i)}$, the diffusing particle may be found in any of these two cells with equal probability. This relation (27.17) also exposes the fact that the notion of Euclidean distance still plays a role in the ultradiffusion considered by Huberman and Kerszberg (1985) since jumps between only the neighboring cells are allowed. The effect of long-range hoppings on the latter model has already been considered by Schreckenberg (1985). Later in this section we shall consider a model of diffusion where the transition matrix is fully ultrametric. Both these models predict an "anomalous" decay of the form $t^{-\alpha}$ (α< 1) for the auto-correlation in the asymptotic regime where w_l is a rapidly decreasing function of l (also see Blumen et al.1986). Indeed, there exists a critical value of $\rho = \rho_c$ above which the decay of the autocorrelation is "normal" random-walk-like $t^{-1/2}$ and a "phase transition" to a regime of "anomalous" dacay takes place at R_c (Teitel and Domany 1986, compare and contrast with Grossmann et al.1985).

Paladin et al.(1985) and Ogielski and Stein (1985) introduced truly ultrametric diffusion which, unlike the Huberman-Kerzberg model, does not use the notion of Euclidean space. The model is defined as follows: consider a system which has N possible states a (a = 1,...,N) and suppose that the characteristic hopping times $1/W_{ab}$ between these states form an ultrametric space (see Fig. 27.7). Suppose, there are R levels in the hierarchy, at each level l the transition rate is W_l and the number of branching is α_l. Therefore, at the l-th level there is a bifurcation with α_l clusters, each of which contains N_l states $N_l = \alpha_1 \cdots \alpha_{l-1}$ and $N = \alpha_1 \cdots \alpha_R$. Let us now characterize the N states by R indices $j_1, j_2, \ldots, j_R$ $(1 \le j_l \le \alpha_l)$(see Fig. 27.7). These indices give the sequence of the branches of the tree one must follow in order to go from the common original root at the R-th level to the end points of the branches at the 1st level of the tree. Suppose $P_a(t)$ is the probability of being in a state a at time t. The corresponding Master equation is given by

$$dP_a(t)/dt = \Sigma\, [W_{ab}\, P_b(t) - W_{ba}\, P_a(t)], \quad a = 1,2,\ldots,N \qquad (27.18)$$

Equation (27.18) can be solved iteratively beginning at the common origin at the highest level. The probability for the state $a \equiv j_1, j_2, \ldots, j_R$ at time t is given by

$$P^{(1)}_{j_1 \ldots jR}(t) = N_1 [\sum_{l=1}^{R} e^{-t/t_l} \{P^{(l)}_{jl \ldots jR}(0)/N_l - P^{(l+1)}_{jl+1 \ldots jR}(0)/N_{l+1}\} + 1/N_{R+1}] \qquad (27.19)$$

with

$$1/t_l = \sum_{r=1}^{R} (W_r - W_{r+1}) N_{r+1}$$

and

$$N_l = \alpha_1 \cdots \alpha_{l-1}$$

where

$$W_{R+1} = 0 \text{ and } N_1 = 1.$$

Depending on the relative weights of the different states and on the choice of the sequences α_l, W_l, one gets the explicit forms of $P^{(1)}(t)$. For example, if all the states are equivalent and if the relaxation times t_l form a sequence where each time scale is much larger than the previous one and if α_l are large, one gets

$$P^{(1)}(t) = \sum_{l=1}^{R} \{\exp(-t/t_l)/N_l\} + (1/N) \qquad (27.20)$$

Note that $P^{(1)}(t)$ in (27.20) is a superposition of exponentials with hierarchical relaxation times, in the same spirit as in the hierarchical model of relaxation of Palmer et al.(1984) described in chapter 19. Similar superposition of exponentials follows also in the special case where there are two branchings per level (uniform bifurcation); in this case the transition matrix can be diagonalized exactly (Ogielski and Stein 1985).

27.7 NEURAL NETWORKS, CONTENT ADDRESSABLE MEMORIES, PATTERN RECOGNITION AND SG

A typical nerve cell consists of the following essential parts: (i) the cell body, (ii) the dendrites and (iii) the axon (see Fig.27.8). The input information is fed through the dendrites, the "processing" of the input information is carried out in the cell body and the output is led to the next cell through the axon. The axon of any cell is connected to the dendrites of some other cell(s) by the so-called synaptic junction(s). The cell "fires" (i.e. is active) only if the sum total of the external stimuli on it exceeds a certain threshold. Therefore, in the simplest model (McCulloch and Pitts 1943) a neuron is represented as a two-state system- "active" and "inactive". Let us denote the state of a neuron at time t by a variable v_i which can have two values: $v_i = 1$ if it is active and $v_i = 0$ if it is inactive. Suppose, the strength of the synaptic junction between two arbitrary cells i and j is C_{ij}. Moreover, let us assume that the sum total of the stimuli at the i-th neuron from all the others be given by $\Sigma C_{ij}v_j$. Let us denote the threshold for the i-th neuron by T_i. Then,

$$v_i = 1 \text{ if } \Sigma C_{ij}v_j > T_i$$

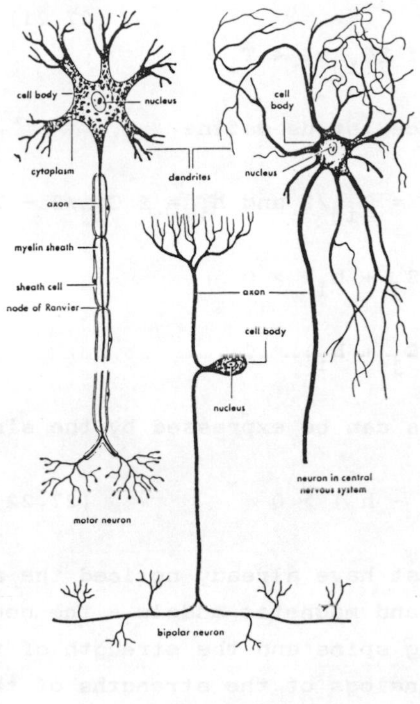

Fig. 27.8. Three neurons of common types (after H.A. Deutsch and D. Deutsch, Physiological Psychology, The Dorsey Press, Homewood, Illinois, 1973).

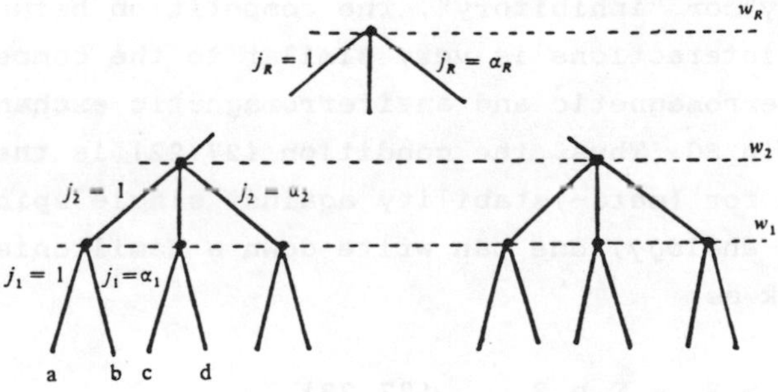

Fig. 27.7 Ultrametric structure of the states: a,b,.. denote the states of the systems. A state is characterized by R numbers j_1, j_2, ..., α_R. The transition rate from one state to another depends only on the closest common node (i.e., the number of successive distinct values $j_1 \neq j_1$, $j_2 \neq j_2$ ... for the two states), e.g., $W_{ab} = W_1$, $W_{bc} = W_2$, ... (after Paladin et. al. 1985).

and (27.21)

$$v_i = 0 \text{ if } \Sigma C_{ij} v_j < T_i$$

For later convenience let us define

$$S_i = 2v_i - 1,\ J_{ij} = C_{ij}/2 \text{ and } h_i = \Sigma\ C_{ij}/2 - T_i$$

$$S_i = +1 \text{ if } (\Sigma\ J_{ij} S_j + h_i) > 0$$

and

$$S_i = -1 \text{ if } (\Sigma\ J_{ij} S_j + h_i) < 0$$

Both these relations can be expressed by the single formula

$$S_i(\Sigma\ J_{ij} S_j - h_i) > 0 \qquad (27.22)$$

The reader must have already noticed the analogy between the neural network and magnetic models - the neurons are the analogs of the Ising spins and the strength of the synaptic junctions are the analogs of the strengths of the exchange interactions. Of course, in a neural network the relation $J_{ij} = J_{ji}$ does not necessarily hold. The synapses in neural systems can be "excitatory" or "inhibitory", the competition between these two types of interactions is very similar to the competition between the ferromagnetic and antiferromagnetic exchange interactions in SG. Thus, the condition (27.22) is the analog of the condition for (meta-)stability against single spin-flips in SG. From this analogy, one can write down a Hamiltonian for the neural network as

$$\mathcal{H} = -\ \Sigma\ J_{ij}\ S_i S_j - \Sigma\ h_i S_i \qquad (27.23)$$

$\{S_i\}$ denotes the state of the whole system at a given instant of time. Both S_i and S_j in (27.23) belong to the same set $\{S_i\}$ in the Hopfield model (1982). The consequences of the latter nature of the hamiltonian (27.23) will be studied later.

A neural network must have capacities for (a) learning

(i.e., storing patterns in the memory of the system) and (b) recollection (i.e., retrieval of the learned patterns). Suppose, the p-patterns

$$\vec{S}^1 = \{S_1^1, \ldots, S_N^1\}$$

$$\vec{S}^2 = \{S_1^2, \ldots, S_N^2\}$$

....

....

$$\vec{S}^p = \{S_1^p, \ldots, S_N^p\}$$

are to be stored in a neural network consisting of N-neurons. Some of the most fundamental questions are:

(i) How to choose the matrix J so that the learned patterns would be stable ? A state is stable if it is invariant in time.

(ii) What is the measure of the error of retrieval of learned patterns ?

(iii) Allowing an upper limit of error, defined in (ii), what is the maximum number of patterns, p, that can be stored in an N-neuron network?

(iv) How is the size of the basins of attraction of the learned states?

The stability criterion (27.22) for a state S of a neural network can be written as follows:

$$J \vec{S} - \vec{h} = A \vec{S}$$

where A is a positive diagonal matrix. Denoting the k-th state to be learned by $\vec{S}^k$, a matrix J guaranteeing the stability of the p states to be learned should satisfy the set of equations:

$$J \vec{S}^k - \vec{h} = A^k \vec{S}^k, \qquad k = 1, \ldots, p$$

where A^k is an arbitrary diagonal matrix with all elements positive or zero. One can always find matrices A^k such that

$$A^k \vec{S}^k + \vec{h} = \vec{S}^k$$

(Personnaz et al.1985, 1986a,b,c,d). Thus, the matrix J must satisfy the relation

$$J \vec{S}^k = \vec{S}^k, \qquad k = 1,\ldots,p.$$

Obviously, taking J as the projection matrix into the subspace spanned by the $\vec{S}^k$ is a simple, exact way of satisfying the above p relations; therefore, the matrix J can be written as

$$J = \Sigma \Sigma^I$$

where Σ is a matrix whose columns are the vectors $\vec{S}^k$ and

$$\Sigma^I = (\Sigma^T \Sigma)^{-1} \Sigma^T$$

(where Σ^T is the transpose of Σ) is the pseudo-inverse of Σ. If the vectors $\vec{S}^k$ to be learned are orthogonal, matrix J takes the form

$$J = (1/N) \Sigma \Sigma^T$$

which is nothing but the Hebb-Cooper prescription (Hebb 1949, Cooper 1974)

$$J_{ij} = (1/N) \sum_{k=1}^{p} \xi_i^k \xi_j^k \qquad (i \neq j) \qquad (27.24)$$

where $\{S_i^k\}$ (k=1,...,p) are the p learned patterns. Note that the model (27.24) is the analog of the Provost-Vallee model (1.12) for SG. The stability analysis is quite straightforward at T = 0. The local field acting on the i-th spin, in the state $S_i = \xi_i^\nu$, is given by

$$h_i = \sum_{j\neq i} J_{ij} S_j = (1/N) \sum_{j\neq i} \sum_{\mu=1}^{p} \xi_i^\mu \xi_j^\mu \xi_j^\nu \simeq \mu_i^\nu (1 + \delta_i)$$

where the noise $\delta_i = (1/N) \sum_{j\neq i} \sum_{\mu\neq\nu} \xi_i^{\mu} \xi_j^{\mu} \xi_i^{\nu} \xi_j^{\nu}$ is a random variable with variance (p-1)/N, which is negligibly small provided $(p/N) \to 0$ as $N \to \infty$ (i.e., p remaining finite). Therefore, the stored patterns would be stable at T = 0 for vanishingly small values of p/N. This observation leads us to the question (iii) above. But the latter question is incomplete, because one must specify the maximum amount of error allowed. This brings us to the question (ii) above, viz., formulation of a quantitative measure of the retrieval error. The retrieval of memory requires specification of a dynamics for the time-evolution of the system.

If the input pattern is one of the learned patterns, the pattern must remain invariant in time according to the definition of learned patterns. However, if the input pattern is not any of the pure learned states, but a superposition of, say, two of them, the network should be able to recollect the "closest" learned pattern. For example, if one of the learned patterns is "J.J. Hopfield" and if the input pattern is "J.J.Hoffield", the desirable characteristic of the evolution dynamics would be that the final pattern is, indeed, "J.J. Hopfield". In other words, each of the learned patterns should act as an attractor in the generalized phase space of the system and should have substantially large basin of attraction around each of them.

So far as the evolution dynamics is concerned, several models have been proposed. In the Hopfield model (1982) only "down-hill" motion in the phase space (equivalent to T=0 Monte-Carlo procedure) is allowed until a local minimum is reached. Suppose $\vec{S}$ be the (noisy) input pattern and $\vec{S}'$ be the final pattern. Then, as a first step towards the measure of retrieval error, one defines the "Hamming distance" d between a pure learned pattern $\vec{S}^k$ and the pattern $\vec{S}'$ as (Hopfield 1982, Kinzel 1985)

$$d(\vec{S}', \vec{S}^k) = [1 - q(\vec{S}', \vec{S}^k)]/2$$

where the overlap (compare with the overlap of the pure states of the SG defined in chapter 9)

$$q(\vec{S}', \vec{S}^k) = (1/N)\vec{S}' \cdot \vec{S}^k$$

between the two patterns $\vec{S}'$ and $\vec{S}^k$ is the number of common bits. The Hamming distance itself can be used as a measure of the retrieval error. At this point it is worth mentioning that the internal noise, which gives rise to occasional spontaneous activity (or inactivity) can be incorporated into the Hopfield model by introducing finite nonzero temperature T (Amit et al.1985a).

Next we shall consider the storage capacity of the networks. First, consider the 'unsaturated' limit, i.e., p remaining finite as $N \to \infty$ (Amit et al.1985a). In this limit, 2p thermodynamically stable Mattis-like states appear below the transition temperature $T_c = 1$. Between $T = 1$ and $T = 0.46$ the Mattis-like states are the only stable states and each of these states is highly correlated with a single learned pattern (see also Kinzel 1985a,b). Therefore, the basins of attraction of the learned patterns are enormously large in this limit. Below $T = 0.46$ additional dynamically stable states appear; but these states are spurious.

From the stability analysis at $T = 0$, presented earlier in this section, one would expect the original learned patterns to get destabilized by increasing p. Indeed, even at $T = 0$, these patterns become unstable for $p > N/(2\ln N)$ (Weisbuch and Fogelman-Soulie 1985, Amit et al. 1985b). However, contrary to naive expectations, very effective retrieval of the patterns is possible even for finite p/N, although the original learned patterns are no longer stable, provided $(p/N) < 0.14$ (Amit et al.1985b, see also Feigelman and Ioffe 1986). Suppose the overlap of a state with a given pattern is q. The average number of errors (i.e., the number of spins which do not align with the embedded patterns) is given by

$$N_e = (N/2)\ (1 - q)$$

Amit et al.(1985b, 1986) showed that N_e/N starts with a value of about 1.5% at $p/N = 0.14$ and goes to zero rapidly with the decrease of p/N. Therefore, if small retrieval errors are allowed then the maximum storage capacity is given by

$$p_{max} = 0.14\ N.$$

The metastable states, which are highly correlated with a single learned pattern, are called "retrieval states". Such retrieval states exist also at finite nonzero temperatures $T < T_M$, where $T_M \sim 1-K\surd(p/N)$ is a decreasing function of p/N (see Amit et al.1985b, 1986). No retrieval is possible between the SG transition temperature $T_g = 1+\surd(p/N)$ and T_M (see Fig.27.9). If $0.05 < (p/N) < 0.138$, the retrieval states remain metastable above a temperature T_S (see Fig.27.9) and become more stable than the SG states below T_S. The retrieval states remain metastable down to $T = 0$ for $0.138 > (p/N) > 0.05$. The replica symmetry breaking, marked by the line T_R, is very weak and therefore, the retrieval of memory is always a fast process.
All the results summarized above have been derived for the long-range model (note the factor N in the denominator of 27.24). Very little is known about the more realistic situation where every neuron would be connected to only a fraction of the other neurons by synaptic junctions (see Kinzel 1985a,b).

Our discussion in this chapter has been based so far on the Hopfield model (Hopfield 1982) defined by (27.23). Any model of associated memory requires a well-defined dynamics so that, given the initial pattern, the system evolves in time so as to relax to a final steady-state pattern. The Hopfield model assumes single-spin-flip Glauber dynamics (see chapter 24 for details of Glauber dynamics) which is equivalent to T=0 Monte-Carlo procedure for spin systems. The generalization of the latter model to T > 0 is usually called the generalized Hopfield model (GHM). Thus, the transition probability W(I|J) from the state J to the next state

I in GHM is given by the usual form

$$W(I|J) = \exp[-\beta\mathcal{H}(I)]/\Sigma \exp[-\beta\mathcal{H}(K)] \qquad (27.25)$$

and the system relaxes to the Gibbs distribution

$$P(\{S_i\}) \sim \exp(-\beta\mathcal{H}\{S_i\}) \qquad (27.26)$$

There exists another simple and (almost) equally powerful model for neural networks; this is called the Little model (Little 1974). The starting point of the latter model is the transition probabilities

$$W(I|J) = \exp[-\beta\mathcal{H}(I|J)]/\Sigma \exp[-\beta\mathcal{H}(K|J)] \qquad (27.27)$$

where

$$\mathcal{H}(I|J) = -\Sigma J_{ij} S_i(I) S_j(J) - \Sigma h_i S_i(I) \qquad (27.28)$$

Notice that the Hopfield hamiltonian (27.23) corresponds to $\mathcal{H}(I|I)$ in the notation of (27.28). Thus, in the Little model, at each time-step all the spins <u>simultaneously</u> check their states against the corresponding local field, and hence such evolution is called <u>synchronous</u> whereas the Hopfield model adopts <u>asynchronous</u> dynamics. Peretto (1985) showed that the Little model leads to a Gibbs-type steady state $\exp(-\beta\bar{\mathcal{H}})$ where the effective Hamiltonian $\bar{\mathcal{H}}$ is given by

$$\bar{\mathcal{H}}(I|I) \equiv H(I) = -(1/\beta)\Sigma \ln [\cosh \{ \beta\Sigma J_{ij} S_j(I)\}].$$

The free energy of the Little model is twice that of the GHM at the extremal points (Amit et al. 1985a, van Hemmen 1986). As a consequence, the nature of the ground states and the metastable states in the two models are identical.

As stated earlier in this chapter, 'learning' corresponds to the specification of the synaptic matrix by the rule (27.24). More specifically, learning in the Hopfield model can be described as an incremental process; a pattern μ^{new} is learned by the increment $\Delta J_{ij} = \mu_i^{new} \mu_j^{new}$. However, it is also possible to 'unlearn' (i.e., forget) a pattern μ^{for} by the reverse procedure, viz., (Hopfield et al. 1983)

$$\Delta J_{ij} = - \epsilon \mu_i^{for} \mu_j^{for}$$

which describes weak unlearning provided $0 < \epsilon << 1$. It is claimed (Crick and Mitchison 1983) that in the rapid-eye-movement sleep one unlearns one's unconscious dreams. In other words, one dreams in order to foget. Hopfield's prescription for unlearning can not only be applicable to such real situations but seems to be an efficient way to get rid of spurious attractors of the network.

From the computation theoretic point of view, what are the special features of neural networks that are distinctly different from those of the conventional computers (the latter have location addressable memories)? Firstly, the memories in the neural networks are content addressable (Kohonen 1980) rather than location addressable. Secondly, the same elements of the networks are involved in more than one pattern, albeit in different combinations so as to correspond to different patterns. Finally, and most crucially, the neural networks are (a) error-correcting devices, i.e., these can retrieve correct informations from partially-incorrect input informations and (b) quite insensitive to defects in the network, provided, of course, the defects are not too large.

For solution of optimization problems by analog processors and anlog-to-digital conversion etc., see the recent works of Hopfield and coworkers (Hopfield 1986, Hopfield and Tank 1985, Hopfield and Tank 1986a,b, Tank and Hopfield 1985, Gelperin et al. 1985). Hopfield model corresponds to the special case x=2 and $\xi_{i\alpha} = \pm 1$ of the general model (van Hemmen 1986)

$$\mathcal{H}_N = -NJ \sum_{\alpha=1}^{p} x^{-1} \left| N^{-1} \sum_{i=1}^{N} \xi_{i\alpha} S_i \right|^x \quad \text{with } x \geq 1.$$

Let us now introduce the concept of the cumulative synaptic

intensity K(p) after learning p patterns. K(p) is defined by

$$K(p) = \langle J_{ij}^2 \rangle - \langle J_{ij} \rangle^2,$$

and treating the number of stored patterns as a continuous "time" variable K'(t) = k(t). Moreover, the threshold intensity k_m for one pattern to be safely stored against a background intensity K is given by

$$k_m = \epsilon^2 K/N.$$

In the standard Hebbs-Cooper learning process, K(t) grows linearly with time, i.e., K(t) = kt, so that the maximum storage capacity $p_{max} = N/\epsilon^2$. The value $p_{max} = 0.138N$ is consistent with the above estimate provided $\epsilon = 2.69$. The latter agrees with the corresponding agreements by alternative methods (Toulouse et al.1986).

In the so-called "marginalist learning" process (Nadal et al. 1986)

$$K(t) = K_0 \exp(\epsilon^2 t/N)$$

In this process the patterns that have been learned earlier (chronologically) are also forgotten earlier. This can be checked by computing the retrieval quality as a function of storage ancestry. This model can always learn new patterns by storing the new ones on top of the older ones. The ability to keep learning is maintained at the cost of storage capacity.

In the so-called "learning within bounds" (Nadal et al. 1986) we have

$$0 \le J_{ij} \le A \text{ or } -A \le J_{ij} \le 0$$

which is a more restricted situation than that considered by Toulouse et al.(1986) where

$$- A \le J_{ij} \le A.$$

Very recently, Parga and Virasoro (1986) have raised objections against the conventional Hebb-Cooper prescription (27.24) for learning. Any two stored patterns, learned according to the Hebb-Cooper prescription, have an average overlap of the order $O(1/N^{1/2})$ and hence such patterns can be assumed to be mutually orthogonal for sufficiently large N. " On the other hand, Parga and Virasoro argue, the human brain learns in a hierarchical fashion and requires a modification of the conventional Hebb-Cooper prescription. When we try to memorize a new pattern we look for all possible relationships with the patterns already stored in our memory and categorize accordingly. In other words, learning by the Hebb-Cooper mechanism is a "thoughtless" activity. This is reasonable as long as completely uncorrelated patterns are being stored. But if the patterns being stored have some relation with one another, as seems to be the case for human brain, a refelction on the part of the learning system leads to a hierarchical (possibly ultrametric) organization of the learned patterns. Parga and Virasoro suggest modifications of the Hebb-Cooper prescription, which, I hope, will be discussed in detail in future editions of this book.

Finally, this field of research seems very promising not only from point of view of basic research in neuroscience and pattern recognition but also for the development of next generation of computer technology. Although our brains have not been able to understand the SG very well we should not be disappointed; let the SG be a tool for understanding the brain!

27.8. Modelling of Prebiotic Evolution and The Origin of Life: Analogy with SG?

The two basic elements essential for the evolution of the biological species are: stability and diversity of the species (Anderson 1983). In the absence of stability one species would mutate into another. Similarly, if there were only one stable structure and no diversity, there would be no possibility to create structures with ever increasing complexity. Analogy with the infinite-ranged models of SG seems promising- the large number of local minima have both these qualitative features, viz., stability and diversity. However, more detailed quantitative works are needed before these interesting ideas can become the foundation of a theory of "prebiotic evolution". We all know that frustration, as defined by sociologists, plays a cruial role in our social life. However, it would be more interesting if frustration, as defined in chapter 1, determines the evolution of life!

CHAPTER 28
CONCLUSION

Finally, what is the present status of our understanding of the physics of SG? The answer is, of course, subjective. An optimist would say, we have not only understood the qualitative nature of the order in a large class of magnetic materials we have also developed mathematically elegant formalism of replica symmetry breaking, the first successful utilization of the special purpose machine for simulating short-ranged SG has opened up a new era in computational science, we have developed formalisms which may, in near future, lead to breakthroughs in computer science and biophysics. On the other hand, a pessimist would, perhaps, say that our present knowledge about SG is no better than the knowledge of the n blind persons about an elephant (Fig. 28.1).

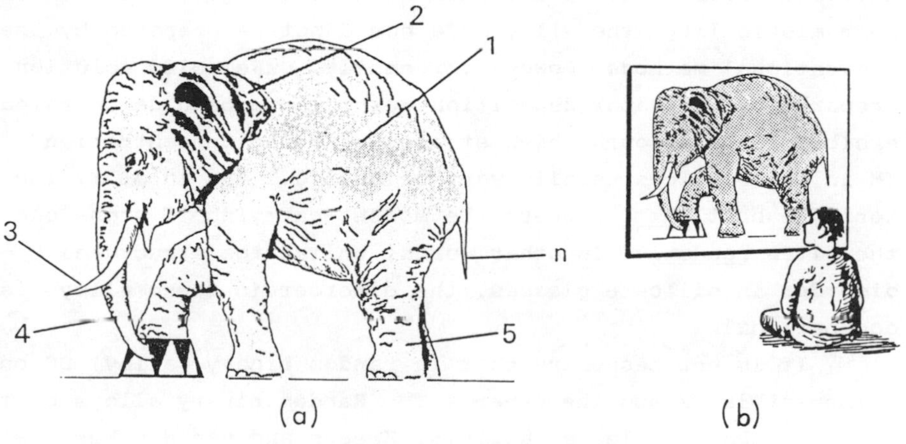

Fig. 28.1. (a) An elephant according to n blind men: 1 wall, 2 fan, 3 spear, 4 snake, 5 tree,..., n rope. (b) Theorists's dream: an elephant model (the figure after E. Shustorovich, Surface Science Reports, 6, 1 (1986).

APPENDIX A

SG SYSTEMS AND THE NATURE OF THE INTERACTIONS

In this appendix we shall list the magnetic systems which have been shown to exhibit SG behaviour. The list is not exhaustive but I shall try to list as many different types of systems as possible to emphasize the diversity. However, I am sure that many more systems will be discovered (even some systems which have been claimed to be SG might turn out not to be true SG) by the time this book appears in print.

The class of SG systems that has been studied most extensively consists of the noble metal - transition metal (NM-TM) alloys, e.g., AuFe, AuMn, CuMn, AgMn, etc. These are usually referred to as the canonical SG or archetypal SG. The alloy systems ZnMn, CdMn and MgMn (Albrecht et al.1982, Murayama et al.1986) should also be included in this category. Because of the poor miscibility, the alloy CuFe could not be prepared by the conventional methods. However, recently, CuFe solid solution prepared by the vapor deposition method have been demonsrated to exhibit SG behaviour (Chien et al.1986). The concentration of the TM in these alloys usually vary between 0.1 and 10 at.%. The NM form the host <u>lattice</u> where the moment-carrying TM atoms occupy the sites <u>randomly</u>. In other words, unlike the structural disorder in silicate glasses, the disorder in these alloys is occupational.

It is not necessary that in random binary (alloy) SG one component is NM and the other a TM. Random binary alloys of two TM, e.g., PdMn (Coles et al.1975, Zweers and van den Berg 1975, Ho et al.1981a,b, 1982), NiMn (Goldfarb and Patton 1981, Schaf et al.1983) which consist of one ferromagnetic and one antiferromagnetic TM, and Fe-Ni invar alloy (Miyazaki et al.1985) which consists of two ferromagnetic TM are well known SG. Moreover, random ternary alloys of three TM, e.g., Fe-Ni-Cr (Majumdar and Blanckenhagen 1984), multi-layered structures of Mn-Ni/Co (Stearns 1984, Vernon et al.1985), Rare-earth (RE) SG alloys, e.g., $La_{1-x}Gd_xAl_2$ (Bennett and Coles 1977, Trainor and McCollum 1974, 1975), $Th_{1-x}Gd_xRu_2$, $Ce_{1-x}Gd_xRu_2$ (Davidov et al.1977), $(La_{1-x}Gd_x)B_6$ (Felsch 1978), $(La_{1-x}Ce_x)Al_2$ (Steglich 1976), $(La_{1-x}Dy_x)B_6$ (Ali and Woods 1983), $La_{1-x}Ce_xRu_2$, $La_{1-x}Th_xCe$ (see Mydosh 1978), $Y_{1-x}Er_x$ (Wendler and Baberschke 1983, Baberschke et al.1984, Bouchiat and Mailly 1985) can also exhibit SG ordering. The crystallinity of the host is also not an essential requirement for the formation of SG. For example, <u>amorphous</u> TM-TM(-metalloid) alloys Fe-Ni, Fe-Mn, Co-Mn, Co-Ni (Salamon et al.1980, Yeshurun et al.1980, 1981a,b, 1982, Rao and Chen 1981, Salamon and Tholence 1982, 1983, Kudo et al.1982, Lynn et al.1981, Coey et al.1981, Chappert et al.1981, Gignoux et al.1982, Goldfarb et al. 1982a,b, Aeppli et al.1982, Hiroyshi and Fukamichi 1981, 1982) as well as FeNiSi (Suran and Gerard 1984), <u>amorphous</u> alloys $CrSnTe_4$, $MnSnTe_4$, $FeSnTe_4$ (Haushalter et al. 1984), amorphous Heusler alloys, e.g., Cu_2MnAl, Cu_2MnIn, Cu_2MnSn (Krusin-Elbaum et al.1983), <u>amorphous</u> RE-Fe alloys (Rhyne 1985), e.g., Tb_xFe_{1-x} (Rhyne and Glinka 1984, Spano and Rhyne 1985, Nd_xFe_{1-x} (Spano et al.1985), Ho_xFe_{1-x} (Pickert et al.1975), amorphous films of GdAl, LaGd (Jamet and Malozemoff 1978, Malozemoff and Hart 1980) are metallic SG.

Note that it is not necessary that the SG have to be metallic. $Eu_xSr_{1-x}S$ (Maletta and Felsch 1979, Maletta and Convert 1979, Eiselt et al.1979), $Eu_xGd_{1-x}S$ (Litterst et al.1982, van Dongen et al.1983) and $Fe_{1-x}Mg_xCl_2$ (Bertrand et al.1982) are insulators whereas $Cd_{1-x}Mn_xTe$ (Galazka et al.1980, Escorne et al.1981, Escorne and Mauger 1982, Giebultowicz et al. 1985, Novak et al.1985, Oseroff and Gandra 1985), $Cd_{1-x}Mn_xSe$ (Oseroff 1982, Oseroff and Acker 1980, Oseroff et al.1979, 1980a,b), $Hg_{1-x}Mn_xTe$ (Nagata et al.1980), $Zn_{1-x}Mn_xTe$ (McAlister et al.1984), $Sn_{1-x}Mn_xTe$ (Escorne et al.1985) are magnetic semiconductors (Furdyna 1982). Random "mixing" of ferromagnetic EuS with diamagnetic SrS leads to the SG properties of $Eu_xSr_{1-x}S$ whereas random "mixing" of ferromagnetic Rb_2CrCl_4 with antiferromagnetic Rb_2MnCl_4 gives rise to SG system $Rb_2Mn_{1-x}Cr_xCl_4$ (Katsumata et al.1982a,b). Other examples of similar systems are $K_2Cu_xMn_{1-x}F_4$ (Yamada 1983), quasi-1D systems $C_6H_{11}NH_3Cu_{1-x}Mn_xCl_3$ (Cheikhourou et al.1983), $(CH_3)_3NHCo_{1-x}Mn_xCl_3.2H_2O$ (Cheikhrouhou et al.1985). Several other quasi 1D systems, e.g., $FeMgBO_4$, $FeMg_2BO_5$ (Wiedenmann and Burlet 1978, Wiedenmann et al.1981a,b), quasi 2D systems, e.g., $LuFeMgO_4$ (Wiedenmann et al.1983) are also SG. Insulating solid solutions with spinel structure form another class of SG materials. For example, the thio-spinels $CdIn_{2-2x}Cr_{2x}S_4$, $ZnAl_{2-2x}Cr_{2x}S_4$ (Alba et al. 1981, 1982, Mery et al.1985), $Cu_{2x}Cr_{2x}Sn_{2-2x}S_4$ (Colombet and Danot 1983), $CdCr_{2x}Ga_{2-2x}S_4$ (Mery et al.1985) and oxi-spinels $ZnCr_{2x}Ga_{2-2x}O_4$ (Fiorani et al.1983, 1984, 1985), $Mn_xZn_{1-x}Cr_2O_4$, $Mn_{0.75}Mg_{0.25}(Cr_{1-y}V_y)_2O_4$ (Le Dang et al.1984) and $Ge_xCu_{1-x}Fe_2O$ (Kulkarni and Baldha 1985) are also SG. Compound of TM, e.g., $(Ti_{1-x}V_x)O_3$ (Dumas and Schlenker 1979a, Dumas et al.1979, Miyako et al.1980, 1981, Chikazawa et al.1980, 1981), Fe_2TiO_5 (Atzmony et al.1979, Yeshurun et al.1984, Yeshurun and Sompolinsky 1985, Tholence et al.1986,; compare with the SG-like properties of $FeTiO_3$-Fe_2O_3 system (Ishikawa et al.1983)), $Fe_xNi_{1-x}Ge$ (Goto and Kanomata 1985) amorphous ionic insulators $PbMnFeF_7$ (Renard et al.1980, Velu et al.1981) and Pb_2MnFeF_9 (Dupas et al.1982), dilute crystalline insulating compund $CsNiFeF_6$ (Pappa et al.1984, 1985a,b, Occio et al.1985), $CsMnFeF_6$ (Kurtz et al.1976, Kurtz and Roth 1977, Dachs and Kurtz 1977, Kurtz 1982), amorphous aluminosilicate glasses (Naegele et al.1978, 1979a,b,c, LeDang et al.1980, Naegele 1981, Chappert et al.1980, Ferre et al.1980a,b, Rajchenbach et al.1980, Bontemps and Rivoal 1982, Morgownick et al.1982, Beauvillain et al.1984), amorphous fluorophosphates $(MnF_2)_x(BaF_2)_y(NaPO_3)_z$ (Dupas et al.1984, Beauvillain et al. 1986), solid solutions of $Mn_{1-x}Ga_xN$ (Garcia et al.1983), Co_xGa_{1-x} (Grover et al.1979, Meisel et al.1982, Zhou et al.1983), $FeCl_3$ intercalated into graphite (Millman and Zimmerman 1983), UGa_xNi_{1-x} (Zeleny et al.1985) have been demonstrated to be SG.

There are systems which are not SG when the composition is stoichiometric but acquire SG properties if the composition is slightly off-stoichiometric. For example, $PrP_{0.9}$ (Yoshizawa et al.1983), which is non-stoichiometric compared to PrP, is a SG. Similarly, MnO is antiferromagnetic at low T; but $MnO_{1.01}$ (Hauser and Waszczak 1984), which is slightly off-stoichiometric, is a SG. Sometimes, magnetic moment is induced in a system by alloying or compound formation; such systems in the pure state have either very small magnetic moment or no moment at all. The moment induced in this way can exhibit SG ordering and such systems are called induced moment SG.

Thus, SG ordering takes place in such diverse class of materials. However, there are at least two characteristic features that are responsible for the "unities" in the "diversities"; these are quenched disorder and frustration (see chapter 1 for the definitions). Let us see how frustration arises in some of the most widely studied SG systems. AuFe is the prototype of the NM-TM alloys. Because of the random occupation of the lattice sites in the Au matrix by the Fe atoms and because of the oscillating nature of the RKKY interaction (see equation 1.7a) a fraction of the spins get frustrated, giving rise to the SG ordering at low T. Note that in order to form a canonical SG the concentration of the magnetic constituent, c, in these systems must be neither too low nor too high. If c is very small the magnetic atoms are effectively isolated from each other so that the interaction among these atoms can be neglected; this is the so-called Kondo regime. On the other hand, if c is very large the magnetic atoms interact via direct exchange interaction and the system exhibits ferromagnetic (or, antiferromagnetic) long-ranged order.

In contrast with the NM-TM alloys, the exchange interactions in the insulating SG alloys are short-ranged. The competition between the ferromagnetic nearset-neighbour interaction and the antiferromagnetic next-nearest-neighbour interaction is responsible for the frustration in $Eu_xSr_{1-x}S$ (see Maletta 1982a,b, for reviews). However, as stressed in chapter 1, competition between ferromagnetic and antiferromagnetic interactions is not necessary for the SG ordering. For example, the frustration in $Cd_{1-x}Mn_xTe$, $Cd_{1-x}Mn_xSe$ and $Hg_{1-x}Mn_xTe$ arise purely from the lattice structure, viz., f.c.c. lattice with only antiferromagnetic exchange interaction. The origin of the frustration in amorphous alloys is somewhat similar to that for NM-TM alloys. However, the randomness in the mutual separation of the spin pairs arises in these systems from the absence of the underlying lattice. From the experimental point of view, the latter systems are claimed to have some additional advantage over the other conventional SG systems. First of all, usually, the composition of the amorphous alloys can be varied over wider range as compared to the crystalline SG materials. Secondly, usually, the amount of short-ranged order in amorphous SG materials is much smaller compared to the crystalline counterparts. One convenient way of controlling the relative strengths of the ferromagnetic and the antiferromagnetic exchange interactions in a SG system is to use a multilayered system, where the thickness of the layers can be varied appropriately.

Note that all the models introduced in chapter 1 assume complete randomness in the site or the bond variables. But even the most carefully prepared samples of real SG materials contain short-ranged spatial correlations (the so-called clustering of the constituents). Such atomic SRO (and the consequent effects on the magnetic properties) have been investigated by various experimental techniques, e.g., X-ray scattering, diffuse neutron scattering, Mossbauer effect, EXAFS etc. (Hayes et al.1980, Bouchiat et al.1981, Bouchiat and Dartyge 1982, Dartyge et al.1982, Dartyge and Fontaine 1984, Whittle and Campbell 1983, Violet and Borg 1983, Morgownik and Mydosh 1983, Cable et al. 1982, 1984, see Fischer 1985 for a brief summary). Moreover, various metallurgical treatments, e.g., annealing etc., can enhance the SRO (see Beck 1978 and Fischer 1985 for reviews).

APPENDIX B

GENERAL FEATURES OF THE EXPERIMENTAL RESULTS

The discussion on the various theories of SG would be incomplete unless their merits and shortcomings are evaluated in the light of the known experimental data. In this appendix we shall briefly summarize the general features of the experimental results to provide an overview of the experimental status of SG-physics for the theoreticians. We shall also refer to the chapters where the experimental data have been compared with the corresponding theories. The latter might help the experimentalists to find the relevant theoretical information without going through the full details of all the theories (For earlier detailed reviews of the experimental status of SG see Fischer 1985, Ford 1982, Huang 1985, Mydosh 1981a, b, 1983. Several other reviews on the specialized topics will be mentioned in the appropriate context. For the history of the experimental developments of SG physics see Coles 1985).

Both the static as well as the dynamic properties of SG have been probed by almost all possible experimental techniques. Almost all these investigations have focussed attention on the (i) temperature (T)-dependence, (ii) field (H)-dependence, (iii) time (t)-dependence and (iv) concentration (c)-dependence of the various thermodynamic and transport properties, etc., of SG materials. So far as the T-dependence is concerned, some of the experiments, e.g., Mossbauer spectroscopy, magnetic susceptibility measurements, etc., exhibit a sharp anomaly at a temperature T_g whereas some of the other experimentally measured quantities, such as, specific heat, vary smoothly over the same range of temperature. A sharp anomaly usually hints at the possibility of a thermodynamic phase transition whereas the smooth variation of the other properties suggests that even if a phase transition takes place at T_g it must be of a very unusual kind.

Measurements of the dynamical properties of SG materials

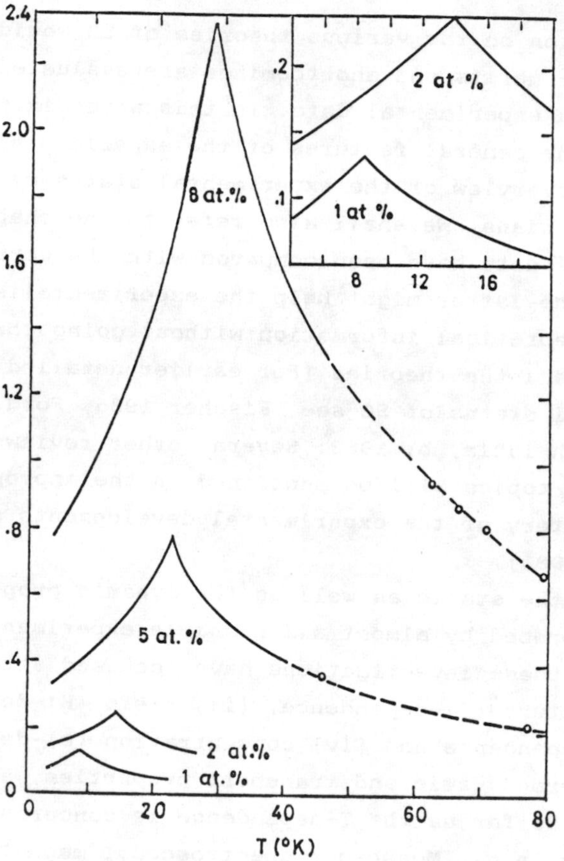

Fig. B.1. Low field $\chi_{ac}(T)$ for AuFe alloys (after Cannella and Mydosh 1972).

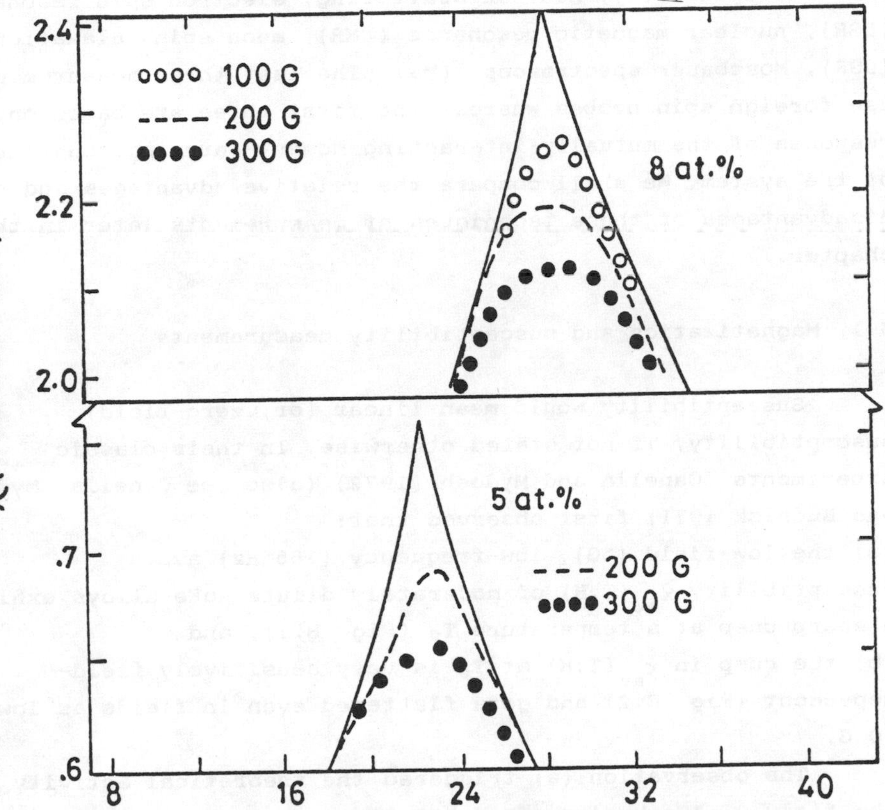

Fig. B.2. $\chi_{ac}(T)$ for AuFe alloys in various applied fields (after Cannella and Mydosh 1972).

have played a crucial role in revealing the peculiarities of the spin-freezing phenomenon in these materials and, therefore, deserve special attention. Such dynamical measurements include a.c. susceptibility, neutron scattering, electron spin resonance (ESR), nuclear magnetic resonance (NMR), muon spin relaxation (μSR), Mossbauer spectroscopy (MS). The last three measurements use foreign spin probes whereas the first three are based on the response of the mutually-interacting moment-carrying constituents of the system. We shall compare the relative advantages and disadvantages of these techniques of measurements later in this chapter.

B.1. Magnetization and susceptibility measurements

Susceptibility would mean linear (or, zero-field) susceptibility, if not stated otherwise. In their classic experiments, Canella and Mydosh (1972) (also see Canella, Mydosh and Budnick 1971) first observed that:

(a) the low-field (5G), low-frequency (155 Hz) a.c. susceptibility $\chi_{ac}(T,H)$ of moderately dilute AuFe alloys exhibits a sharp cusp at a temperature T_g (Fig. B.1), and

(b) the cusp in $\chi_{ac}(T,H)$ at T_g is very sensitively field-dependent (Fig. B.2) and gets flattened even in fields as low as 50 G.

The observation (a) triggered the theoretical activity in the field of SG-physics. This has now become one of the main characteristics by which the experimentalists identify a material as a SG. The failure of all the earlier measurements to observe the sharp cusp can be explained by the observation (b) above; all the earlier measurements used higher fields which was unavoidable for the limitations of accuracy. This is an example of how modern technological advances can open up new areas of research in basic sciences. Breakthroughs in the understanding of the results (a) and (b) above was made by Edwards and Anderson (1975) (see chapter 3).

The temperature, T_g, corresponding to the susceptibility

cusp is an increasing function of the concentration of the magnetic constituent in canonical SG alloys, viz., $T_g \propto c^x$, where x varies between about 0.6 and 0.8 depending on the sample. The latter variation of T_g with c has been analyzed in chapter 3.

What is the significance of using an a.c. technique instead of a d.c. measurement ? Is χ_{ac} at a given temperature T for a given field H identical with χ_{dc} at the same T and H ? The answers to these questions are closely related to the t-dependences of magnetization and susceptibility of SG materials. Susceptibility calculated from the d.c. magnetization measurements exhibits a different behavior depending on the experimental conditions. One must distinguish between the two types of remanent magnetizations in SG (Tholence and Tournier 1974):

(i) Isothermal remanent magnetization (IRM) is observed when the external field H, applied at a temperature $T < T_g$ after cooling in zero field, is suppressed,

(ii) Thermo remanent magnetization (TRM) is observed when the sample is cooled in the presence of an external magnetic field from a temperature above T_g to a temperature below T_g and then the field is switched off. The total d.c. susceptibility consists of the reversible and irreversible contributions; the irreversible contribution $\chi_{ir} = [TRM(H)/H]_{H\to 0}$ comes from the TRM. The a.c. susceptibility χ_{ac} is identical with the reversible part χ_r of χ_{dc} because only the reversible part can follow the alternation of the magnetic field. Using external field much smaller than those used by Tholence and Tournier, Guy (1975, 1977, 1978) not only confirmed the qualitative conclusions of these authors, but also showed that the field cooled (FC) magnetization is almost independent of time. Therefore, he claimed that equilibrium magnetization state can be alternatively achieved by field-cooling. Unfortunately, that is not true! It has been demonstrated that

(i) the magnetization produced by field-cooling depends on the cooling rate; below T_g, the slower is the cooling rate the higher is the magnetization (Wenger and Mydosh 1984),

(ii) Immediately after a small change of the temperature, the FCM attains a value different from the corresponding equilibrium value, and then very slowly relaxes towards the latter (Lundgren et al. 1983a),

(iii) the rate of magnetization relaxation depends on the waiting time t_w (the time interval for which the sample is kept in the field-cooled state before turning the field off); the longer is t_w, the slower is the relaxation. Therefore, the field-cooled state cannot be considered to be the true equilibrium state.

The field-dependences of IRM and TRM, shown in Fig. B.3 (Tholence and Tournier 1974, Bouchiat and Monod 1982), have been compared with the corresponding results of computer simulations in chapter 3.

Guy (1975, 1977, 1978, see also Salamon and Tholence 1982, Tholence and Salamon 1983, Wendler and Baberscke 1983, Yeshurun et al. 1982) observed that the IRM as well as TRM decay logarithmically with time, i.e.,

$$M(t) = M_0 - \Delta m\ (t)$$

and for intermediate time-scales (B.1)

$$\Delta m\ (t) = S(T) \ln t$$

where S(T) depends only on T but not on t, and it is a measure of the "magnetic viscosity" that slows down the relaxation at all $T < T_g$. Moreover, not only the decay but also the growth of magnetization in time seems to be logarithmic. It has been realized for a long time (Prejean and Souletie 1980, Berton et al. 1980, Ferre et al. 1981) that the magnetization relaxation data can also be fitted to a power-law decay of the form

$$M(t) = M_0\ t^{-a} \qquad (B.2)$$

provided a is very small. A stretched exponential decay of the form (see Ngai 1979, 1980)

$$M(t) = M_0 \exp\left[-(t/\tau)^{\alpha}\right] \quad (\alpha < 1) \qquad (B.3a)$$

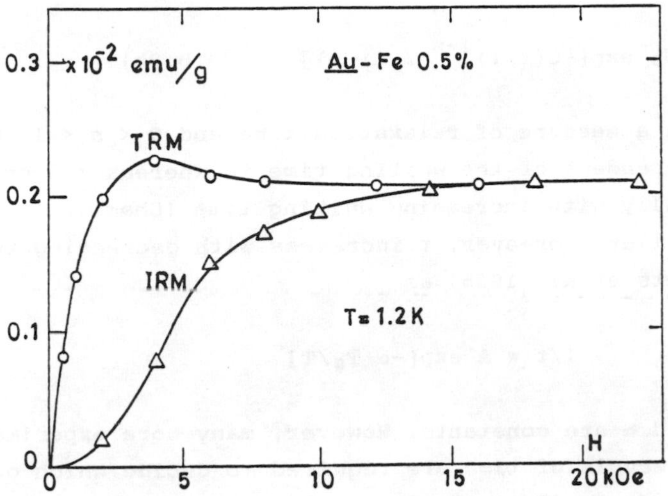

Fig. B.3. Field-dependence of the IRM and TRM of AuFe (0.5 at.%) (after Tholence and Tournier 1974).

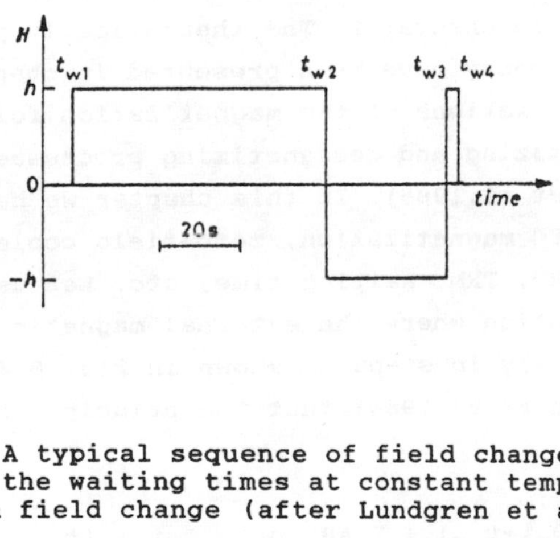

Fig. B.4. A typical sequence of field changes where t_{wi}'s refer to the waiting times at constant temperature prior to the i-th field change (after Lundgren et al. 1986).

has also been reported by Chamberlin et al. (1984). These authors suggested the empirical relation

$$M(t) = M_0 \exp[-C(t/\tau)^{1-n} / (1-n)] \qquad \text{(B.3b)}$$

where τ is a measure of relaxation time and $0 < n < 1$. M_0, n and C are independent of the waiting time t_w whereas τ increases exponentially with increasing waiting time (Chamberlin 1984), as stated earlier. Moreover, τ increases with decreasing temperature (Hoogerbeets et al. 1985) as

$$1/\tau = A \exp[-\alpha\, T_g/T]$$

where A and α are constants. However, many more experiments over longer intervals of time are required to decide which of the functional forms (B.1), (B.2) and (B.3) is the best possible fit to the experimental data (Nordblad et al. 1986). The slow relaxation of magnetization is argued (Prejean and Souletie 1980, Omari et al.1984, Souletie 1983, 1985, also see Wassermann and Herlach 1984) to be consistent with the phenomenological theory of TLS discussed in chapter 2. The theoretically predicted forms of magnetization decay have been presented in chapter 18. The non-monotonic relaxations of the magnetization following successive magnetizing and demagnetizing processes have also been observed (Carre et al.1986). In this chapter we have introduced the concepts of FC magnetization, zero-field cooled (ZFC) magnetization, IRM, TRM, waiting time, etc. Let us now consider the general situation where the external magnetic field is changed sequentially in steps as shown in Fig. B.4. It has been claimed (Lundgren et al.1986) that the principle of superposition applies, so that

$$M(t) = H\, \chi_{FC}(0, t+t_{wn}) + \Sigma\, \Delta H_i\, p(t_{wi}, t + (t_{wn} - t_{wi}))$$

where $p(t_w, t)$ is the response function at time t corresponding to the waiting time t_w and the first term drops out in the case of ZFC samples.

In the a.c. susceptibility measurements one can measure both the in-phase and out-of-phase components χ' and χ'', respectively, which are related via the Kramers-Kronig relations. It is χ' that exhibits a sharp cusp at T_g. On the other hand, χ'' exhibits a broad maximum at a temperature just below T_g, but the point of inflection in χ'' coincides with T_g as shown in Fig. B.5 (Mulder et al. 1981a,b, 1982a,b, Tholence 1981, Lundgren et al. 1981). The frequency ω of the a.c. measurement can be varied to probe the dynamics of the spin-freezing process over varying time-scales. It has been observed that not only χ' but also χ'' depend on ω (Huser et al. 1983a); the ω-dependence is larger in insulating SG than that in metallic SG. The most important feature is that the higher is the frequency ω, higher is the corresponding cusp temperature $T_g(\omega)$ (see Fig. B.6) (Tholence 1980, 1981a,b, Tholence et al. 1978, Loehneysen et al. 1978, Dahlberg et al.1978, 1979, Maletta et al. 1978, Fiorani et al. 1981, Holtzberg et al. 1979, 1982, Lecomte and Loehneysen 1983, Loehneysen and Tholence 1980, Aarts et al.1980, Hauser and Felder 1983, Mulder et al. 1981, Guyot et al. 1980, Yoshizawa et al. 1983, Hasanain et al. 1984, Lundgren et al. 1981, 1982a, Sanchez et al.1984, Beauvillain et al.1984 a,b, Zibold 1978, 1979, Zibold and Korn 1980). Quantitatively speaking, how does T_g vary with $\omega(=1/\tau)$? Most of the experimental data could be fitted better to the Vogel-Fulcher law (Tholence 1980,a,b, 1981)

$$\tau = \tau_0 \exp[\text{constant}/\{T_g(\omega)-T_0\}] \qquad (B.4)$$

than to the Arrhenius law

$$\tau = \tau_0 \exp[\text{constant}/T_g(\omega)] \qquad (B.5)$$

However, at the accuracy attainable at present, it is difficult to decide whether the Vogel-Fulcher law or a power law

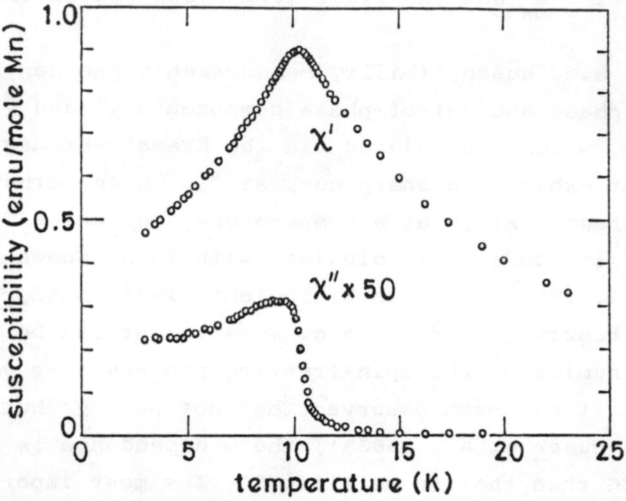

Fig. B.5. Temperature-dependence of χ' and χ'' for AuMn (2.98 at.% Mn) (after Mulder et al. 1982a).

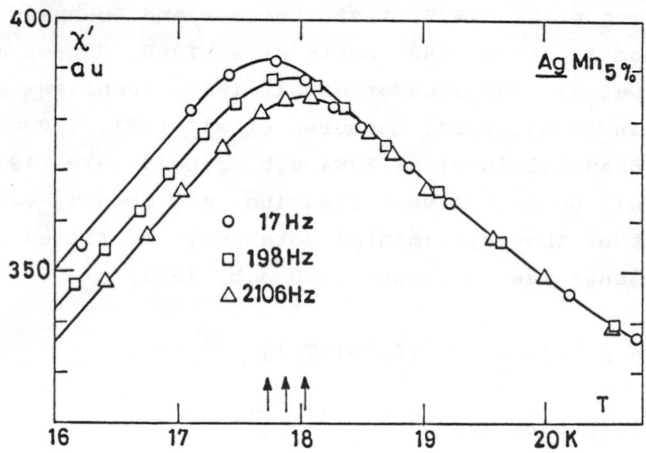

Fig. B.6. $\chi_{ac}(T)$ for AgMn (5 at.% Mn) for various frequencies of measurement (after Tholence 1981a).

$$\tau \sim [T-T_g(\omega)]^{-x} \qquad (B.6)$$

is most appropriate for all SG (Souletie and Tholence 1985, 1986) although the latter seems to fit the data better than the former. The latter observation has been interpreted by different authors by different theories (chapters 8,9,18,20,21,22 and 23). The χ_{ac} measurement (Ocio 1985) has also demonstrated 1/f noise spectrum (see Datta and Horn 1980 for introduction to 1/f noise) in SG.

The magnetization of SG at $T < T_g$ is strongly history-dependent, as already indicated in this appendix. The most striking feature of the hysteresis curve is that it is laterally displaced from the origin as shown in Fig. B.7 (Monod et al. 1979)(compare with Guy 1982, Felten and Schwink 1984). This observation has been analyzed in the light of the anisotropic interactions in SG in chapters 14 and 17.

Measurement of the high-temperature susceptibility is no less important than that near T_g. The susceptibility at sufficiently high temperatures ($T > 5T_g$ in most of the systems) is given by the Curie-Weiss law $\chi(T) = C/(T - \Theta)$ where C is the Curie constant and Θ is the Curie temperature. Θ can be estimated by extrapolation from the high temperature susceptibility. Gradual deviations from the Curie-Weiss behavior begins at rather high temperatures (Morgownick and Mydosh 1981a,b, Morgownick et al.1984, Rao et al. 1983, Majumdar et al. 1983) and hence the need for truly high temperature susceptibilities. Wide varieties of concentration-dependence of Θ have been observed (Morgownick and Mydosh 1981a,b, Morgownick 1983). Moreover, Θ can be positive as well as negative depending on the SG system studied.

Concepts of universality and scaling (see Stanley 1971 for an introduction) are intimately related to phase transitions and critical phenomena. Does the paramagnet (P)- SG transition exhibit any universality and scaling? We shall consider scaling with respect to temperature (T), field (H) and concentration (c) of the magnetic constituent. Yeshurun et al. (1981a,b) fitted their magnetization data to the scaling form
$M = t^{\beta} m^*(h/t^{\beta\delta})$ along the paramagnet-ferromagnet (P-F)

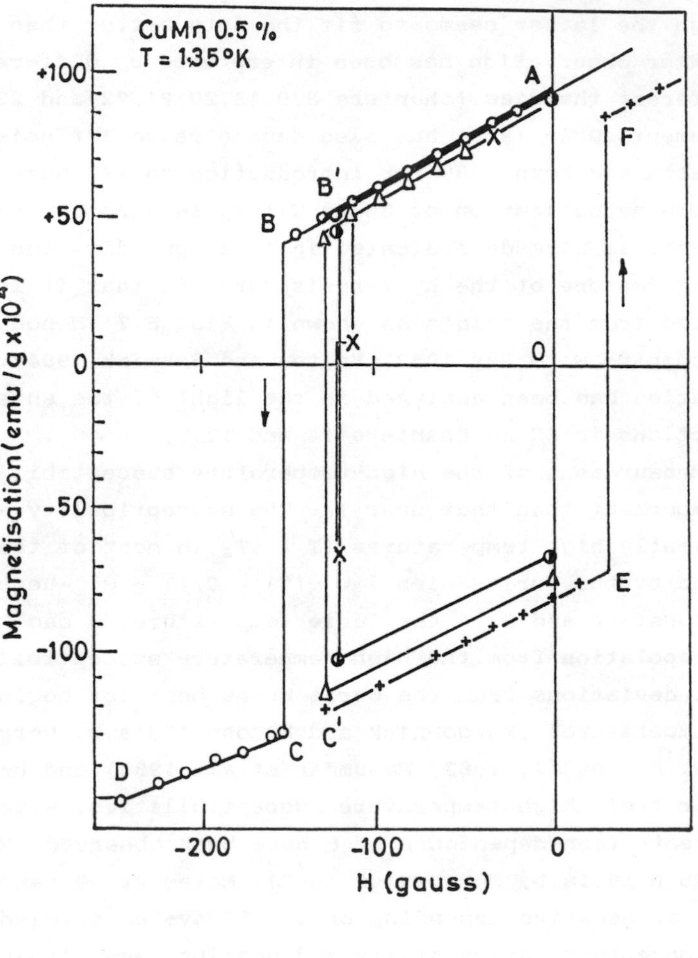

Fig. B.7. Hysteresis of CuMn (0.5 at.% Mn) at 1.35 °K. Different symbols represent different runs (after Monod et al. 1979).

transition line, where $t = (T/T_{pf}-1)$ is the reduced temperature and m^* is the scaling function, to estimate the effective exponents β and δ. Similarly, the corresponding exponents β and δ along the F-SG line were also extracted from the experimental data. The P-F line and the F-SG line meet at a multicritical point. The corresponding crossover exponent ϕ was determined by fitting the experimental χ_{ac} data to the corresponding scaling function. Depending on the system, the "effective exponents" β and δ vary and the agreement with the mean-field exponents (chapter 17) is neither too good nor too bad (see also Dublon and Yeshurun 1982). Several groups have estimated the critical exponents for the P-SG transition (Barbara et al.1981, Berton et al.1982, Omari et al.1983, Beauvillain et al.1984, beauvillain et al.1986a,b, Taniguchi et al.1985). One of the most recent estimates of this set of exponents is that by Bouchiat (1986). From the magnetization measurements, she obtained $\delta = 3.1 \pm 0.2$, $\gamma = 2.2 \pm 0.2$ and $\beta = 1 \pm 0.1$. However, these values should be considered as tentative because even the existence of a truly thermodynamic phase transition has not been established conclusively.

The non-linear susceptibility is related to the order parameter susceptibility of SG (for theoretical justification see chapter 16) and, therefore, expected to diverge at T_g if SG transition is a true phase transition. How does one measure nonlinear susceptibility? In the d.c. measurements, one measures the magnetization M as a function of the external field H. Then the nonlinear susceptibility can be extracted from these data following the steps described below:

(i) A field expansion of the magnetization

$$M(H)/H = \chi_0(T) - \chi_2(T)H^{\alpha(T)} \qquad (B.7)$$

and $\chi_0(T)$ is obtained by a linear extrapolation to zero field from the M/H versus H plots.

(ii) Then, a log-log plot of χ_0 - M/H against H gives $\chi_2(T)$ and $\alpha(T)$ (Barbara et al. 1981, Malozemoff et al. 1982a, Novak et al.

1986).
The data, indeed, show that χ_0 is almost independent of temperature for $T < T_g$ whereas χ_2 increases very sharply near T_g where it exhibits a maximum. On the other hand, nonlinear susceptibility measured by a.c. technique has only logarithmic divergence (Miyako et al. 1980, 1981, Chikazawa et al. 1980, 1981) for some systems whereas for other systems it exhibits power law divergence (Taniguchi et al. 1985)! Is this difference related to the difference in the measurement techniques, viz. d.c. and a.c. measurements, or does this difference have deeper implications? We have discuss the various possible reasons in chapter 16. Moreover, Omari et al. (1983a,b) used the expansion

$$M/\chi_0 H = a_1 - (1/15)a_3 x^2 + (2/305)a_5 x^4 - .....$$

where $x=\mu H/k(T-\theta)$ and θ is the Curie-Weiss temperature (recall that $L_1=1/3$, $L_3=-1/45$, $L_5=2/915$,..etc. are the numerical coefficients which enter the series expansion of the Langevin function and for paramagnets $a_1=a_3=a_5=...=a_{2n+1}=1$). χ_0 is obtained by linear extrapolation in a M/H versus H^2 plot. Then a_3 and a_5 are extracted from the intercept and slope, respectively, of the plot $(1-M/\chi_0 H)/H^2$ versus H^2. Both a_3 and a_5 have been observed to increase by three and six orders of magnitude, respectively, in the vicinity of T_g. In case of superparamagnets (see chapter 21 for an introduction to superparamagnetism), the appropriate expansion seems to be (Fiorani et al.1986)

$$M = \chi_0 H - b_3(\chi_0 H)^3 + b_5(\chi_0 H)^5 + ...$$

where the variation of χ_0 with temperature is taken into account. In contrast with the SG, the coefficients b_3, b_5 are independent of temperature.

Let us investigate the effect of external magnetic field on the nature of the transition as well as on the susceptibility. Does a finite external field H destroy SG order (remember the smearing of the cusp in χ_{ac} by small H)? If not, how does T_g vary with H? There are various different ways of identifying T_g for a given H:

(i) Chamberlin et al. (1982) assumed that $T_g(H)$ "marks the end of a plateau region in dM/dT for given H; M being the d.c. magnetization.

(ii) Monod and Bouchiat (1982) identified $T_g(H)$ as the temperature T at which the field-cooled magnetization M(T) departs from the low-T ($T \rightarrow 0$) magnetization by 3%.

(iii) From the observation of relaxation of ZFC magnetization, Wendler and Baberscke (1983) derived the temperature corresponding to the vanishing of the magnetic viscosity (S=0 in (B.1)), for a given H, and identified it with $T_g(H)$. The same criterion, viz. temperature corresponding to vanishing magnetic viscosity, has also been used by Salamon and Tholence (1982, 1983) for identifying $T_g(H)$. For comparison, they also plotted the temperature T(H) at which, for given H, the magnetic viscosity S attains a maximum. The latter line also looks qualitatively somewhat similar to the line $T_g(H)$.

(iv)Yeshurun et al. (1982) extracted the $T_g(H)$ from the vanishing of the magnetic viscosity of field-cooled as well as zero-field cooled magnetization

(v)Salamon and Tholence (1983) identified $T_g(H)$ with the temperature where the imaginary part of the susceptibility χ'' vanishes. Interestingly, this line was found to shift towards lower temperature with the increase of the time scale of observation (Fig.B.8).

(vi) Bontemps et al.(1983) measured the change in the magnetization ΔM as a result of the change ΔH of the external field in the presence of another larger static field H. Suppose, t_m is the time interval between changing the field and measuring the consequent magnetization change. For a given H and t_m, the temperature T where ΔM vanishes was plotted as a function of the field H. Although all the lines corresponding to the different t_m appear very similar there was a clear shift to lower tempeerature with increasing t_m in agreement with Salamon and Tholence (1983) (Also see Rajchenbach and Bontemps 1983 and 1984).

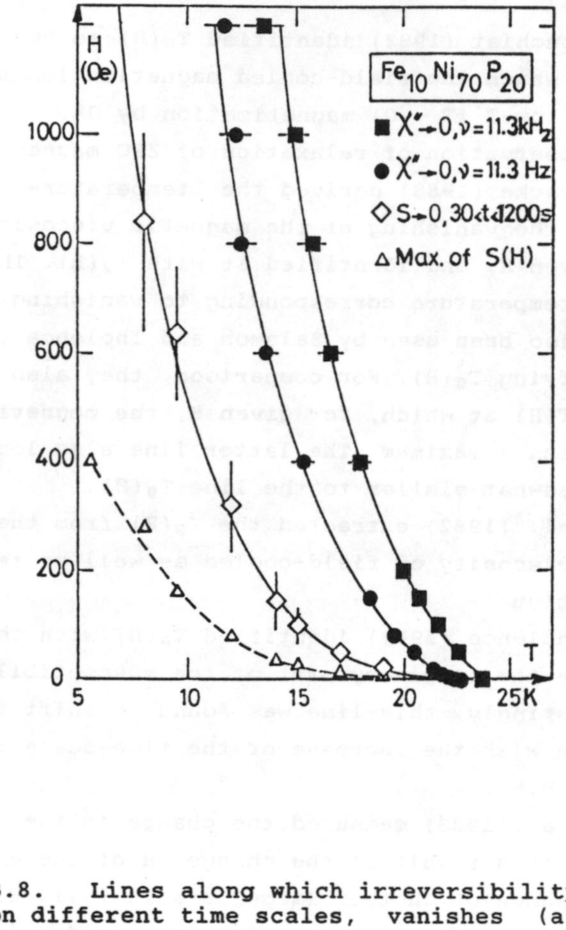

Fig. B.8. Lines along which irreversibility, as determined on different time scales, vanishes (after Salamon and Tholence).

(vii) Paulsen et al. (1984) identified the temperature corresponding to the half-maximum value of the imaginary part of the susceptibility χ'' with $T_g(H)$.
(viii) Campbell et al. (1983) (see also de Courtenay et al.1984) identified the line $T_g(H)$ as the contour of the temperature at which the anisotropy torque (the torque developed as a consequence of changed direction of the external magnetic field) vanishes for the given field H.

Surprisingly, all these criteria led to

$$\tilde{t} \sim H^x \qquad (B.8)$$

where $\tilde{t}$ is the reduced temperature $(1-T/T_g)$ and x is approximately 2/3. The line in the H-T plane defined by (B.8) is called the AT line for reasons explained in chapter 5. It is important to realize that AT-like line has been observed even for superparamagnets where $\tilde{t} = (T - \langle T_B\rangle/\langle T_B\rangle)$, with $\langle T_B\rangle$ corresponding to the maximum in χ_0 (Fiorani et al.1986).

As discussed in chapters 15 and 23, dynamical scaling form depends on whether T_g=0 or $T_g \neq 0$. Measuring χ' and χ'' and using appropriate criterion for defining $T_g(H,\omega)$, one can, in principle, test the applicability of these scaling forms to the real SG materials. Unfortunately, in practice, it is difficult to distinguish between the goodness of fit between the scaling forms for T_g=0 and $T_g \neq 0$ (Bontemps et al. 1984).

How does susceptibility scale with the field H? The following field-scaling law was first suggested (Barbara et al. 1981, Malozemoff et al. 1982a) for the nonlinear susceptibility :

$$\chi_0 - \chi = H^{2/\delta} f(t/H^{2/\phi})$$

where the scaling function f(y) is such that
$f(y) \to$ constant as $y \to 0$
$f(y) \to y^{-\gamma}$ as $y \to \infty$
and $\phi = \gamma\delta/(\delta-1)$
However, the treatment was not self-consistent because χ_0 was

determined by (B.7). Later, a more general scaling (that allows for a self-consistent treatment) for the total susceptibility was proposed (Malozemoff et al.1982b, Barbara et al.1984):

$$\chi(H,T) = \chi_0 + \chi_n$$

where the linear susceptibility is given by

$$\chi_0 = \chi_p - At^{\beta} \qquad (B.9)$$

and the nonlinear susceptibility is given by

$$\chi_n = 1/[B(t^{\gamma}/H^2) + CH^{-2/\delta}] \qquad (B.10)$$

where the seven constants χ_p, A, B, C, β, γ and δ are independent of T and could be estimated by fitting the above forms with the experimental data (also see Beauvillain et al.1983). The agreements with the corresponding mean-field exponents (chapter 17) are rather poor. Setting the two terms in (B.10) equal, one gets a line

$$H^{2/\phi} = (B/C)^{1/\gamma} t \text{ where } \phi = \gamma\delta/(\delta-1)$$

describing the crossover between T-dependent H^2 behavior at low H and T-independent $H^{2/\delta}$ behaviour at high H. Experimental value $\phi \simeq 3$ implies the crossover line $t \sim H^{2/3}$ (this is no extra information, but demonstrates consistency with (B.8) above. What do we learn from these scaling forms? We have discussed these issues in chapter 16.

B.2. Specific Heat Measurements

By the term "specific heat" we shall always mean the "magnetic contribution to the specific heat", if not otherwise stated. The magnetic contribution to the specific heat is defined by the difference between the total specific heat of the SG sample and the specific heat that the system would have if the moment carrying atoms were non-magnetic (e.g., Fe in AuFe). Specific heat measurement belongs to the category of calorimetric measurements where the excited states of the system are probed by supplying heat energy to the system in a controlled manner. Unfortunately, such measurements do not directly yield the density of states. However, the latter can often be inferred indirectly from the nature of the temperature-dependence of the specific heat.

So far as the experiments are concerned, it is quite difficult to subtract the true electronic contribution to the specific heat from the experimental data because the latter as well as the former vary linearly with temperature T. The separation of the phonon contribution is simpler because the latter varies as T^3. In most of the experimental investigations, the electronic specific heat (as well as the phonon contribution) is subtracted, effectively, by subtracting the specific heat of the pure host from that of the alloy under investigation.

From the measurements on the metallic SG over the last decade following general features emerged (Wenger and Keesom 1975, 1976, Martin 1979, Fogle et al. 1983 Sato and Miyako 1982):
(i) the specific heat C exhibits a broad hump whose maximum usually occurs at a temperature higher than that corresponding to the cusp in the susceptibility (Fig. B.9a),
(ii) $d(C/T)/dT$ and $d^2(C/T)/dT^2$ exhibit structure (Figs. B.9b,c,d) although C/T does not exhibit any anomaly (Fig. B.9a),
(iii) the T-dependence of C is given by

$$C = A_1 T + A_2 T^2$$

The same data can often be fitted to the form $C \sim T^{3/2}$ (Thomson and Thompson 1981),
(iv) only about 30% of the total entropy is recovered upto T_g, the remaining 70% develops above T_g,
(v) the maximum in C(T) shifts monotonically to higher T with increasing external field H (Brodale et al. 1983), and

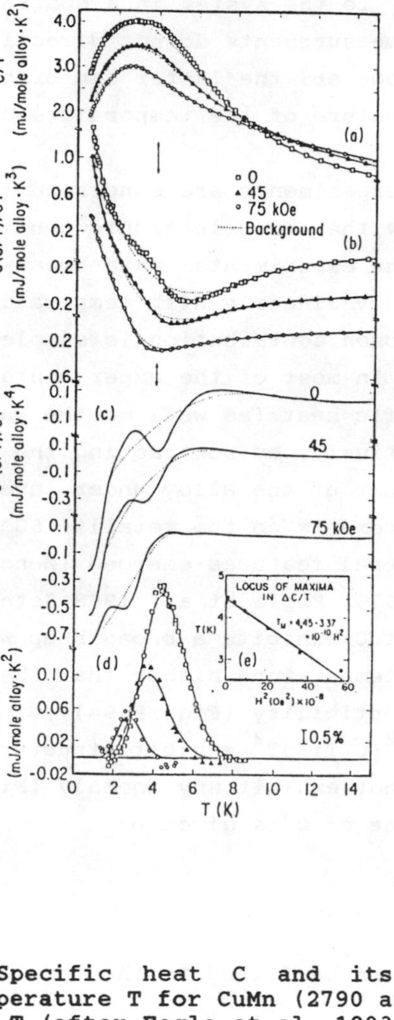

Fig. B.9. Specific heat C and its derivatives with respect to temperature T for CuMn (2790 at. ppm) are shown as function of T (after Fogle et al. 1983).

$$(\partial C/\partial H)_T < 0 \text{ at } T << T_g$$

and

$$(\partial C/\partial H)_T > 0 \text{ at } T >> T_g$$

(see Fig. B.10). For H < 1 KG, $C/T = A + B\,H^2$ where B is negative for almost all T and has a minimum at an intermediate T (Fig. B.11).

The fall of specific heat at low temperature is very fast (almost exponential) in the presence of very high magnetic field (Loehneysen et al. 1985, Marcenat et al. 1985). Possible interpretation of the latter observations has been reviewed in chapter 17.

We have stressed the various aspects of the magnetization relaxation in SG systems in the preceeding section. In the energy relaxation experiments thermal energy is supplied to the system from an external heater in a controlled manner. The sample is thermally connected to a helium bath and the temperature of the sample can be measured and maintained at a fixed value quite stably. The heat flowing from the heater to the sample is equal to the heat flowing from the sample to the heat bath provided the temperature of the sample remains steady. Thus, the energy flux from the heater into the sample, which can be measured with high accuracy, is a measure of the energy relaxation. It has been observed (Berton et al.1979, 1980, 1981a,b) that energy relaxation takes place whenever magnetization relaxaes in a SG sample. Simultaneous relaxation of magnetization and energy has been argued to be consistent with the phenomenologival TLS theory of SG (chapter 2).

B.3 Neutron Scattering :

Neutron scattering is one of the most powerful probes for the studies of magnetic ordering in solids (see Axe and Nicklow 1985, Lander and Price 1985, Pynn and Fender 1985 for elementary introduction to neutron scattering technique and Marshall and

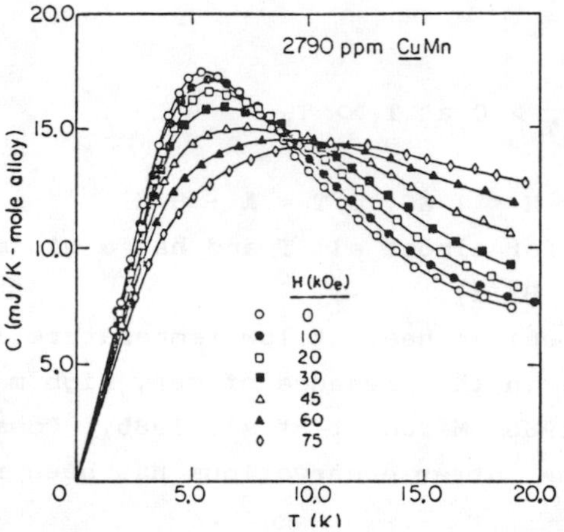

Fig. B.10. Specific heat for CuMn (2790 at. ppm) for various applied fields H (after Brodale et al. 1983).

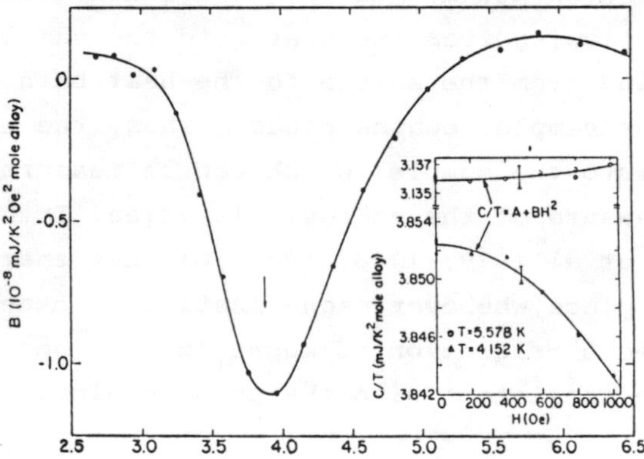

Fig. B.11. The field-dependence of $C/T = A + BH^2$ (after Brodale et al. 1983).

Lovesey 1971 for details). By "neutron scattering" we shall mean magnetic contribution to neutron scattering, if not otherwise stated. The neutron scattering cross section for a spin system is given by (Marshall and Lovesey 1971)

$$(d^2\sigma/d\Omega\ d\omega) = N(\gamma e^2/mc^2)^2(k'/k)S(q,\omega) \quad (B.11)$$

where N is the number of spins, $(\gamma e^2/mc^2)^2$ can be considered as a coupling constant between the local spin and the spin of the neutron, and k and k' are the initial and final wave vectors of the neutrons, respectively. $S(q,\omega)$ is the scattering function given by

$$S(q,\omega) = (1/N\pi)F^2\,\Sigma \int \exp[i\{q\cdot(r_i-r_j)-\omega t\}]\cdot\langle S_i(0)S_j(t)\rangle dt \quad (B.12)$$

where $F(q)$ is the form factor of the spins. Using the trivial identity

$$\langle S_i(0)S_j(t)\rangle = \{\langle S_i(0)S_j(t) - \langle S_i\rangle\langle S_j\rangle\} + \langle S_i\rangle\langle S_j\rangle$$

we get nontrivial simplicity in the analysis of neutron scattering. The total cross-section (B.11) can now be decomposed into static and dynamic parts

$$S_s(q,\omega) = (2/N)\ F^2(q)\ \delta(\omega)\ \Sigma\ \exp[iq\cdot(r_i-r_j)]\ \langle S_i\rangle\langle S_j\rangle \quad (B.13)$$

and

$$S_d(q,\omega) = 2[\omega/\{1-\exp(-\omega/kT)\}]\ F^2\ \chi(q)\ f(q,\omega)/g^2\mu^2 \quad (B.14)$$

where $\chi(q)$ is the wavevector-dependent susceptibility and $f(q,\omega)$, for sufficiently small q, is given by the Lorentzian form

$$f(q,\omega) = (1/\pi)[\Gamma/(\Gamma^2 + \omega^2)] \quad (B.15)$$

where Γ is, in general, a function of q. The static contribution to the cross section is elastic and is also called the Bragg scattering cross section. $S_s(q)$ is often referred to as the static structure factor. Assuming that there is no short-range order in SG and the spin-orientations are frozen completely randomly, we get

$$S_s(q) = 2 F^2(q) Q \quad (B.16)$$

where Q is the EA order parameter. Thus, the elastic scattering cross section yields a direct measure of the EA order in SG.

The dynamic contribution to the neutron scattering cross section is also called diffuse scattering. The diffuse scattering gives three types of important informations in the three different regions of temperature. At very low temperatures ($T \ll T_g$), inelastic scattering by magnons is the dominant contribution to the diffuse scattering cross section and hence the latter yields the magnon (spin wave excitation) spectrum. At temperatures T near T_g diffuse scattering is often referred to as the critical scattering for obvious reasons. The latter provides informations about the spin-correlation function. Finally, at high temperatures ($T \gg T_g$), diffuse scattering cross section is proportional to the magnetic form factor.

In the quasi-static approximation ($\omega \ll kT$) and provided that the incident energy is much larger than ω, the total scattering cross section is given by

$$(d\sigma/d\Omega)_{total} = 2 N (\gamma e^2/mc^2)^2 kT \chi(q) F^2(q)/(g^2\mu^2)$$

and, hence, is a direct measure of $\chi(q)$. In summary, the spatial growth of the SG-ordering can be probed by varying the momentum transfer q in neutron scattering experiments and the temporal growth can be probed by varying the energy resolution of the measurement.

Murani's (1976) small angle neutron scattering data for metallic SG alloys (Fig. B.12) looks very similar to the corresponding data for ferromagnets near the Curie temperature except for the fact that the temperature T_g corresponding to the maximum of the cross-section depends on q. Two opposing interpretations of the latter observations have been reviewed in

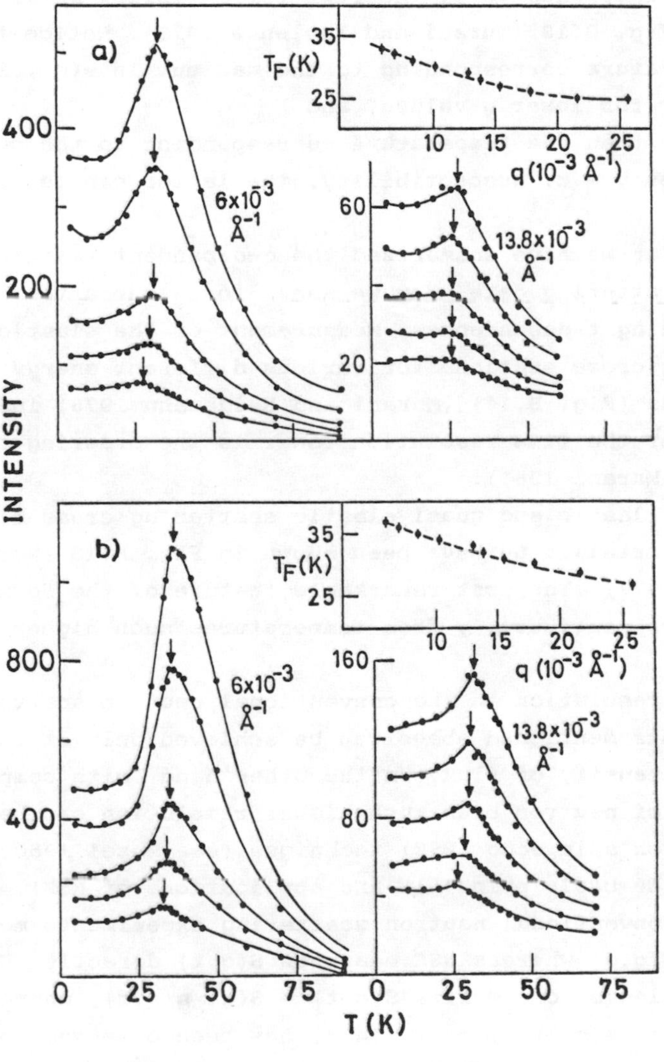

Fig. B.12. Forward scattering intensity as a function of temperature for a series of q values for AuFe alloys containing (a) 10 at.% Fe and (b) 13 at.% Fe (after Murani 1976).

chapters 3,20,21 and 22.

The T-dependence of the $\chi(T)$ for a typical metallic SG is shown in Fig. B.13 (Murani and Tholence 1977). Notice that $T_g(q)$, the temperature corresponding to the maximum in $\chi(q)$, is
(a) higher for lower q-values, and
(b) higher than the temperature corresponding to the cusp in the low-frequency a.c. susceptibility, the latter can be looked as $\chi(0)$.

So far we have summarized the q-dependent features of the neutron scattering data. Now we shall focus our attention on the corresponding t-dependences. Measurement of the elastic scattering cross sections for various different energy resolutions (Fig. B.14) (Murani and Heidemann 1978) indicates that higher the time resolution lower is the ordering temperature (see also Murani 1981).

The elastic and quasi-elastic scattering cross sections for a typical metallic SG have been shown in Fig. B.15 (Murani and Tholence 1977). The most remarkable feature of the former is that it develops continuously from temperatures much higher than $T_g(q)$.

The resolution in the conventional neutron scattering measurements mentioned above can be achieved only at the high cost of intensity of flux. On the other hand, with comparable intensity of neutron beam much higher resolution can be achieved with neutron spin echo (NSE) technique (see Mezei 1980, 1982, 1983 for the basic principle and applications of NSE). As stated earlier, conventional neutron scattering experiments measure the quantity $S(q,\omega)$ whereas NSE measures $S(q,t)$ directly. Usually, for convenience, one writes $S(q,t) = S(q)\ s(q,t)$, where $S(q) = S(q,t{=}0)$ and $s(q, t{=}0) = 1$. $s(q,t)$ has been observed to relax very slowly (logarithmically ?).

B.4 Electron Spin Resonance (ESR)

The spin dynamics in a system of interacting spins is determined crucially by the nature of the spin-spin interaction

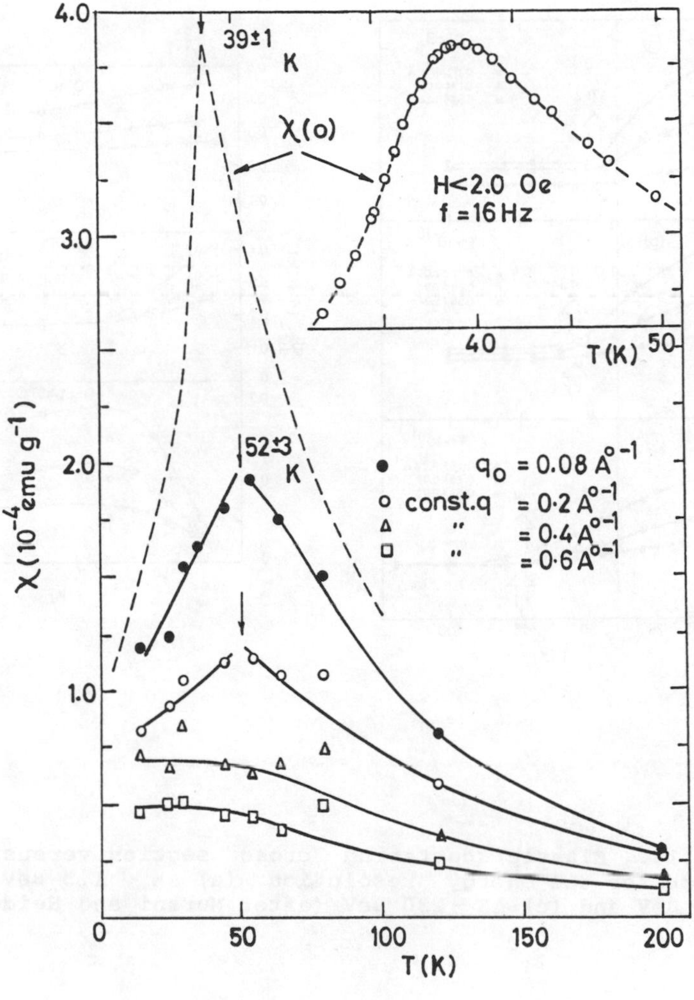

Fig. B.13. The susceptibility $\chi(q)$ versus T for CuMn (8 at.% Mn) alloy. The dashed curve gives the χ_{ac} for the same sample (after Murani and Tholence 1977).

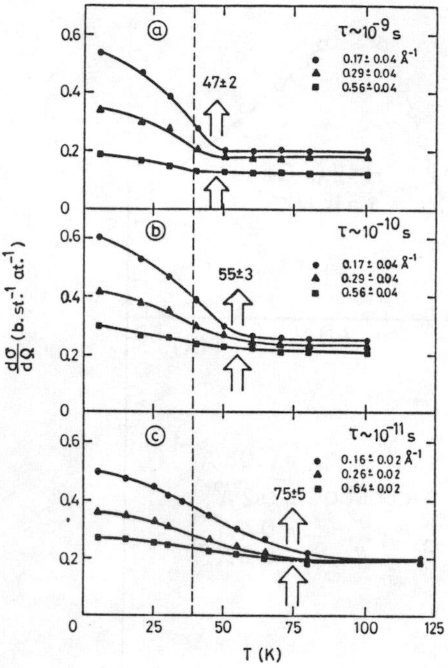

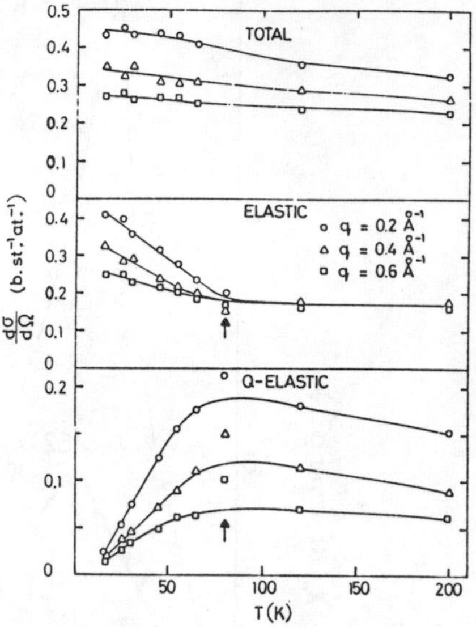

Fig. B.14. Elastic scattering cross section versus temperature for the energy resolution (a) $\Delta E \sim 1.5\ \mu eV$, (b) $\Delta E \sim 25\ \mu eV$ and (c) $\Delta E \sim 230\ \mu eV$ (after Murani and Heidemann 1978).

Fig. B.15. The neutron scattering cross sections for CuMn (8 at.% Mn) alloy. The total (a), the elastic (b) and the quasi-elastic (c) scattering cross sections are shown separately (after Murani and Tholence 1977).

and the coupling of the spin system with the 'heat bath'. Since the resonant absorption of the electromagnetic energy depends on the spin dynamics, it provides a wealth of informations about various types of interactions in the system. In the conventional experimental set-up for ESR measurements, a static magnetic field H_0 polarizes the system and a periodically oscillating magnetic field perpendicular to H_0 provides the electromagnetic energy to be absorbed by the system. In the conventional experimental set up it is the resonance field that is varied while keeping the frequency fixed in order to achieve the resonance. Not only the magnitude of the resonance frequency of absorption, but also the intensity and shape of the absorption line yield important informations about the system. So far as the SG are concerned, most of the ESR measurements focussed on the following quantities:

(a) the resonance frequency, ω_{res}, identified as that corresponding to the maximum of the absoption spectrum,

(b) the resonance line-width, ΔH, usually expressed as proportional to the width at the half-maximum, and

(c) the resonance line shift, δH, defined by $\delta H = H_2 - H_r$, where H_2 is the field corresponding to a gyromagnetic factor g=2 and H_r is the observed resonance field.

Now, we should summarize the main experimental results (see Malozemoff and Jamet 1977, Jamet and Malozemoff 1978, Malozemoff and Hart 1980, Dahlberg et al. 1979 for earlier works):

(i) when cooled in strong magnetic field so as to produce saturated thermoremanent magnetization (Monod and Berthier 1980), the resonant frequency for the so-called ω^+ mode is given by

$$\omega_+/\gamma = \alpha H_r + H_i \qquad \text{(B.17)}$$

where $\alpha \simeq 1$, H_r is the resonance field and H_i is related to the macroscopic anisotropy energy. On the other hand, for SG cooled in the absence of external magnetic field (Schultz et al. 1980, da Silva and Abe 1983), the resonance condition is of the same form (B.17) except that the slope $\alpha \simeq 0.5$ (the intercept H_i is, of course, system-dependent).

(ii) The so-called ω^- mode, with a slope $\alpha \simeq -0.5$, has also been observed in some experiments (Schultz et al. 1980), but has not been observed in some other experiments (Hoekstra et al. 1984).

(iii) If an ESR mode is independent of the field and does not couple with any of the other field-dependent modes, conventional experimental set up for ESR measurement would fail to detect this mode. There are experimental as well as theoretical claims that a field-independent longitudinal mode, ω_L, exists in SG which manifests itself in conventional ESR measurements through mixing with one of the transverse modes (Gullikson et al. 1983). Recently, a novel transmission swept-frequency spectroscopic set up (Gullikson et al. 1985) has enabled direct observation of the longitudinal mode.

(iv) The time-dependences of ΔH and δH at $T < T_g$ are similar to that of the remanent magnetization (Oseroff et al. 1983); also see Dahlbarg et al.(1979) for opposite views.

(v) If the ESR line width would exhibit a power law divergence at T_g, it would strongly support the existence of a phase transition at T_g in the SG material under investigation. Unfortunately, the situation is not so clear cut; the line width does, indeed, increase quite sharply with the decrease of T just above T_g (Jamet et al. 1980, Oseroff 1982), but ultimately saturates, instead of diverging (Owen et al. 1957, Okuda and Date 1969, Salamon 1979, Hou et al. 1984, Mozurkewich et al. 1984)! However, the critical part of the line width $(\Delta H)_c$ (the total line width above T_g = residual width + bT + $(\Delta H)_c$ where b is the thermal broadening coefficient) obeys the dynamic scaling form

$$(\Delta H)_c \sim t^{-g}\, G(\omega/t^{\nu z})$$

where

$$t = (T - T_g)/T_g \text{ and } g = 1.5,\ \nu z = 2.5$$

The same data could also be fitted to the field-scaling form

$$(\Delta H)_c \sim t^{-g}\, G(H/t^{(\beta + \gamma)/2})$$

with $g = 1.5$ and $(\beta + \gamma_s)/2 = 2.5$. A recent interpretation of the latter scaling forms has been discussed in chapter 18.

B.5. Nuclear Magnetic Resonance (NMR)

The earlier NMR measurements showed a minimum of relaxation time T_2 as a function of temperature (Levitt and Walstedt 1977, MacLaughlin and Alloul 1976, 1977) which was interpreted as an evidence of random freezing of the spins in SG. However, these conventional experimental setups used finite, albeit small, magnetic fields to polarize the system. As ststed earlier in this chapter external fields smear, for example, the susceptibility cusp. Therefore, a more desirable technique would be to perform a zero-field NMR. Indeed, such zero-field measurements (Alloul 1979a,b, 1983, Chen and Slichter 1983) seem to support the earlier conclusions (for detailed description of the basic principle and results of these experiments see the original papers; also see Roshen (1982, 1983) for a mean-field theory of NMR in SG).

ESR measurements (together with simultaneous measurement of the magnetization of the same sample) (Schultz et al. 1980) and NMR measurements (together with transverse susceptibility measurements) (Alloul and Hippert 1980, Hippert and Alloul 1982) have provided important informations about the nature of the anisotropy in SG materials (see Alloul and Hippert 1983 for reviews). The most remarkable feature of this anisotropic energy are:

(i) The anisotropy energy is independent of the magnitude of the remanent magnetization. Thus, unlike that of ferromagnets, it is independent of the crystal axes. In other words, the anisotropy energy is isotropic, a phenomenon which is often referred to as the "isotropy of the anisotropy". The latter conclusion has also been confirmed by torque measurements (Fert and Hippert 1982, Fert et al. 1984, Campbell et al. 1983). Moreover, this anisotropy can be enhanced by nonmagnetic impurities with strong spin-orbit coupling (Prejean et al. 1980). The macroscopic theories of spin dynamics incorporating the latter type of anisotropy energy has been reviewed in chapter 17 and the microscopic origin of this anisotropy has been presented in chapter 14.

(ii) the anisotropy can be varied by appropriate combination of cooling and heating in external field (see Fig. B.16). In fact, "the remanent magnetization and the anisotropy of SG turned out to be "two similar but independent properties associated with different types of correlations; the remanent magnetization represents correlation between the spins and the direction of the applied field and the anisotropy represents correlations between the spins and the directions of local anisotropy fields" (Fert et al. 1983). Therefore, in analogy with remanent magnetization, one can also define Isothermal Remanent Anisotropy (IRA) and Thermo Remanent Anisotropy (TRA) (see Fig. B.17). Some plausible justifications of the remanent nature of the anisotropy in SG has been presented in chapter 18.

B.6. Muon Spin Relaxation (μSR)

The basic principle of transverse-field μSR (TF-μSR) is very similar to that of NMR (see Heffner and Fleming 1984 for an elementary introduction, also see Schenk 1978 and Karlsson 1982 for more details). Positive muons are produced with their spins polarized along the beam as a consequence of the parity violation in pion decay. In the TF-μSR set up an external magnetic field is applied perpendicular to the initial polarization vector so that the stopped muon starts precessing. The intensity of the positron beam emitted during the decay of such precessing muons monotored at a given position in the plane of precession exhibits damped oscillations with a decaying amplitude profile. The observed damping rate is a measure of the spread of the (local-) field. In the first μSR measurement of SG (Murnick et al. 1976) observed a rapid change of the depolarization rate Λ around T_g, the latter temperature being known from χ_{ac} measurement (also see Emmerich

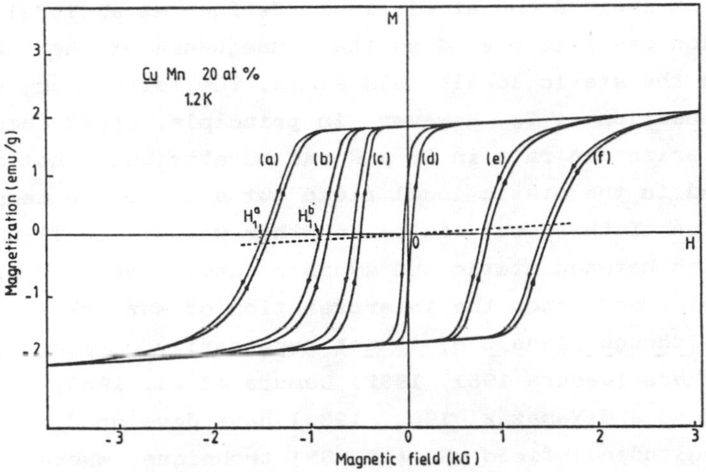

Fig. B.16. Series of hysteresis cycles of CuMn (20 at.% Mn) at 1.2 °K under the following conditions: after cooling the sample down to 1.2K in 9.1 kG, the cycle is recorded by simply sweeping the field between ±9.1 kG (curve a). Curves (b), (c), (d), (e) and (f): stopping the cycle (a) at -9.1kG, the sample was heated up to T ((b) : T = 11K, (c): T = 24K, (d): T = 31K, (e) T = 60K, (f) T = 88K) and cooled back to 1.2K. Then the cycle was recorded by sweeping between ±9.1kG (after Fert et al. 1983).

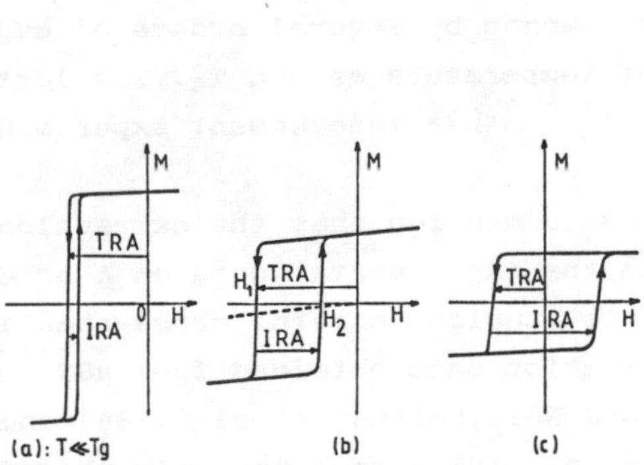

Fig. B.17. Typical hysteresis curve of CuMn (at T < T_g) showing contributions from thermo-remanent anisotropy (TRA) and isothermal remanent anisotropy (IRA) (after Fert et al. 1983).

and Schwink 1981, Brown et al. 1981, Heffner et al.1982). This observation was interpreted as the consequence of the onset of spread in the static local field at T_g, the latter suggesting a phase transition at T_g. However, in principle, rapid change in the depolarization rate in TF-μSR can be attributed not only to the spread in the static local field but also to the temporal fluctuation of the local field. In other words, the TF-μSR cannot distinguish between static and dynamic inhomogeneities of the local field, and hence the interpretation of Murnick et al. (1976), although plausible, is not necessarily correct. Uemura and coworkers (Uemura 1981, 1982, Uemura et al. 1980, T. Yamazaki 1982, Uemura and Yamazaki 1982, 1983) have developed, and used, a novel longitudinal-field μSR (LF-μSR) technique, where both the forward and backward emitted positron fluxes, EF and EB respectively, are measured. The ratio (EF-EB)/(EF+EB) directly yields the longitudinal muon relaxation function $G_z(t)$. This technique can distinguish between the static and dynamic origins of changes in Λ. Since the latter measurement can be performed even in zero field, no smearing of the transition is caused by the field. Using the latter method, Uemura and coworkers (Uemura 1980, 1981, 1982, Uemura and Yamazaki 1982, 1983, Uemura et al. 1980, 1981a,b, 1983) observed that the correlation time, τ_c, of an impurity spin changes by several orders of magnitude within a small interval of temperature around T_g. The latter observation has been supported by other independent experiments (Heffner et al. 1982).

I would like to mention that the extraction of useful informations from the experimental data on Λ or $G_z(t)$ is model-dependent and hence requires careful cross-checking. Comparison of the spin-correlation data obtained from μSR and χ_{ac} (Uemura 1984), from μSR and NSE (Heffner et al. 1984) and from μSR, χ_{ac} and NSE (Uemura et al. 1984) have not only showed compatibility of the data from these different measurements, but also demonstrated the existence of long-time tail over several decades of time scale at $T < T_g$ whereas relaxation is very fast at $T > T_g$.

B.7. Mossbauer spectroscopy

The Mossbauer spectrum of SG materials were probably the first demonstration of the existence of the (apparent?) phase transition in these systems (Violet and Borg 1966, 1967). Unfortunately, these works did not received as much attention as they deserved until 1972 ! (See Window 1969, 1970, Window et al. 1970). The basic principle is that if there is a time-averaged hyperfine-magnetic field at a site where decay of the foreign probe spin takes place, a splitting of the spectrum (the so-called six-finger pattern for **) appears (see Fig. B.18). Observation of such a splitting at a sharply defined temperature T_g strongly indicated the possibility of a spin-freezing, at least over the time-scale of observation.

B.8. Resistivity and Magnetoresistance:

By the term "resistivity" we shall always mean electrical resistivity, if not otherwise stated. Resistivity arises from the conduction electron scattering and hence variation of the resistivity with temperature, field, concentration of the magnetic constituent etc. provide indirect information on the mechanisms of the scatterings involved. In the study of SG, one is mainly intersted in the two sources of scattering, viz.,
(a) the random spatial fluctuation of the local potential (which arises from the formation of the random alloy), and
(b)the magnetic contribution (which arises from the s-d spin interaction).
Since one is not interested in the other sources of scattering, e.g., the phonon contribution, one simply subtracts the resistivity of the pure host from that of the alloy at the same T (compare with the measurement of specific heat) to get the desired quantity

$$\rho(T) = \rho_{alloy}(T) - \rho_{pure\ host}(T)$$

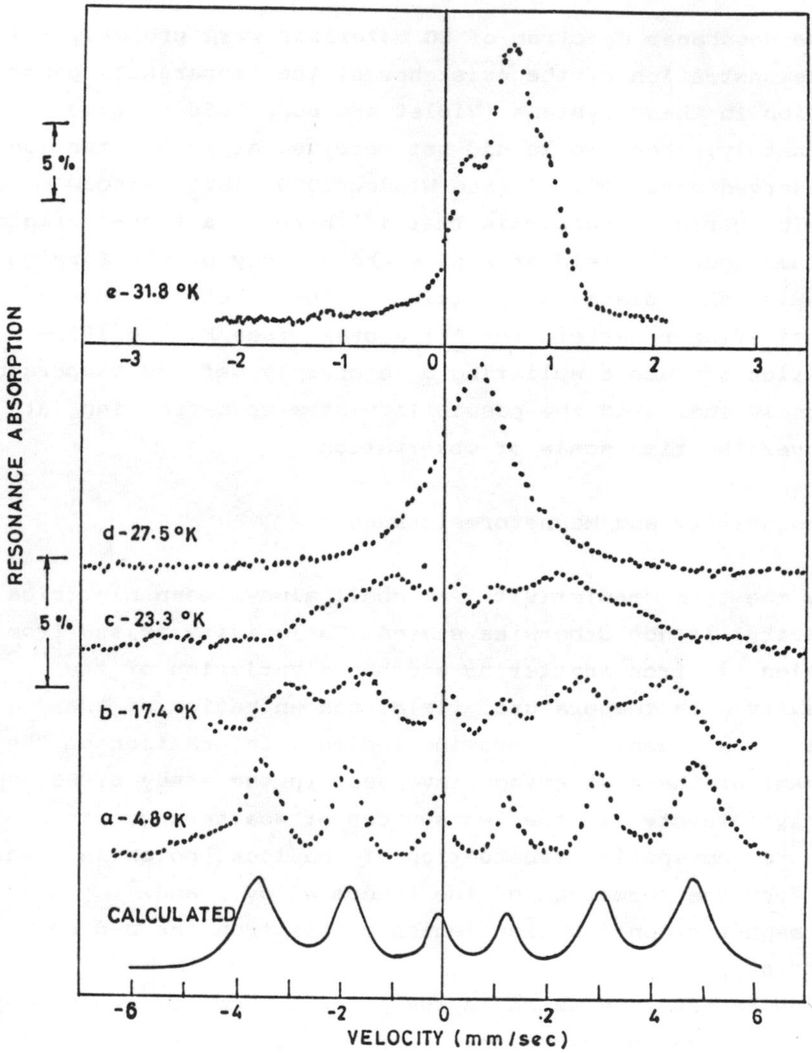

Fig. B.18. Resonance absorption spectra of AuFe (6.7 at.% Fe) (after Violet and Borg 1966).

However, such simple-minded method of measuring $\rho(T)$ becomes increasingly erroneous with the increasing deviation from Matheisen's rule. The magnetoresistance is given by

$$\Delta\rho(H,T) = \rho(H,T) - \rho(0,T)$$

The main results on the resistivity (Ford and Mydosh 1976, Tari 1976) can be summarized as follows:
(a) $\rho(T)$ exhibits a very broad maximum at a temperature $T_{max} >> T_g$ (Fig. B.19),
(b) the best fit to the data is obtained with

$$\rho(T) = \rho_0 + A\, T^{3/2}$$

where ρ_0 is the residual resistivity and A is a constant
(c) T_{max} as well as ρ_0 are increasing functions of c.

The main results of the magnetoresistance measurements (Nigam and Majumdar 1983, Senoussi 1980, Majumdar 1983, Majumdar and Oestreich 1984) are as follows:
(a) $\Delta\rho$ is isotropic, i.e., the longitudinal magnetoresistance is identical with the corresponding transverse magnetoresistance,
(b) $[\rho(H) - \rho(0)]/\rho(0) \propto M^2$ over a wide range of H and T (see Fig. B.20).

A crucial feature of every experimental probe is the corresponding characteristic time-scale of observation. A list of the latter time-scale for some of the most conventional techniques of measurements is presented at the end of this appendix. Most of the dynamical experiments very emphatically demonstrate the existence of a distribution of relaxation times (DRT) in SG. Various consequences of the DRT in SG have been considered in chapters 8, 19, 20 and 22.

One special feature of the real SG materials is the role of anisotropy (see chapters 15 and 18). The strength of the anisotropy can be controlled by varying the composition of the material. For example, the anisotropy strength $D > 0$ in ZnMn, $D <$

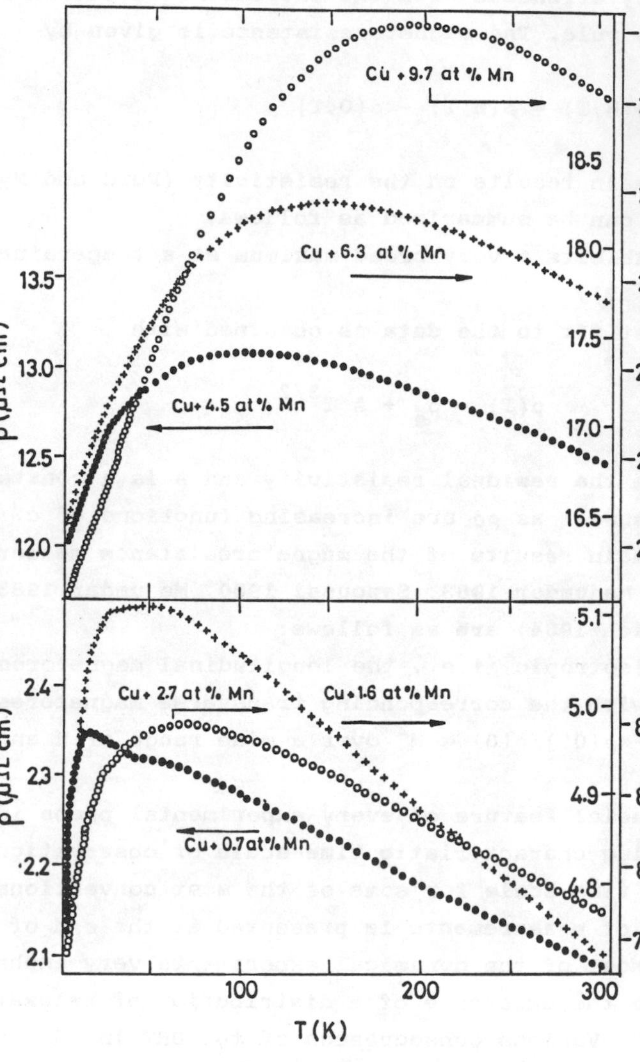

Fig. B.19. Magnetic contribution to the resistivity of CuMn alloys (after Ford and Mydosh 1976).

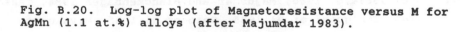

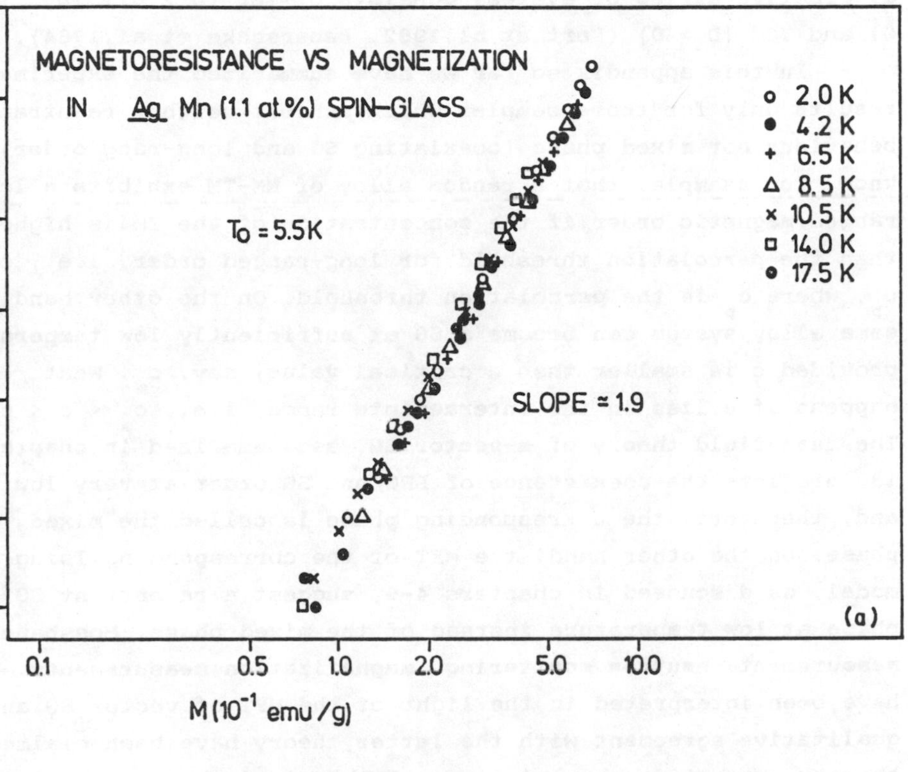

Fig. B.20. Log-log plot of Magnetoresistance versus M for AgMn (1.1 at.%) alloys (after Majumdar 1983).

0 in CdMn and D is almost vanishingly small in MgMn (Albrecht et al.1982). However, one disadvantage of using this alloy system for a comparative study of the Ising-like , XY-like and Heisenberg-like SG is that nonmagnetic hosts are different. An alternative series of systems consists of YEr ($D > 0$), YDy ($D < 0$) and YGd ($D \simeq 0$) (Fert et al.1982, Baberschke et al.1984).

In this appendix so far we have summarized the experimental results only for those samples which exhibit neither reentrant behaviour nor mixed phase (coexisting SG and long-rang order). We know, for example, that a random alloy of NM-TM exhibits a long-ranged magnetic order if the concentration of the TM is higher than the percolation threshold for long-ranged order, i.e., $c > c_p$, where c_p is the percolation threshold. On the other hand, the same alloy system can become a SG at sufficiently low temperature provided c is smaller than a critical value, say, c_m. What happens if c lies in the intermediate range, i.e., $c_m < c < c_p$? The mean-field theory of m-vector SG, as summarized in chapter 13, predicts the coexistence of LRO and SG order at very low T and, therefore, the corresponding phase is called the mixed phase. On the other hand, the MFT of the corresponding Ising model, as discussed in chapters 4-9, suggest a re-entrant SG phase at low temperature instead of the mixed phase. Mossbauer measurement, neutron scattering, magnetization measurements, etc. have been interpreted in the light of the MFT of vector SG and qualitative agreement with the latter theory have been claimed, thereby supporting the existence of the mixed phase (Varret et al. 1982, Campbell et al.1983, Lauer and Keune 1982, Brand et al. 1983, Hennion et al.1983, 1984, Ketelsen and Salamon 1984, Razzaq and Kouvel 1984a,b, 1985, Majumdar and Blanckenhagen 1984, Takeda et al.1985, Taniguchi et al.1985, Sato et al.1985, Miyako et al.1986). On the other hand, several other groups claim that the results could be explained as a re-entrant phenomenon where the low-temperature phase is a pure SG without co-existing LRO (Maletta 1983, Aeppli et al.1984, Dillon et al.1985, Erwin et al.1985, Shapiro et al.1985, Mazumdar et al.1984, 1985, Shapiro et al.1985). If the latter interpretation is correct, would it imply the real SG to be Ising like? Not necessarily; an alternative mechanism (Aeppli et al.1984) suggests the destruction of the LRO at low temperatures by random magnetic fields which could exist naturally in random magnetic systems. Co-existence of antiferromagnetic ordering and SG ordering has also been reported (Datta et al.1984, Lecomte et al.1984, Wong et al.1985, Chowdhury et al.1986). (Also see Borg and Violet 1984, Senoussi and Oner 1985, O'Shea and Sellmeyer 1985, Goldfarb et al.1985 for related works).

We shall not discuss the results of several other experimental investigations employing, as for example, Faraday rotation technique (Ayadi and Ferre 1983, 1984, Dillon et al. 1984), elastic light scattering (Geschwind et al. 1984), optical microscopy (Dillon et al. 1984), spin-flip Raman scattering (Carlson et al. 1984), ultrasonic measurements (Moran et al. 1973, Hawkins and Thomas 1978, Hawkins et al.1979), thermal conductivity measurements (Herlach et al. 1983a,b, Wassermann and Herlach 1984), thermal expansion coefficient measurement (Simpson et al. 1981, Geerken and Nieuwenhuys 1982), effect of pressure on SG properties (Razavi and Scilling 1983), X-ray scattering (Dartyge et al. 1982, Alexander and Dorenbusch 1982), Point contact spectroscopy (d'Ambrumenil et al. 1983), Perturbed angular correlation spectroscopy (Rots et al. 1984a,b) etc.

While summarizing the experimental results we have focussed mainly on the most general qualitative features and did not mention the type of the sample studied. Readers interested in the specific informations, e.g., the composition, lattice structure, etc., of these samples studied are requested to refer to the original papers.

We have mentioned on several different occasions in this book the different time scales of observation associated with different experimental probes. Now we present a table of the typical time scales for some of the most conventional experimental techniques (Chowdhury and Mookerjee 1984a, Fischer 1985):

neutron scattering	$10^{-12} - 10^{-8}$ **sec.**
μSR	$10^{-11} - 10^{-5}$ **sec.**
Mossbauer	$10^{-9} - 10^{-8}$ **sec.**
A.C. susceptibility	$10^{-3} - 10^{-1}$ **sec.**
remanence	$10 - 10^{3}$ **sec.**

Finally, I would like to remind the experimentalists that experiments with SG are difficult for various reasons, e.g, history-dependence, long-tails of relaxation, etc. Nevertheless, the experimentalists should not be biased by the findings of the theorists; the aim of an experimentalist is to find the "truth", nothing but the "truth"!

"Some are to be read, some to be studied, and some may be neglected entirely, not only without detriment, but with advantage". (Anonymous)

REFERENCES:

Aarts J, Felsch W, Loehneysen H V and Steglich F 1980 Z. Phys. B40, 127

Abrikosov A A 1978 J. Low Temp. Phys. 33, 505

Abrikosov A A 1980 in: Lec. Notes in Phys. vol. 115

Abrikosov A A and Moukhin S I 1978 J. Low Temp. Phys. 33, 207

Aeppli G and Bhanot G 1981 J. Phys. C14, L593

Aeppli G, Shapiro S M, Birgeneau R J and Chen H S 1982 Phys. Rev. 25, 4882

Aeppli G, Shapiro S M, Maletta H, Birgenau R J and Chen H S 1984 J. Appl. Phys. 55, 1628

Aharony A 1975 Phys. Rev. B12, 1028

Aharony A 1978a J. Magn. Mag. Mat. 7, 198

Aharony A 1978b J. Phys. C11, L457

Aharony 1983 J. Magn. Mag. Mat. 31-34 1432

Aharony A 1986 J. Magn. Mag. Mat. 54-57, 27

Aharony A and Imry Y 1976 Solid State Commun. 20, 899

Aharony A and Pfeuty P 1979 J. Phys. C12, L125

Aharony A and Pytte E 1980 Phys. Rev. Lett. 45, 1583

Aho A V, Hopcroft J E and Ullman J D 1975 The design and analysis of computer algorithms, Addison-Wesley

Akhiezer I A and Spolnik A I 1983 Sov. Phys. Sol. State 25, 81

Alba M, Hammann J and Nogues M 1981 Physica 107B, 627

Alba M, Hammann J and Nogues M 1982 J.Phys. C15, 5441

Albrecht H, Wassermann E F, Hedgecock F T and Monod P 1982 Phys. Rev. Lett. 48, 819

Alexander R B and Dorenbusch W E 1982 Solid State Commun. 42, 223

Alexander S and Pincus P 1980 J. Phys. A13, 263

Ali N and Woods S B 1983 Solid State Commun. 45, 471

Alloul H 1979a Phys. Rev. Lett. 42, 603

Alloul H 1979b J. Appl. Phys.50, 7330

Alloul H 1983 in: Lec. Notes in Phys. vol. 192

Alloul H and Hippert F 1980 J. de Phys.Lett. (Paris) 41, L201

Alloul H and Hippert F 1983 J. Magn. Mag. Mat. 31-34, 1321

Alloul H and Mendels P 1985 Phys. Rev. Lett. 54, 1313

Alloul H, Mendels P, Beauvillain P and Chappert C 1986 Europhys. Lett. 1, 595

Amit D J 1978 Field Theory, Renormalization Group and Critical Phenomena (McGraw-Hill)

Amit D J 1986 Preprint

Amit D J, Gutfreund H and Sompolisnky H 1985a Phys. Rev. B32, 1007

Amit D J, Gutfreund H and Sompolinsky H 1985b Phys. Rev. Lett. 55, 1530

Amit D J, Gutfreund H and Sompolinsky H 1986 Preprint

Anderson P W 1970 Mat. Res. Bull. 5, 319

Anderson P W 1973 in: Amorphous magnetism I, eds. Hooper H O and deGraff A M (Plenum)

Anderson P W 1977 in: Amorphous magnetism II, eds. Levy R A and Hasegawa R (Plenum)

Anderson P W 1978a J. Appl. Phys.49, 1599

Anderson P W 1978b J. Less Comm. Met. 62, 291

Anderson P W 1979 in: Ill Condensed Matter ed. Balian R, Maynard R and Toulouse G (North Holland)

Anderson P W 1983 Proc. Natl. Acad. Sci. USA 80, 3386

Anderson P W 1984 Basic Notions of Condensed Matter Physics", Benjamin

Anderson P W, Halperin B I and Varma C M 1972 Phil. Mag. 25, 1

Anderson P W, Thouless D J and Palmer R G 1977 Phys. Lett. 62A, 456

Anderson P W and Pond C M 1978 Phys. Rev. Lett. 40, 903

Andre G, Bidaux R, Carton J P, Conte R and deSeze L 1979 J. de Phys. (Paris) 40, 479

Andreev A F 1978 Sov. Phys. JETP 47,411

Arai M, Ishikawa Y, Saito N and Takei H 1985, J. Phys. Soc. Jap. 54, 795

Arai M and Ishikawa Y 1985 J. Phys. Soc. Jap. 54, 795

Ariosa D, Droz M and Malaspinas A 1982 Helv. Phys. Acta 55, 29

Atzmony U, Gurewitz E, Melamud M, Pinto H, Shaked H, Gorodesky G, Hermon E, Hornreich R M. Strikman S and Wanklyn B 1979 Phys. Rev. Lett. 43, 782

Axe J D and Nicklow R M 1985 Phys. Today 38, no.1, 26

Ayadi M and Ferre J 1983 Phys. Rev. Lett. 50, 274

Ayadi M, Nordblad P, Ferre F, Mauger A and Triboulet R 1986 J. Magn. Mag. Mat. 54-57 1223

Baberschke K, Pureur P, Fert A. Wendler R and Senoussi S 1984 Phys. Rev. B29, 4999

Bachas C P 1984 J. Phys. A17, L709

Bachas C P 1985 Phys. Rev. Lett. 54, 53

Banavar J R and Cieplak M 1982a Phys. Rev. Lett. 48, 832

Banavar J R and Cieplak M 1982b Phys. Rev. B26, 2662

Banavar J R and Cieplak M 1983 J. Phys. C16, L755

Banavar J R and Cieplak M 1984 Phys. Rev. B28, 3813

Bantilan F T and Palmer R G 1981 J. Phys. F11, 261

Barahona F 1982 J. Phys. A15, 3241

Barahona F, Maynard R, Rammal R and Uhry J P 1982 J. Phys. A15, 673

Barbara B, Malozemoff A P and Imry Y 1981 Phys. Rev. Lett. 47, 1852

Barbara B, Malozemoff A P and Barnes S E 1984 J. Appl. Phys. 55, 1655

Barber M N 1983 in: Phase Transition and Critical Phenomena, ed. Domb C and Lebowitz J L, vol. 8

Barnes S E 1981a J. Phys. F11, L249

Barnes S E 1982b Physica 108 B, 771

Barnes S E 1981c Adv. Phys. 30, 801

Barnes S E 1984 Phys. Rev. B30, 3994

Barnes S E, Malozemoff A P and Barbara B 1984 Phys. Rev. B30, 2765

Beauvillain P, Dupas C, Renard J P and Veillet P 1983 J. Magn. Mag. Mat. 31-34, 1377

Beauvillain P, Dupas C, Renard J P and Veillet P 1984a Phys. Rev. 29, 4086

Beauvillain P, Chappert C and Renard J P 1984b J. de Phys. Lett. (Paris) 45, L665

Beauvillain P, Matecki M, Prejean J J and Renard J P 1986 Europhys.Lett.

Beauvillain P, Chappert C, Renard J P and Seiden J 1986 J. Magn. Mag. Mat. 54-57, 127

Beck P A 1978 Prog. Mat. Sci. 23, 1

Becker K W 1982a Phys. Rev. B26, 2394

Becker K W 1982b Phys. Rev. B26, 2409

Beckman O 1985 in: Festkorperprobleme 25

Benamira F, Provost J P and Vallee G 1985 J. de Phys. 46, 1269

Bennet M H and Coles B R 1977 Physica 86-88B, 844

Benyoussef A and Boccara N 1981 Phys. Lett. 86A, 181

Benyoussef A and Boccara N 1983 J. Phys. C16, 1901

Benyoussef A and Boccara N 1984 J. Phys. C17, 285

Berker A N and Kadanoff L P 1980 J. Phys. A13, L259

Berton A, Chaussy J, Odin J, Rammal R, Souletie J and Tournier R 1979 J de Phys. Lett. (Paris) 40, L391

Berton A, Chaussy J, Odin J, Prejean J J, Rammal R, Souletie J and Tournier R 1980 J. Magn. Mag. Mat. 15-18, 203

Berton A, Chaussy J, Odin J. Peyirard J, Prejean J and Soletie J 1981a Solid State Comm. 37, 241

Berton A, Chaussy J, Odin J. Rammal R, Souletie J, Tholence J L, Tournier R, Holtzberg F and von Molnar S 1981b J. Appl. Phys. 52, 1763

Bertrand D, Fert A R, Schmidt M C, Bensamka F and Legrand S 1982 J. Phys. C15, L883

Beton P H 1985 J. Phys. C18, 1225

Beton P H and Moore M A 1983 J. Phys. C16, 1245

Beton P H and Moore M A 1984 J. Phys. C17, 2157

Beton P H and Moore M A 1985 J. Phys. C18, L37

Beton P H and Moore M A 1985 J. Phys. C18, L145

Bhanot G and Creutz M 1980 Phys. Rev. B22, 3370

Bhatt R N and Young A P 1984 Phys. Rev. Lett. 54, 924

Bhargava R and Kumar D 1977 Solid State Comm. 22, 545

Bhargava R and Kumar D 1979 Phys.Rev. B19, 2764

Bhat R N and Young A P 1985 Phys Rev. Lett. 54, 924
Bhattacharya S, Nagel S R, Fleishman L and Susman S 1982 Phys.Rev.Lett. 48, 1267
Bien K and Usadel K D 1986 J. Magn. Mag. Mat. 58, 117
Binder K 1977 in: Festkoerperprobleme, vol. 17, ed. Treusch J (Vieweg, Braunschweig) p. 55
Binder K 1977a Z. Phys. B26, 339
Binder K 1978 J. dePhys. (Paris) 39, C6, 1527
Binder K 1979 ed: Monte Carlo methods in statistical physics, Springer Verlag
Binder K 1980a in: Fundamental problems in statistical mechanics V, ed. Cohen E G D (North Holland)
Binder K 1980b in: Ordering in strongly fluctuating condensed matter systems, ed. Riste T (Plenum)
Binder K 1986 "Recent trends in the development and applications of the Monte Carlo method" (Preprint)
Binder K and Stauffer D 1976 Phys.Lett. 57A, 177
Binder K and Stauffer D 1984 in: Applications of the Monte Carlo Method in Statistical Physics, ed. Binder K (Springer-Verlag)
Binder K and Schroeder K 1976 Phys. Rev. B14, 2142
Binder K and Kinzel W 1981 in: Lecture Notes in Phys. vol. 149 (Springer Verlag)
Binder K and Kinzel W 1983 in: Lec.Notes in Phys. vol. 192 (Springer Verlag)
Binder K, Kinzel W and Stauffer D 1979 Z. Phys. B36, 161
Binder K and Young A P 1984 Phys. Rev. B29, 2864
Birge N O, Jeong Y H. Nagel S R, Bhattacharya S and Susman S 1984 Phys. Rev. B30, 2306
Bishop A R and Schneider T 1978 Solitons and Condensed Matter Phys. Springer Series in Solid State Sciences 8 (Springer-Verlag)
Blanckenhagen P V and Scheerer B 1983 J. Magn. Mag. Mat. 31-34, 1349
Blandin A 1961 Ph.D. Thesis (unpublished)
Blandin A 1978 J. de Phys. (Paris), 39, C6-1499

Blandin A, Gabay M and Garel T 1980 J. Phys. C13, 403
Blumen A, Klafter J and Zumofen G 1986 J. Phys.A19, L77
Bontemps N and Rivoal J C 1982 J. Phys. C15, 1301
Bontemps N, Rajchenbach J and Orbach R 1983 J.de Phys. Lett. (Paris) 44, L47
Bontemps N, Rajchenbach J, Chamberlin R V and Orbach R 1984 Phys Rev. B30, 6514
Borg R J and Violet C E 1984 J. Appl. Phys. 55, 1700
Bouchiat H 1983 J. Phys. C16, L145
Bouchiat H 1984 Phys. Rev. B30, 3963
Bouchiat H 1986 J. de Phys. (Paris) 47, 71
Bouchiat H, Dartyge E, Monod P and Lambert M 1981 Phys. Rev. B23, 1375
Bouchiat H and Monod P 1982 J. Magn. Mag. Mat. 30, 175
Bouchiat H and Dartyge E 1982 J. de Phys.(Paris) 43, 1699
Bouchiat H and Mailly D 1985 J. Appl. Phys. 57, 3453
Bowman D R and Levin K 1982a Phys. Rev. Lett. 49, 240
Bowman D R and Levin K 1982b Phys.Rev. B25, 3438
Bowman D R and Halley J W 1982 Phys.Rev.B25, 1892
Brand R A, Lauer J and Keune W 1983 in:Lec. Notes in Phys. vol. 192
Brand R A and Keune W 1984 Phys.Rev.Lett. 52, 2097
Bray A J 1982 J. Phys. C15, L57
Bray A J and Moore M A 1977 J. Phys. F7, L333
Bray A J and Moore M A 1978 Phys. Rev. Lett. 41, 1068
Bray A J and Moore M A 1979a J. Phys. C12, L441
Bray A J and Moore M A 1979b J. Phys. C12, 79
Bray A J and Moore M A 1980a J. Phys. C13, L469
Bray A J and Moore M A 1980b J. Phys. C13, L655
Bray A J and Moore M A 1980c J. Phys. C13, 419
Bray A J and Moore M A 1981a J. Phys. C14, 1313
Bray A J and Moore M A 1981b J. Phys.A14, L377
Bray A J and Moore M A 1981c Phys. Rev. Lett. 47, 120
Bray A J and Moore M A 1981d J. Phys. C14, 2629
Bray A J and Moore M A 1982 J. Phys. C15, 3897
Bray A J and Moore M A 1982a J. Phys. C15, L765

Bray A J and Moore M A 1984a J. Phys. C17, L463
Bray A J and Moore M A 1984b J. Phys. C17, L613
Bray A J and Moore M A 1985a Phys. Rev. B31, 631
Bray A J and Moore M A 1985b J. Phys. C18, L139
Bray A J and Moore M A 1985c J. Phys. A18, L683
Bray A J, Moore M A and Reed P 1978 J. Phys. C11, 1187
Bray A J, Moore M A and Reed P 1979 J. Phys. C12, L477
Bray A J and Roberts S A 1980 J. Phys. C13, 5405
Bray A J and Viana L 1983 J. Phys. C16, 4679
Bray A J, Grzonka R B and Moore M A 1983 J. Magn. Mag. Mat.31-34, 1293
Bray A J, Moore M A and Young A P 1984 J. Phys. C17, L155
Bray A J, Moore M A and Young A P 1986 Phys. Rev. Lett. 56, 2641
Brieskorn G and Usadel K D 1986 Preprint
Brodale G E, Fischer R A, Fogle W E, Phillips N E and van Curen J 1983 J. Magn. Mag. Mat. 31-34
Brout R 1959 Phys. Rev. 115, 824
Brout R 1965 Phase Transitions (Benjamin)
Brown J A, Heffner R H, Kitchens T A, Leon M, Olsen C E, Schillaci M E, Dodds S A and MacLaughlin D E 1981 J.Appl. Phys. 52, 1766
Bruinsma R and Aeppli G 1984 Phys. Rev. B29, 2644
Bulaevskii L N, Buzdin A I, Kulic M L and Panjukov S V 1985 Adv. Phys. 34, 175
Bullough R K and Caudrey P J 1980 "Solitons", Topics in Current Phys. 17 (Springer-Verlag)
Burkhardt T W 1982 in: Real space renormalization, ed. Burkhardt T W and van Leeuwen J M J
Buzdin A I, Bulaevskii L N, Kulic M L and Panjukov S V 1984 Sov. Phys. Usp. 27, 927
Cable J W, Werner S A, Felcher G P and Wakabayashi N 1982 Phys. Rev. Lett. 49, 829
Cable J W, Werner S A, Felcher G P and Wakabayashi 1984 Phys.Rev. B29, 1268
Caflisch R G and Banavar J R 1985 Phys. Rev. B32, 7617
Caflisch R G, Banavar J R and Cieplak M 1985 J.Phys. C18, L991

Caliri R, Mattis D and Read R C 1985 Phys. Lett. 109A, 282
Callaway D J E 1985 Contemp. Phys 26,23 and 95
Campbell I A 1981 Phys. Rev. Lett. 47, 1473
Campbell I A 1986 Phys. Rev. B33, 3587
Campbell I A, Ford P J and Hamzic A 1982 Phys. Rev. B26, 5195
Campbell I A, Hurdequint H and Hippert F 1986 Phys. Rev. B33, 3540
Campbell I A, Arvanitis D and Fert A 1983 Phys. Rev. Lett. 51, 57
Campbell I A, Senoussi S, Varret F, Teillet J and Hamzic A 1983a Phys. Rev. Lett. 50, 1615
Cannella V, Mydosh J A and Budnick J I 1971 J. Appl. Phys. 42, 1689
Cannella V and Mydash J A 1972 Phys. Rev. B6, 4220
Canisius J and van Hemmen J L 1981 Phys. Rev. Lett. 47, 1487
Canisius J and van Hemmen J L 1986 Europhys. Lett. 1, 319
Carlson N W, Geschwind S, Devlin G E, Batlogg B, Dillon J F, Rupp L W and Maletta 1984 J. Appl. Phys. 55, 1679
Carre E, Prejean J J and Souletie J 1986 J. Magn. Mag. Mat. 54-57,205
Catchings R M, Borg R J and Violet C E 1985 J. Appl. Phys. 57, 3435
Ceccatto H A 1985 preprint
Chakrabarti A and Dasgupta C 1986 Phys. Rev. Lett. 56, 1404
Chakrabarti B K 1981 Phys. Rev. B24, 4062
Chakrabarti B K 1986 J. Phys. A19, 1273
Chalupa J 1977 Solid State Commun. 22, 315
Chamberlin R V 1984 Phys. Rev. B30, 5393
Chamberlin R V 1985 J. Appl. Phys. 57, 3377
Chamberlin R, Hardiman M, Turkevich L A and Orbach R 1982 Phys. Rev. B25, 6720
Chamberlin R V, Mozurkewich G. and Orbach R 1984 Phys. Rev. Lett. 52, 867
Chang M C 1983 Solid State Comm. 47, 529
Chang M C 1983a Phys. Rev. B28, 6592
Chang M C and Sak J 1984 Phys. Rev. B29, 2652
Chappert C, Beauvillain P, Renard J P and Knorr K 1980 J. Magn.

Mag. Mat. 15-18, 117
Chappert C, Coey J M D, Lienard A and Ribouillat J P 1981 J. Phys. F11, 2727
Chayes J, Chayes L, Sethna J P and Thouless D J 1986 Preprint
Cheikhrohou A, Dupas C and Renard J P 1983 J. de Phys. Lett. (Paris) 44, L777
Cheikhrouhou A, Dupas C, Renard J P and Veillet P 1985 J. Magn. Mag. Mat. 49, 201
Chen J H and Lubensky T C 1977 Phys. Rev. B16, 2106
Chen M C and Slichter C P 1983 Phys. Rev. B27, 278
Cheung H F and McMillan W L 1983a J. Phys.C16, 7027,
Cheung H F and McMillan W L 1983b J. Phys.C16, 7033
Cherry R J and Domb C 1978 J. Phys. A11, L5
Chien C L, Liou S H, Kofalt D, Yu W, Egami T and McGuire T R 1986 Phys.Rev. B33, 3247 Mag. Mat. 49, 201
Chikazawa S, Youchunus Y G and Miyako Y 1980 J. Phys. Soc. Jap. 49, 1276
Chikazawa S, Sandberg C J and Miyako Y 1981 J. Phys. Soc. Jap. 50, 2884
Ching W Y, Leung K M and Huber D L 1977 Phys. Rev. Lett. 39, 729
Ching W Y and Huber D L 1978 J. Appl. Phys. 49, 1633
Ching W Y and Huber D L 1979 Phys. Rev. B20, 4721
Ching W Y, Huber D L and Verbeek B H 1979 J. Appl. Phys. 50, 1715
Ching W Y, Huber D L and Leung K M 1980 Phys. Rev. B21, 3708
Ching W Y, Huber D L and Lueng K M 1981 Phys. Rev. B23, 6126
Ching W Y and Huber D L 1984 Phys. Rev. B30, 179
Choi M Y and Huberman B A 1983a Phys. Rev. B28, 2547
Choi M Y and Huberman B A 1983b Phys. Rev. B29, 2796
Choi M Y and Stroud D 1985 Phys. Rev. B32, 7173
Choi M Y, Chung J S and Stroud D 1986 Preprint
Chowdhury D 1984 Phys. Lett. 100A, 370
Chowdhury D 1985 J. Phys. A18, L95
Chowdhury D 1986 Comments on Solid State Phys. 12, 69
Chowdhury D and Mookerjee A 1983a J. Phys. F13, L19
Chowdhury D and Mookerjee A 1983b Solid State Commun. 48, 887
Chowdhury D and Mookerjee A 1984a Phys. Rep. 114, 1

Chowdhury D and Mookerjee A 1984b J. Phys. F14, 245
Chowdhury D and Mookerjee A 1984c Physica 124B, 255
Chowdhury D and Mookerjee A 1984e Unpublished
Chowdhury D and Banerjee K 1984 Phys. Lett. 101A, 273
Chowdhury D and Bhattacharjee J K 1984 Phys. Lett. 104A, 100
Chowdhury D, Harders T M, Mookerjee A and Gibbs P 1986 Solid State Commun. 57, 603
Choy T C and Sherrington D 1984 J. Phys. C17, 739
Chudnovsky E M, Saslow W M and Serota R A 1986 Phys. Rev. B33, 251
Cieplak M and Banavar J R 1983 Phys. Rev.B27, 293
Cieplak M and Banavar J R 1984a Phys. Rev.B29, 469
Cieplak M and Banavar J R 1984b in: Elementary excitations and Fluctuations in magnetic systems", ed. Lovesey S W, Balucani U, Borsa F and Tognetti V (Springer)
Cieplak M and Cieplak M Z 1985 J. Phys. C18, 1481
Cieplak M, Cieplak P and Kotur M A 1986 Preprint
Cochran W T, Gaines J R, McCall R P, Sokol P E and Patton R B 1980 Phys. Rev. Lett. 45, 1576
Cochrane R W, Harris R and Zuckermann M J 1978 Phys. Rep. 48, 1
Coey J M D 1978 J. Appl. Phys. 49, 1646
Coey J M D, Givord D, Lienard A and Ribouillat 1981 J. Phys. F11, 2707
Coles B R 1980 J. Magn. Mag. Mat. 15-18, 157
Coles B R 1985 Ann. Phys. Fr. 10, 63
Coles B R, Jamieson H, Taylor R H and Tari A 1975 J. Phys. F5, 565
Collet P and Eckmann J P 1984 Commun. Math. Phys. 93, 379
Collet P, Eckmann J P, Glaser V and Martin A 1984 J. Stat. Phys. 36, 89
Colombet P and Danot M 1983 Solid State Commun. 45, 311
Condon J H and Ogielski A T 1985 Rev. Sci. Instrum. 56, 1691
Continentino M A and Malozemoff A P Phys. Rev.B33, 3591
Continentino M A and Malozemoff A P Preprint
Cooper B R 1969 J. Appl.Phys. 40, 1344
Cooper L N 1974 in: Proc. nobel Symp. on Cllective Prop. of Phys.

Systems, ed. Lundquist B and Lundquist S (Academic Press)
Corbelli G and Morandi G 1979 Solid State Commun. 29, 589
Courtens E 1982 J de Phys. Lett. 43, L199
Courtens E 1983 Helv. Phys. Acta 56, 705
Courtens E 1984 Phys. Rev. Lett. 52, 69
Courtens E, Rosenbaum T F, Nagler S E and Horn P M 1984 Phys. Rev. B29, 515
Cragg D M and Sherrington D 1982 Phys. Rev. Lett. 49, 1190
Cragg D M, Sherrington D and Gabay M 1982 Phys. Rev. Lett. 49, 158
Creutz M 1982 in: Melting, localization and Chaos, eds. Kalia R and Vashista P (Elsevier)
Crick and Mitchison 1983 Nature 304, 111
Crisan M 1986 J. Magn. Mag. Mat. 59, 191
Cyrot M 1979 Phys. Rev. Lett. 43, 173
Cyrot M 1981 Solid State Commun. 39, 1009
Dachs H and Kurtz W 1977 J. Magn. Mag. Mat. 4, 262
Dahlberg E D, Hardiman M, Orbach R and Souletie J 1979 Phys. Rev. Lett. 42, 401
Dahlberg E D, Hardiman M and Souletie J 1978 J. de Phys. Lett. (paris) 39, L389
D'Ambrumenil N, Duif A M, Jansen A G M and Wijder P 1983 J. Magn. Mag. Mat. 31-34, 1415
Danielian A 1964 Phys. Rev. 133, 1345
Dartyge E, Bouchiat H and Monod P 1982 Phys. Rev. B25, 6995
Dartyge E and Fontaine A 1984 J. Phys. F14, 721
Das S K, Tripathi R S and Joshi S K 1982 Phys. Rev. B25, 1921
DasGupta C, Ma S K and Hu C K 1979 Phys. Rev. B20, 3837
Dasgupta C and Sompolinsky H 1983 Phys. Rev. B27, 4511
Da Silva J M M and Abe H 1983 J. Magn. Mag. Mat. 31-34, 1351
Datta P and Horn P M 1981 Rev. Mod. Phys. 53, 497
Datta T, Thornberry D, Jones E R and Ledbetter H M 1984 Solid State Comm. 52, 515
Davidov D, Baberschke K, Mydosh J A and Nieuwenhuys G J 1977 J. Phys. F7, L47
Davies J H, Lee P A and Rice T M 1982 Phys. Rev. Lett. 49, 758

Davies J H, Lee P A and Rice T M 1984 Phys. Rev. B29, 4260
de Almeida J R L and Thouless D J 1978 J. Phys. A11, 983
de Almeida J R L, Jones R C, Kosterlitz J M and Thouless D J 1978 J.Phys. C11, L871
de Almeida J R L and Lage E J S 1983 J. Phys. C16, 939
deChatel P F 1981 J. Mag. Mag. Mat. 23, 28
de Courtenay N, Fert A and Campbell I A 1984 Phys. Rev. B30, 6791
DeDominicis C 1978 Phys. Rev. B18, 4913
DeDominicis C 1979 in: Lec. Notes in Phys. vol.104
DeDominicis C 1980 Phys. Rep. 67, 37
DeDominicis C 1983 in: Lec. Notes in Phys. vol.192
DeDominicis C 1984 in: Lec. Notes in Phys. vol.216
DeDominicis C and Garel T 1979 J.de Phys. Lett. (Paris) 22, L576
DeDominicis C, Gabay M, Garel T and Orland H 1980 J. de Phys. (Paris) 41, 923
DeDominicis C, Gabay M and Orland H 1981 J. de Phys Lett.(Paris) 42, L523
DeDominicis C, Gabay M and Duplantier B 1982 J. Phys. A15, L47
DeDominicis C and Young A P 1983a J. Phys. C16, L641
DeDominicis C and Young A P 1983b J. Phys. A16, 2063
DeDominicis C and Hilhorst H J 1985 J. de Phys. Lett. (Paris) 46, L909
DeDominicis C, Orland H and Lainee F 1985 J. de Phys. Lett. (Paris) 46, L463
DeDominicis C and Kondor 1983 Phys. Rev. B27, 606
DeDominicis C and Kondor I 1984a J.de Phys. Lett. (Paris) 45, L205
DeDominicis C and Kondor I 1984b in: Lec. Notes in Phys. vol.216
DeDominicis C and Kondor I 1985 J.de Phys. Lett. 46, L1037
deGennes P G 1962 J. Phys. Radium 23, 630
Deo P and Mishra S 1984 Phys. Rev. B29, 2811
Derrida B 1980a Phys. Rev. Lett. 45, 79
Derrida B 1980b Phys. Rep. 67, 29
Derrida B 1981 Phys. Rev. B24, 2613
Derrida B 1985 J. de Phys. Lett. (Paris)
Derrida B, Pomeau Y, Toulouse G and Vannimenus J 1979 J. de Phys.

(Paris) 40, 617
Derrida B, Pomeau Y, Toulouse G and Vannimenus J 1980 J. de Phys. (Paris) 41, 213
Derrida B and Vannimenus J 1980 J. Phys. C13, 3261
Derrida B and Hilhort H 1981 J. Phys. C14, L539
Derrida B and Toulouse G 1985 J. de Phys. Lett. (Paris) 46, L223
Derrida B and Pomeau Y 1986 Europhys. Lett. 1, 45
Derrida B and Gardner E 1986 J. Phys. C19, 2253
deSeze L 1977 J. Phys. C10, L353
Devoret M, Esteve D and Sullivan N S 1982 J.Phys. C15, 5455
Devoret M and Esteve D 1983 J.Phys. C16, 1827
deYoro J J, Meissner M, Pohl R O, Rowe J M, Rush J J and Susman S 1983 Phys. Rev. Lett. 51, 1050
Dhar D in: Lec. Notes in Phys. vol. 184 (Springer)
Dieny B and Barbara B 1985 J. de Phys. 46, 293
Diep H T, Lallemand P and Nagai O 1985a J. Phys. C18, 1067
Diep H T, Ghazali A and Lallemand P 1985b J. Phys.C18, 5881
Dillon J F, Albiston S D, Batlogg B, Schreiber H and Maletta H 1984, J. Appl. Phys. 55, 1673
Dillon J F, Rupp L W, Batlogg B and Maletta H 1985 J. Appl. Phys. 57, 3488
Ditzian R V and Kadanoff L P 1979 Phys. Rev. B19, 4631
Domb C 1960 Adv. Phys. 9, 245
Dormann J L, Fiorani D, Tholence J L and Sella C 1983 J. Magn. Mag. Mat. 35, 117
Dos Santos R R, dos Santos R Z and Kischinhevsky M 1985 Phys. Rev. B31, 4694
Dotsenko V S 1985 J. Phys. C18, 6023
Dotshenko V S 1985b J. Phys.C18, L1017
Dotshenko V S and Feigelman M V 1981 J. Phys. C14, L823
Dotshenko V S and Feigelman M V 1982 J. Phys. C16, L565
Dotshenko V S and Feigelman M V 1982 Sov. Phys. JETP 56, 189
Dotshenko V S and Feigelman M V 1984 Sov. Phys. JETP 59, 904
Dublon D and Yeshurun Y 1982 Phys. Rev. B25, 4899
Dumas J and Schlenker C 1979 J. Phys. C12, 2381
Dumas J, Schlenker C, Tholence J L and Tournier R 1979 Phys. Rev.

20, 3913
Dunlop M W and Sherrington D 1985 J. Phys. C18, 1465
Dupas C, Renard J P, Fonteneau G and Lucas J 1982 J. Magn. Mag. Mat. 27, 152
Dupas C, Renard J P and Matecki M 1984 J. Magn. Mag. Mat. 43, 283
Duplantier B 1981 J. Phys. A14, 283
Dzyaloshinskii I E 1979 in: Lec. Notes in Phys. vol.115
Dzyaloshinskii I E 1983 JETP Lett. 37, 227
Dzyaloshinskii I E and Obukhov S P 1982 Sov. Phys. JETP 56, 456
Dzyaloshinskii I E and Volovik G E 1978 J. de Phys. (Paris) 39, 693
Ebner C and Stroud D 1985 Phys. Rev. B31, 165
Edwards S F 1970 in: Proc. Third Int. Conf. on Amorphous Mat.
Edwards S F 1976 J. Phys. F6, 1923
Edwards S F and Anderson P W 1975 J. Phys. F5, 965
Edwards S F and Anderson P W 1976 J. Phys. F6, 1927
Ehrenreich H, Seitz F and Turnbull D 1980 Solid State Phys. vol. 35 (Academic Press)
Eilenberger G 1981 Solitons Mathematical methods in Phys. Springer series in solid state sciences 19, (Springer Verlag)
Eiselt G, Koetzler J, Maletta H, Stauffer D and Binder K 1979 Phys. Rev. B19, 2664
Elderfield D 1983 J. Phys. A16, L439
Elderfield D 1984a J. Phys. A17, L307
Elderfield D 1984b J. Phys. A17, L517
Elderfield D and McKane A J 1978 Phys. Rev. B18, 3730
Elderfield D and Sherrington D 1982 J. Phys. A15, L437
Elderfield D and Sherrington D 1982a J. Phys. A15, L513
Elderfield D and Sherrington D 1982b J. Phys. C15, L783
Elderfield D and Sherrington D 1983a J. Phys.C16, L497
Elderfield D and Sherrington D 1983b J. Phys.C16, L971
Elderfield D and Sherrington D 1983c J. Phys. C16, L1169
Elderfield D and Sherrington 1984 J. Phys. C17, 5595
Emery V J 1975 Phys. Rev. B11, 239
Emmerich K and Schwink Ch 1981 Hyp. Int. 8, 767

Erwin R W, Lynn J W, Rhyne J J and Chen H S 1985 J. Appl. Phys. 57, 3473
Erzan A 1984 Portgal. Phys. 15, 9
Erzan A and Lage E J S 1983 J. Phys. C16, L555
Escorne M, Mauger A, Triboulet R and Tholence J L 1981 Physica 107B, 309
Escorne M and Mauger A 1982 Phys. Rev. B25, 4674
Escorne M, Godinho M, Tholence J L and Mauger A 1985 J. Appl. Phys. 57, 3424
Ettelaie R and Moore M A 1985 J. de Phys. Lett. (Paris) 46, L893
Fahnle M and Egami T 1982a Solid State Commun. 44, 533
Fahnle M and Egami T 1982b J. Appl. Phys. 53, 7693
Farquhar I E 1964 Ergodic Theory in Statistical Mechanics (John Wiley)
Feigelman M V 1983 J. Phys. C16, 6275
Feigelmann M V and Tsvelik A M 1979 Sov. Phys. JETP 50, 1222
Feigelman M V and Ioffe L B 1983 Z. Phys. B51, 237
Feigleman M V and Ioffe L B 1985 J. de Phys. Lett. (Paris) 46, L695
Feigelman M V and Ioffe L B 1986 Europhys. Lett. 1, 197
Feile R, Loidl A and Knorr K 1982 Phys. Rev. B26, 6875
Felsch W 1978 Z. Phys. B29, 203
Felten G and Schwink C 1984 Solid State Commun. 49, 233
Fernandez J F and Medina R 1979 Phys. Rev. B19, 3561
Fernandez J F and Streit T S J 1982 Phys. Rev. B25, 6910
Ferre J, Pommier J, Renard J P and Knorr K 1980a J.Phys. C13, 3697
Ferre J, Pommier J, Jahn I R and Ranabe Y 1980b Solid State Commun. 33, 27
Ferre J, Rajchenbach J and Maletta 1981 J. Appl. Phys. 52, 1697
Ferre J and Ayadi M 1984 J. Appl. Phys. 55, 1720
Fert A and Hippert F 1982 Phys. Rev. Lett. 49, 1508
Fert A, Pureur P, Hippert F, Baberschke K and Bruss F 1982 Phys. Rev. B26, 5300
Fert A, Arvanitis D and Senoussi S 1983 J.de Phys. Lett. (Paris)

44, L345
Fert A, Arvanitis D and Hippert F 1984 J. Appl. Phys. 55, 1640
Fert A, de Courtnay N and Campbell I A 1985 J. Appl. Phys. 57, 3398
Fiorani D, Tholence J L and Dorman J L 1981 Physica 107B, 643
Fiorani D, Nogues M and Viticoli S 1982 Solid State Commun. 41, 537
Fiorani D, Viticoli S, Dormann J L, Nogues M, Tholence J L, Hammann J and Murani A P 1983 J. Magn. Mag. Mat. 31-34, 947
Fiorani D, Viticoli S, Dormann J L, Tholence J L and Murani A P 1984 Phys. Rev. B30, 2776
Fiorani D, Dormann J L, Tholence J L and Soubeyroux J L 1985 J. Phys. C18, 3053
Fiorani D, Dormann J L, Tholence J L, Bessais L and Villers D 1986 J. Magn. Mag. Mat. 54-57, 173
Fisch R and Harris A B 1977 Phys. Rev. Lett. 38, 785 and 1981 Phys. Rev. Lett. 47, 620
Fischer B and Klein M W 1975 Phys. Rev. B11, 2025
Fischer K H 1975 Phys. Rev. Lett. 34, 1438
Fischer K H 1976 Solid State Commun. 18, 1515
Fischer K H 1977 Physica 86-88B, 813
Fischer K H 1979 Z. Phys. B34, 45
Fischer K H 1981a Z. Phys. B42, 27
Fischer K H 1981b Z. Phys. B42, 245
Fischer K H 1981c Z. Phys. B43, 291
Fischer K H 1983 Phys. Stat. Sol. 116b, 357
Fischer K H 1983a Solid State Commun. 46, 309
Fischer K H 1983b J. Phys. Soc. Jap. Suppl. 52S, 235
Fischer K H 1983c Z. Phys. B50, 107
Fischer K H 1985a Phys. Stat. Sol. 130b, 13
Fischer K H 1985a Z. Phys. B60, 151
Fischer K H and Kinzel W 1984 J. Phys. C17, 4479
Fischer K H and Zippelius A 1985 J. Phys. C18, L1139
Fisher D S and Sompolinsky H 1985 Phys. Rev. Lett. 54, 1063
Fisher D S and Huse D A 1986 Phys. Rev. Lett. 56, 1601
Fisher M E 1982 in: Lec.Notes in Phys. vol. 186 (Springer-Verlag)

Fisher M E and Selke W 1980 Phys. Rev. Lett. 44, 1502
Fogle W E, Boyer J D, Phillips N E and Curen J V 1981 Phys. Rev. Lett. 47, 352
Fogle W E, Boyer J D, Fischer R A and Phillips N E 1983 Phys. Rev. Lett. 50, 1815
Ford P J 1982 Contemp. Phys. 23, 141
Ford P J and Mydosh J A 1976 Phys. Rev. B14, 2057
Forster D 1975 Hydrodynamic Fluctuations, Broken Symmetry, and Correlation Functions (Benjamin)
Forti M C, Kishore R and daCunha Lima I C 1982 Phys. Lett. 89A, 96
Fradkin E, Huberman B A and Shenker S H 1978 Phys. Rev. B18, 4789
Fu Y and Anderson P W 1986 J. Phys. A19, 1605
Fujiki S and Katsura S 1981 Prog. Th. Phys. 65, 1130
Furdyna J K 1982 J. Appl. Phys. 53, 7637
Gabay M and Toulouse G 1981 Phys. Rev. Lett. 47, 201
Gabay M, Garel T and DeDominicis C 1982 J. Phys. C15, 7165
Gaines J R and Sokol P E 1983 in: Quantum fluids and solids-1983 ed. Adams E D and Ihas G G, American Institute of Physics
Galazka R R, Nagata S and Keesom P H 1980 Phys. Rev. B22, 3344
Garcia J, Rojo J A, Navarro R, Bartolome J and Gonzalez D 1983 J. Magn. Mag. Mat. 31-34, 1401
Gardner E and Derrida B 1985 J. Stat. Phys. 39, 367
Gardner E 1984 J. de Phys. (Paris) 45, 1755
Gardner E 1985 Nucl. Phys. B257 [FS14] 747
Garel T 1980 J. Phys. C13, 4385
Garland C W, Kwiecien J Z and Damien J C 1982 Phys. Rev. B25, 5818
Gary M R and Johnson D S 1979 Computers and Intractability: A guide to the theory of NP-completeness, Freeman, San Francisco
Gaunt D S and Guttman A J 1974 in: Phase transitions and critical phenomena, ed. Domb C and Green M S
Gautier F 1982 in: Magnetism in metals and alloys ed. M. Cyrot (North Holland)
Gawlinski E T, Grant M, Gunton J D and Kaski K 1985 Phys. Rev.

B31, 281
Geerken B M and Nieuwenhuys G J 1982 Physica 115b, 5
Gelperin A, Hopfield J J and Tank D W 1986 in: Model Neural Networks and Behaviour, ed: Selverston A I (Plenum)
Georges A, Hansel D and Doussal P L 1985 J.de Phys.(Paris) 46, 1309
Geschwind S, Devlin G E, Dillon J F, Batlogg B and Maletta H 1984 J. Appl. Phys. 55, 1676
Ghatak S K 1976 Phys. Lett. 58A, 279
Ghatak S K 1979 Phys. Lett. 74A, 135
Ghatak S K and Moorjani 1976 J. Phys. C9, L923
Ghatak S K and Sherrington D 1977 J. Phys. C10, 3149
Ghazali A and Diep H T 1985 J. Appl. Phys. 57, 3427
Gibb T C 1976 Principles of Mossbaue spectroscopy, Chapman and Hall, London
Giebultowicz T M, Rhyne J J, Ching W Y and Huber D L 1985 J. Appl. Phys. 57, 3415
Gignoux D, Givord D and Lienard A 1982 J. Appl. Phys. 53, 2321
Ginzberg S L 1981a Sov. Phys. JETP 53, 124
Ginzberg S L 1981b Sov. Phys. JETP 54, 737
Ginsberg S L 1983 Sov. Phys. JETP 57, 225
Giri M R and Stephen M J 1978 J. Phys. C11, L541
Goetze W and Sjoegren L 1984 J. Phys. C17, 5597
Goldbart P M 1985 J. Phys. C18, 2183
Goldbart P and Elderfield D 1985 J. Phys. C18, L229
Goldberg S M and Levy P M 1986 Phys. Rev. B33, 291
Goldberg S M, Levy P M and Fert A 1986 Phys. Rev. B33, 276
Goldfarb R B and Patton C E 1981 Phys. Rev. B24, 1360
Goldfarb R B, Fickett F R, Rao K V and Chen H S 1982 J. Appl. Phys. 53, 7687
Goldfarb R B, Rao K V, Chen H S and Patton C E 1982 J. Appl. Phys. 53, 2217
Goldfarb R B, Rao K V and Chen H S 1985 Solid State Comm. 54, 799
Goldman A M and Wolf S A 1984 "Percolation, Localization and Superconductivity", Plenum, New York
Goldscmidt Y Y 1983 Nucl. Phys. B225, 123

Goldschmidt Y Y 1984 Phys. Rev. B30, 1632
Goltsev A V 1984a J. Phys. C17 L241
Goltsev A V 1984b J. Phys. A17, 237
Goltsev A V 1986 J.Phys. C19, L215
Gotaas J A, Rhyne J J and Werner S A 1985 J. Appl. Phys. 57, 3404
Goto T and Kanomata T 1985 J. Appl. Phys. 57, 3450
Green J E 1985 J. Phys. A17, L43
Green J E, Bray A J and Moore M A 1983 J. Phys. A15, 2307
Green J E, Moore M A and Bray A J 1982 J. Phys. C16, L815
Grest G S, Soukoulis C M and Levin K 1984 J. Appl. Phys. 55, 1634
Grest G S and Soukoulis C M 1983 Phys. Rev. B28, 2886
Grest G S, Soukoulis C M and Levin K 1985 Preprint
Griffith G, Volkening F A and Klaus H 1985 J. Appl. Phys. 57, 3392
Grinstein G 1985 in: Fundamental Problems in Statistical Mechanics VI, ed. Cohen E G D
Grinstein G and Luther A H 1976 Phys. Rev. B13, 1329
Gross D and Mezard M 1984 Nucl. Phys. B240 [FS] 431
Gross D J, Kanter I and Sompolinsky H 1985 Phys. Rev. Lett. 55, 304
Grossmann S, Wegner F and Hoffmann K H 1985 J. de Phys. Lett. (Paris)46, L575
Grover A K, Gupta L C, Vijayaraghavan R, Matsumura M, Nakano M and Asayama K 1979
Gruenewald M, Pohlmann B, Schweitzer L and Wuertz D 1982, J. Phys. C15, L1153;
Gruenewald M, Pohlmann B, Schweitzer L and Wuertz D 1983 J. Non. Cryst. Sol. 59 & 60, 77
Grzonka R B and Moore M A 1984 J. Phys. C17, 2785
Gubser D U, Francavilla T L, Wolf A and Leibowitz J R 1980 "Inhomogeneous Superconductors" (AIP, New York)
Gulacsi M, Gulacsi Zs and Crisan M 1983 J.Low Temp. Phys. 50, 371
Gulacsi M and Gulacsi Zs 1986 Phys. Rev. B33, 3483
Gullikson E M and Schultz S 1982 Phys. Rev. Lett. 49, 238
Gullikson E M, Fredkin D R and Schultz S 1983 Phys. Rev. Lett. 50, 537

Gullikson E M, Dalichaouch R and Schultz S 1985 Phys. Rev. B32, 507
Guy C N 1975 J. Phys. F5, L242
Guy C N 1977 J. Phys. F7, 1505
Guy C N 1978 J. Phys. F8, 1309
Guy C N 1982 J. Phys. F12, 1453
Guy C N and Park J G 1983 J. Phys. F13, 1955
Gyorgy E M and Walker L R 1985 J. Appl. Phys. 57, 3395
Guyot M, Foner S, Hasanain S K, Guertin R P and Westerholt K 1980 Phys.Lett. 79a, 339
Haase D G and Perrel L R 1983 in: Quantum Fluids and Solids 1983 ed. Adamas E D and Ihas G G, American Inst. Phys.
Halperin B I and Hohenberg P C 1969 Phys. Rev. 188, 898
Halperin B I and Saslow W M 1977 Phys. Rev. B16, 2154
Halsey T C 1985 Phys. Rev. Lett. 55, 1018
Harris A B, Lubensky T C and Chen J H 1976 Phys. Rev. Lett 36, 415
Harris A B, Washburn S and Meyer H 1983 J. Low Temp. Phys. 50, 151
Harris A B and Meyer H 1985 Can. J. Phys. 63, 3
Harris A B, Caflisch R G and Banavar J R 1986 Preprint
Harris R 1983 Solid State Commun. 45, 711
Harris R and Zobin D 1977 J. Phys. F7, 337
Harris W F 1977 Sci. Am. 237, (Dec.) 130
Hasanain S K, Guertin R P, Westerholt K, Guyot M and Foner S 1984 Phys. Rev. B24, 5165
Hauser J J and Hsu F S L 1981 Phys. Rev. B24, 1550
Hauser J J and Felder R J 1983 Phys. Rev. B27, 6999
Hauser J J and Waszczak J V 1984 Phys. Rev. B30, 5167
Hauser J J, Felder R J and Blitzer L D 1986 Solid State Commun. 57, 881
Haushalter R C, O'Connor C J, Umarji A M, Shenoy G K and Saw C K 1984 Solid State Commun. 49, 929
Hawkins G F and Thomas R L 1978 J. Appl. Phys. 49, 1627
Hawkins G F, Thomas R L and De Graaf A M 1979 J. Appl. Phys.50, 1709

Hayase S, Futamura T, Sakashita H and Terauchi H 1985 J. Phys. Soc Jap. 54, 812

Hayes T M, Allen J W, Boyce J B and Hauser J J 1980 Phys. Rev. B22, 4503

Hebb D O 1949 "The Organization of Behaviour" (Wiley, New York)

Heffner R H, Leon M, Schillaci M E, MacLaughlin D E and Dodds S A 1982 J. Appl. Phys. 53, 2174

Heffner R H, Leon M and MacLaughlin D E 1984 Hyp. Int. 17-19, 457

Heffner R H and Fleming D G 1984 Physics Today, December, p. 38

Held C and Klein M W 1975 Phys. Rev. Lett. 35, 170

Henley, C L, Sompolinsky H and Halperin B I 1982 Phys. Rev. B25, 5849

Henley C L 1983 Ph.D. Thesis, Harvard University (Unpublished)

Henley C L 1984a Ann. Phys. 156, 324

Henley C L 1984b Ann. Phys. 156, 368

Henley C L 1985 Phys. Rev. Lett. 55, 1653

Hennion B, Hennion M, Hippert F and Murani A P 1983 Phys. Rev. B28, 5365

Hennion B, Hennion M, Hippert F and Murani A P 1984 J. Appl. Phys. 55, 1694

Herlach D M, Wassermann E F and Willnecker 1983a Phys. Rev. Lett. 50, 529

Herlach D M, Wassermann E F and Willnecker 1983b, J. Magn. Mag. Mat. 31-34, 1404

Hertz J A 1978 Phys. Rev. B18, 4875

Hertz J A 1979 Phys. Rev. B19, 4796

Hertz J A 1980 in: Proc. Taniguchi Int. Symposium on Electron Correlation in Narrow Bands, Springer

Hertz J A 1983a J. Phys. C16, 1219

Hertz J A 1983b J. Phys. C16, 1233

Hertz J A 1983c Phys. Lett. Lett. 51, 1880

Hertz J A 1985 J. Appl. Phys. 57, 3366

Hertz J A and Klemm R A 1978 Phys. Rev. Lett. 40, 1397

Hertz J A and Klemm R A 1979 Phys. Rev. B20, 316

Hertz J A, Fleishman L, Anderson P W 1979 Phys. Rev. Lett. 43, 942

Hertz J A, Khurana A and Klemm R A 1981 Phys. Rev. Lett. 46, 496
Hertz J A and Klemm R A 1983a Phys.Rev. B28, 2877
Hertz J A and Klemm R A 1983b Phys.Rev. B28, 3849
Hippert F and Alloul H 1982 J. de Phys. (Paris) 43, 691
Hiroyoshi H and Fukamichi K 1981 Phys. Lett. 85A, 242
Hiroyoshi H and Fukamichi K 1982 J. Appl. Phys. 53, 2226
Ho S C, Maartense I and Williams G 1981a J. Phys. F11, 699
Ho S C, Maartense I and Williams G 1981b J. Phys. F11, 1107
Ho S C, Maartense I and Williams G 1981c Phys. Rev. B24, 5174
Ho S C, Maartense I and Williams G 1982 J. Appl. Phys. 53, 2235
Hoechli U T 1982 Phys. Rev. Lett. 48, 1494
Hoekstra F R, Niewenhuys, Baberschke and Barnes S E 1984 Phys. Rev. B29, 1292
Hoever P and Zittartz J 1981 Z. Phys. B44, 129
Hoever P, Wolff W F and Zittartz J 1981 Z. Phys. B41, 43
Hohenberg P C and Halperin B I 1977 Rev. Mod. Phys. 49, 435
Holtzberg F, Tholence J L, Godfrin H and Tournier R 1979 J. Appl. Phys. 50, 1717
Holtzberg F, Francavilla T L, Huang C Y and Tholence J L 1982 J. Appl. Phys. 53, 2229
Honda K and Nakano H 1981 Prog. Th. Phys. 65, 95
Hoogerbeets R, Luo W L and Orbach R 1985 Phys. Rev. Lett. 55, 111
Hopfield J J 1982 Proc. Natl. Acad. Sci. USA 79, 2554
Hopfield J J 1984 Proc. Natl. Acad. Sci. USA. 81, 3088
Hopfield J J 1986 in: The Lesson of Quantum Theory, ed: de Boer D, Dahl E and Ulfbeck O
Hopfield.J J, Feinstein D I and Palmer R G 1983 Nature, 304, 158
Hopfield J J and Tank D W 1985 Biol. Cybern. 52, 141
Hopfield J J, Tank D W 1986a in: Disordered Systems and Biological Organization (Springer)
Hopfield J J and Tank D W 1986b Science
Horner H 1984 Z. Phys. B57, 29
Horner H 1984 Z. Phys. B57, 39
Hornreich R M and Schuster H G 1982 Phys. Rev. B26, 3929
Hou M K, salamon M B and Ziman T A L 1984 J. Appl. Phys. 55, 1723
Hou M K, Salamon M B and Pechan M J 1985 J. Appl. Phys. 57, 3482

Houghton A, Jain S and Young A P 1983a J. Phys. C16, L375
Houghton A, Jain S and Young A P 1983b Phys. Rev. B28, 2630
Huang C Y 1985 J. Magn. Mag. Mat. 51, 1
Huber D L 1985 Phys. Rev. B31, 4420
Huber D L and Ching W Y 1980 J. Phys. C13, 5579
Huber D L, Ching W Y and Fibich M 1979 J. Phys. C12, 3535
Huberman B A and Kerszberg M 1985 J. Phys. A18, L331
Hudak O 1983 J. Phys. C16, 5203
Hurd C M 1982 Contemp. Phys. 23, 469
Huse D A and Morgenstern I 1985 Phys.Rev. B32, 3032
Huser D, Wenger L E, van Duyneveldt A J 1983a Phys. Rev. B27, 3100
Huser D, Rewiersma M J F M, Mydosh J A and Nieuwenhuys G J 1983b Phys. Rev. Lett. 51, 1290
Ihm J 1985 Phys. Rev. B31, 1674
Iida S and Terauchi 1983 J. Phys. 4044
Imry Y 1984 J. Stat. Phys. 34, 849
Indekeu J O, Smedt P de and Dekeyser R 1984 Phys. Rev. B30, 495
Ioffe L B 1983 Sov. Phys. JETP 57, 220
Ioffe L B and Feigelman M V 1983 J. de Phys. Lett. (Paris) 44, L971
Ioffe L B and Feigelman M V 1983 Sov. Phys. JETP 58, 1047
Ishibashi Y and Suzuki I 1985 J. Phys. Soc. Jap. 54, 1443
Ishii H and Yamamoto T 1985 J. Phys. C18, 6225
Ishikawa Y, Arai M, Saito N, Kohgi M and Takei H 1983 J. Magn. Mag. Mat. 31-34, 1381
Ishikawa M 1982 Contemp. Phys. 23, 443
Ishikawa Y, Saito N, Arai M and Watanabe Y 1985 J. Phys. Soc. Jap. 54, 312
Ishimoto H, Nagamine K and Kumura Y 1976 J. Phys. Soc. Jap. 40, 312
Izuyama T 1980 Prog. Th. Phys. Suppl. 69, 69
Jackiw R 1980 Rev. Mod. Phys. 52, 661
Jaeckle J 1984 Physica 127B, 79
Jagannathan A, Schaub B and Kosterlitz J M 1985 Preprint
Jaggi N K 1980 J. Phys. C13, L623

Jain S and Young A P 1986 Preprint
Jamet J P and Malozemoff A P 1978 Phys.Rev. B18, 75
Jamet J P, Dumais J C, Seiden T and Knorr K 1980 J. Magn. Mag. Mat. 15-18, 197
Jayaprakash C and Kirkpatrick S 1979 Phys. Rev. B21, 4072
John S and Lubensky T C 1985 Phys. Rev. Lett. 55, 1014
John S and Lubensky T C 1986 Preprint
Johnston R and Sherrington D 1982 J. Phys. C15, 3757
Jones T E, Kwak J F, Chock E P and Chaikin P M 1978 Solid State Commun. 27, 209
Jonsson T 1982 Phys. Lett. 91A, 185
Kalos M 1985 Preprint
Kaneyoshi T 1975 J. Phys. F5, 1014
Kaneyoshi T 1979 Phys. Lett. 71A, 287
Kanter I and Sompolinsky H 1986 Phys. Rev. B33, 2073
Kaplan T A 1981 Phys. Rev. B24, 319
Kaplan T A 1986 Phys. Rev. B33, 2848
Karder M and Berker A N 1982 Phys. Rev. B26, 219
Karlsson E 1982 Phys. Rep. 82, 271
Kasuya T 1956 Prog. Th. Phys. 16, 45, 58
Katori M and Suzuki M 1985 Prog. Th. Phys. 74, 1175
Katsumata K, Nire T, Tanimoto M and Yoshizawa H 1982a Phys. Rev. B25, 428
Katsumata K, Nire T and Tanimoto M 1982b Solid State Commun. 43,711
Katsura S 1976a J. Phys. C9, L619
Katsura S 1976b Prog. Th. Phys. 55, 1049
Kawamura H and Tanemura M 1985 J. Phys. Soc. Jap. 54, 4479
Ketelsen L J P and Salamon M B 1984 Phys. Rev. B30, 2718
Ketelsen L J P and Salamon M B 1986 Phys. Rev. B33, 3610
Khurana A 1982 Phys.Rev. B25, 452
Khurana A and Hertz J A 1980 J. Phys. C13, 2715
Khurana A, Jagannathan A and Kosterlitz J M 1984 Nucl.Phys. B240 1
Kinzel W 1979 Phys. Rev. B19, 4595
Kinzel W 1982a Phys.Rev. B26, 6303

Kinzel W 1982b Z. Phys. B46, 59
Kinzel W 1984 in: Lec. Notes in Phys. vol. 206
Kinzel W 1985 Z. Phys.B60, 205, Erratum 1986 Z. Phys. B62, 267
Kinzel W and Fischer K H 1977a J. Phys. F7, 2163
Kinzel W and Fischer K H 1977b Solid State Commun. 23, 687
Kinzel W and Fischer K H 1978 J. Phys. C11, 2115
Kinzel W and Binder K 1983 Phys. Rev. Lett. 50, 1509
Kinzel W and Binder K 1984 Phys. Rev. B29, 1300
Kirkpatrick S 1977a Phys. Rev. B15, 1533
Kirkpatrick S 1977b Phys. Rev. B16, 4630
Kirkpatrick S 1979 in:Ill Condensed Matter, ed.Balian R, Maynard R and Toulouse G (North-Holland)
Kirkpatrick S 1984 J. Stat. Phys. 34, 975
Kirkpatrick S and Sherrington D 1978 Phys. Rev. B17, 4384
Kirkpatrick S and Young A P 1981 J. Appl. Phys.52, 1712
Kirkpatrick S, Gelatt C D and Vecchi M P 1983 Science 220, 671
Kirkpatrick S and Toulouse G 1985 J. de Phys. (Paris) 46, 1277
Klein M W 1964 Phys.Rev. 136, A1156
Klein M W 1969 Phys.Rev. 188, 933
Klein M W 1976 Phys.Rev. B14, 5008
Klein M W and Brout R 1963 Phys. Rev. 132, 2412
Klein M W and Shen L 1972 Phys. Rev. B5, 1174
Klein M W, Schowalter L J and Shukla P 1979 Phys. Rev. B19, 1492
Kleman M 1985 J. de Phys. Lett. (Paris) 46, L723
Klemm R A 1979 J. Phys. C12, L735
Klenin M A 1979 Phys. Rev. Lett. 42, 1549
Klenin M A 1982 Phys. Rev. B26, 3969
Klenin M A 1983a Phys. Rev. B28, 5199
Klenin M A 1983b in: Quantum fluids and solids - 1983 Ed. Adams E D and Ihas G G, American Institute of Physics
Klenin M A and Pate S F 1981 Physica 107B, 185
Kogut J B 1979 Rev. Mod. Phys. 51, 659
Kogut J B 1984 J. Stat. Phys. 34, 941
Kohonen T 1980 Content addressable memories (Springer)
Kokshenev V B 1985a Solid State Commun. 55, 143
Kokshenev V B 1985b Sov. J. Low Temp. Phys. 11, 881

Kondor I 1983 J.Phys. A16, L127
Kondor I 1985 in: Festkorperprobleme 25
Kondor I and DeDominicis C 1983 J. Phys. A16, L73
Kosterlitz J M, Thouless D J and Jones R C 1976 Phys. Rev. Lett. 36, 1217
Kotliar G, Anderson P W and Stein D L 1983, Phys.Rev. B27, 602
Kotliar G and Sompolinsky H 1984 Phys. Rev. Lett. 53, 1751
Krey U 1980 Z. Phys. B38, 243
Kery U 1981 Z. Phys. B42, 231
Krey U 1982 J. Magn. Mag. Mat. 28, 231
Krey U 1983 in: Lec. Notes in Phys. vol. 192
Krey U 1985 J. de Phys. Lett. (Paris) 46, L845
Krusin-Elbaum D, Malozemoff A P and Taylor R C 1983 Phys.Rev.B27, 562
Kudo T, Egami T and Rao K V 1982 J. Appl. Phys. 53, 2214
Kulkarni R G and Baldha G J 1985 Solid State Comm. 53, 1001
Kumar A A, Saslow W M and Henley C L 1986 Phys. Rev. B33, 305
Kumar A A and Saslow W M 1986 Phys. Rev. B33, 313
Kumar D and Barma M 1978 Phys. Lett. 67, 217
Kumar D and Stein J 1980 J. Phys. C13, 3011
Kumar D 1986 Private Communication
Kurtz W 1982 Solid State Commun. 42, 871
Kurtz W, Geller R, Dachs H and Convert P 1976 Solid State Commun. 18, 1479
Kurtz W and Roth S 1977 Physica 86-88B, 715
Lage E J S and deAlmeida J R L 1982 J. Phys. C15, L1187
Lage E J S and Erzan A 1983 J. Phys. C16, L873
Lallemand P, Diep H T, Ghazali A and Toulouse G 1985 J. de Phys. Lett. (Paris) 46, L1087
Lander G H and Price D L 1985 Phys. Today 38, no.1, 38
Lange R V 1965 Phys. Rev. Lett. 14, 3
Lange R V 1966 Phys. Rev. 146, 301
Larkin A I and Khmelnitskii D E 1970 Sov. Phys. JETP 31, 958
Larsen U 1976 Phys. Rev. B14, 4356
Larsen U 1977 Solid State Commun. 22, 311
Larsen U 1981 Phys. Lett. A85, 471

Larsen U 1983 Phys. Lett. A97, 147
Larsen U 1984 Phys. Lett. 105A, 307
Larsen U 1985 J. Phys. F15, 101
Larsen 1986 Phys. Rev. B33, 4803
Lauer J and Keune W 1982 Phys. Rev. Lett. 48, 1850
Lecomte G V and Loehneysen H V 1983 Z. Phys. B50, 239
Lecomte G V, Loehneysen H V, Baufhofer W and Guentherodt G 1984 Solild State Comm. 52, 535
Le Dang K, Mery M C and Veillet P 1984 J. Magn. Mag. Mat. 43, 161
Levin K, Soukoulis C M and Grest G S 1979 J. Appl. Phys. 50, 1695
Levin K, Soukoulis C M and Grest G S 1980 Phys. Rev. B22, 3500
Levitt D A and Walstedt R E 1977 Phys. Rev. Lett. 38, 178
Levy P M and Fert A 1981a Phys. Rev. B23, 4667
Levy P M and Fert A 1981b J. Appl. Phys. 52, 1718
Levy P M, Morgan Pond C M and Fert A 1982 J. Appl. Phys. 53, 2168
Levy P M, Morgan-Pond C G and Raghavan R 1983 Phys. Rev. Lett. 50, 1160
Levy P M, Morgan-Pond C G and Raghavan R 1984 Phys. Rev. B30, 2358
Levy P M and Zhang Q 1986 Phys. Rev. B33, 665
Litterst F J, Fried J M, Tholence J L and Holtzberg F 1982 J. Phys. C15, 1049
Little W A 1974 Math. Biosc. 19, 101
Liu S H 1967 Phys. Rev. 157, 411
Loehneysen H V, Tholence J L and Tournier R 1978 J.de Phys.Coll (Paris) 39, C6-922
Loehneysen H V and Tholence J L 1980 J. Magn. Mag. Mat. 15-18, 171
Loehneysen H V, van den Berg R and Lecomte G V 1985 Phys. Rev. B31, 2920
Loidl A, Feile R, Knorr K, Renker B, Daubert J, Durand D and Suck J B 1980 Z. Phys. B38, 253
Loidl A, Feile R and Knorr K 1982 Phys. Rev. Lett. 48, 1263
Loidl A, Knorr K, Feile R and Kjems J K 1983 Phys. Rev. Lett. 51, 1054
Loidl A, Feile R, Knorr K and Kjems J K 1984 Phys. Rev. B29, 6052

Loidl A, Muller M, McIntyre G F, Knorr K and Jex H 1985 Solid State Comm. 54, 367

Loidl A, Schraeder T, Boehmer R, Knorr K, Kjems J K, Born R 1986 Preprint

Lubensky T C 1979 in: Ill Condensed Matter, ed: Balian R, Maynard R and Toulouse G (North-Holland)

Lundgren L, Svedlindh P and Beckman O 1981 J. Magn. Mag. Mat. 25, 33

Lundgren L, Svedlindh P and Beckman O 1982a J. Phys F12, 2663

Lundgren L, Svedlindh P and Beckman O 1982b Phys. Rev. B26, 3990

Lundgren L, Svedlindh P and Beckman O 1983a Phys. Rev. Lett. 51, 911

Lundgren L, Svedlindh P and Beckman O 1983b J. Magn. Mag. Mat. 31-34, 1349

Lundgren L, Nordblad P, Svedlindh P and Beckman O 1985 J. Appl. Phys. 57, 3371

Lundgren L, Nordblad P and Sandlund L 1986 Europhys. Lett. 1, 529

Luttinger J M 1976 Phys. Rev. Lett. 37, 778

Luttinger J M Phys Rev. B26, 3990

Luttinger J M 1983a, Phys. Rev. Lett. 51, 911

Luttinger J M 1983b, J. Magn. Mag. Mat. 31-34, 1349

Luty F and Ortiz-Lopez J 1983 Phys. Rev. Lett. 50, 1289

Lynn J W, Erwin R W, Rhyne J J and Chen H S 1981 J. Appl. Phys. 52, 1738

Ma S K 1980 Phys. Rev. B22, 4484

Ma S K 1981 J. Stat. Phys. 26, 221

Ma S K and Rudnick J 1978 Phys. Rev. Lett. 40, 589

Ma S K and Payne M 1981 Phys. Rev. B24, 3984

Mackenzie N D and Young A P 1982 Phys. Rev. Lett. 49, 301

MacLaughlin D E and Alloul H 1976 Phys. Rev. Lett. 36, 1158

MacLaughlin D E and Alloul H 1977 Phys. Rev. Lett. 38, 181

Majumdar A K 1983 Phys. Rev. B28, 2750

Majumdar A K, Oestereih V and Weschenfelder D 1983 Solid State Commun 45, 907

Majumdar A K and Blanckenhagen P v 1984 Phys. Rev. B29, 4079

Majumdar A K and Oestreich V 1984 Phys. Rev. B30, 5342
Maletta H 1983 in: Lec.Notes in Phys. vol.192 (Springer-Verlag)
Maletta H and Felsch W 1979 Phys. Rev. B20, 1245
Maletta H and Convert P 1979 Phys. Rev. Lett. 42, 108
Maletta H, Zinn W, Scheuer H and Shapiro S M 1981 J. Appl. Phys. 52, 1735
Maletta H, Felsch W and Tholence J L 1978 J. Magn. Mag. Mat. 9, 41
Malozemoff A P and Jamet J P 1977 Phys. Rev. Lett. 39, 1293
Malozemoff A P and Hart S C 1980 Phys. Rev. B21, 29
Malozemoff A P, Barbara B and Imry Y 1982a, J. Appl. Phys.53, 2205
Malozemoff A P, Imry Y and Barbara B 1982b J. Appl. Phys. 53, 7672
Malozemoff A P, Barnes S E and Barbara B 1983 Phys. Rev. Lett. 51, 1704
Mandelbrot B B "The Fractal Geometry of Nature", (Freeman)
Manheimer M A, Bhagat S M and Webb D J 1985 J. Appl. Phys. 57, 3476
Marcenat C. Benoit A, Briggs A, Arzoumanian C, deGoer A M and Holtzberg F 1985 J. de Phys. Lett. (Paris) 46, L569
Maritan A and Stella A 1986 J.Phys.A19, L269
Marshall W 1960 Phys. Rev. 118, 1519
Marshall W and Lovesey S W 1971 Theory of Neutron Scattering, (Oxford University Press)
Martin D L 1979 Phys. Rev. B20. 368
Martin P C, Siggia E and Rose H 1973 Phys. Rev. A8, 423
Matsushita E and Matsubara T 1985 J. Phys. Soc. Jap. 54, 1161
Matsubara F and Sakata M Prog. Th. Phys. 55, 1049
Matsubara F and Katsumata K 1984 Solid State Comm. 49, 1165
Matsui M, Malozemoff A P, Gambino R J and Krusin-Elbaum L 1985 J. Appl. Phys. 57, 3389
Mattis D C 1976 Phys. Lett. 56A, 421
Mattis D C 1981 The Theory of Magnetism I, Springer Verlag
Mattis D C 1985 The Theory of Magnetism II, Springer Verlag
Mazumdar P, Bhagat S M, Manheimer M A and Chen H S 1984

J. Appl. Phys. 55, 1685
Mazumdar P, Bhagat S M and Manheimer M A 1985 J. Appl. Phys. 57, 3479
McCulloch W S and Pitts W A 1943 Bull. Math. Biophys. 5, 115
McKay S R, Berker A N and Kirkpatrick S 1982a Phys. Rev. Lett. 48, 767
McKay S R, Berker A N and Kirkpatrick S 1982b J. Appl. Phys. 53, 7974
McKay S R and Berker A N 1984 J. Appl. Phys. 55, 1646
McMillan W L 1983 Phys. Rev. B28, 5216
McMillan W L 1984a Phys. Rev. B30, 476
McMillan W L 1984b Phys. Rev. B29, 4026
Medina R, Fernandez J F and Sherrington D 1980 Phys. Rev. B21, 2915
Medvedev M V 1979a Phys. Stat. Solidi 91b, 713
Medvedev M V 1979b Fiz. Tverd. Tela 21, 3356
Meisel M W, Zhou W S, Owers-Bradley J R, Ociai Y, Brittain J O and Halperin W P 1982 J. Phys. F12, 317
Meissner M, Knaak W, Sethna J P, Chow K S, De Yoreo J J and Pohl R O 1985 Phys. Rev. B32, 6091
Mermin N D 1979 Rev. Mod. Phys. 51, 591
Mertz B and Loidl A 1985 J. Phys. C18, 2843
Mery M C, Veillet P and Le Dang K 1985 Phys. Rev. B31, 2656
Meschde D, Steglich F, Felsch W. Maletta H and Zinn W 1980 Phys. Rev. Lett. 44, 102
Meyer H and Washburn S 1984 J. Low Temp. Phys. 57, 31
Mezard M, Parisi G, Sourlas N, Toulouse G and Virasoro M 1984 Phys. Rev. Lett. 52, 1156
Mezard M, Parisi G, Sourlas N, Toulouse G and Vilrasoro M l984 J. de Phys. (Paris) 45, 843
Mezard M and Virasoro M A 1985 J. de Phys.(Paris) 46, 1293
Mezard M and Parisi G 1985 J. de Phys. Lett. (Paris) 46, L771
Mezard M, Parisi G and Virasoro M A 1985 J. de Phys. Lett.(Paris) 46, L217
Mezard M, Parisi G and Virasoro M A 1986 Europhys. Lett. 1, 77
Mezei F 1980 in: Lec. Notes in Phys. vol. 128 (Springer Verlag)

Mezei F 1982 J. Appl. Phys. 53, 7654
Mezei F 1983 Physica 120B, 51
Mezei F and Murani A P 1979 J. Magn. Mag. Mat. 14, 211
Mezei F, Murani A P and Tholence J L 1983 Solid State Commun. 45, 411
Michel K H and Rowe J M 1980 Phys. Rev. B22, 1417
Millman S E and Zimmerman G O 1983 J. Phys. C16, L89
Miyako Y, Chikazawa S, Sato T and Saito T 1980 J. Magn. Mag. Mat. 15-18, 139
Miyako Y, Chikazawa S, Saito T and Yuochunus Y G 1981 J. Appl. Phys. 52, 1779
Miyako Y, Nishioka T, Sato T, Takeda Y, Morimoto S and Ito A 1986 J. Magn. Mag. Mat. 54-57, 149
Miyazaki T, Ando Y and Takahashi M 1985 J. Appl. Phys. 57, 3456
Mody A and Rangwala A A 1981 Physica 106B, 68
Monod P, Prejean J J and Tissier B 1979 J. Appl. Phys. 50, 7324
Monod P and Berthier Y 1980 J. Mag. Magn. Mat. 15-18, 149
Monod P and Bouchiat H 1982 J. de Phys.Lett. (Paris) 43, L45
Mookerjee A 1978 Pramana 11, 223
Mookerjee A 1980 Pramana 14, 11
Mookerjee A 1980a J. Phys. F10, 1559
Mookerjee A and Chowdhury D 1983a J. Phys.F13, 365
Mookerjee A and Chowdhury D 1983b J. Phys.F13, 413
Moore M A 1986 J.Phys. A19, L211
Moore M A and Bray A J 1982 J.Phys. C15, L301
Moore M A and Bray A J 1985 J. Phys. C18, L699
Moorjani K and Coey J M D 1984 Magnetic Glasses (Elsevier)
Moorjani K, Poehler T O and Satkiewicz F G 1985 J. Appl. Phys. 57, 3444
Moran T J and Thomas R L 1973 Phys. Lett. 45A, 413
Morgan Pond C G 1982 Phys. Lett. A92, 461
Morgan Pond C G 1983 Phys. Rev. Lett. 51, 490
Morgenstern I 1983a in: Lec. Notes in Phys. vol. 192
Morgenstern I 1983b Phys. Rev. B27, 4522
Morgenstern I and Binder K 1979 Phys. Rev. Lett. 43, 1615
Morgenstern I and Binder K 1980a, Phys Rev B22, 288

Morgenstern I and Binder K 1980b Z. Phys. B39, 227
Morgenstern I, Binder K and Hornreich R M 1981 Phys. Rev. B23, 287
Morgenstern I and van Hemmen J L 1984 Phys. Rev. B30, 2934
Morgownick A F J and Mydosh J A 1981a Phys. Rev. B24, 5277
Morgownick A F J and Mydosh J A 1981b Physica 107B, 305
Morgownick A F J, Mydosh J A and Wenger L E 1982 J. Appl. Phys. 53, 2211
Morgownick A F J and Mydosh J A 1983 Solid State Commun. 47, 325
Morgownick A F J, Mydosh J A and Foiles C L 1984 Phys. Rev. B29, 4144
Moriyasu K 1983 An elementary primer for gauge theory, World Scientific (Singapore)
Morris B W 1985 Phys. Lett. 110A, 415
Morris B W and Bray A J 1984 J.Phys. C17, 1717
Morris B W, Colborne S G, Moore M A, Bray A J and Canisius J 1986 J. Phys. C19, 1157
Morrish A H 1965 "The Physical Principles of Magnetism" (Wiley), p.360
Mottishaw P J 1986 Europhys. Lett. 1, 409
Mottishaw P J and Sherrington D 1985 J. Phys. C18, 5201
Moy D, Dobbs J N and Anderson A C 1984 Phys. Rev. B29, 2160
Mozurkewich G, Elliot J H, Hardiman M and Orbach R 1984 Phys. Rev. B29, 278
Mulder C A M, van Duyneveldt A J and Mydosh J A 1981a Phys. Rev. B23, 1384
Mulder C A M, van Dyneveldt A J, van der Linden H W M, Verbeek B H, van Dongen J C M, Niewenhuys G J and Mydosh J A 1981b, Phys. Lett. 83A, 74
Mulder C A M, van Duyeneveldt A J and Mydosh J A 1982a, Phys. Rev. B25, 515
Mulder C A M and van Dyneveldt A J 1982b Physica 113b, 123
Murani A P and Coles B R 1970 J. Phys. C3, S159
Murani A P 1976 Phys. Rev. Lett. 37, 450
Murani A P 1977 J. Magn. Mag. Mat. 5, 95
Murani A P 1978a J. Appl. Phys. 49, 1604

Murani A P 1978b J. de Phys. (Paris) 39, C6-1517
Murani A P 1978 Phys. Rev. Lett. 41, 1406
Murani A P 1980 Solid State Commun. 34, 705
Murani A P 1980a Phys. Rev. B22, 3495
Murani A P 1981 J. Magn. Mag. Mat. 25, 68
Murani A P and Coles B R 1970 J. Phys. C3, S159
Murani A P and Heidenmann A 1978 Phys. Rev. Lett. 41, 1402
Murani A P and Tholence J L 1977 Solid State Commun. 22, 25
Murani A P, Mezei F and Tholence J L 1981 Physica 108B, 1283
Murayama S, Yokosawa K, Miyako Y, Wassermann E F 1986
J. Magn. Mag. Mat.
Murnick D E, Fiory A T and Kossler W J 1976 Phys. Rev. Lett. 36, 100
Mydosh J A 1978 J. Magn. Mag. Mat. 7, 237
Mydosh J A 1980 J. Magn. Mag. Mat. 15-18, 99
Mydosh J A 1981a in: Lec.Notes in Phys. vol.149,(Springer Verlag)
Mydosh J A 1981b in: Magnetism in solids - some current topics ed. Cracknell A P and Vaughan R A p. 85
Mydosh J A 1983 in: Lec.Notes in Phys. vol 192 (Springer Verlag)
Nadal J P, Toulouse G, Changeux J P and Dehaene S 1986 Europhys. Lett. 1, 535
Naegele W 1981 Z. Phys. B42, 135
Naegele W, Knorr K, Prandl W, Convert P and Bueroz J L 1978 J. Phys. C11, 3295
Naegele W, Prandl W and Knorr K 1979a J. Magn. Mag. Mat. 13, 141
Naegele W, Blanckenhagen P v, Knorr K and Suck J B 1979b Z. Phys. B33, 251
Naegele W, Blanckenhagen P v and Heidemann A 1979c J. Magn. Mag. Mat. 13, 149
Nagata S, Galazka R R, Mullin D P, barzadeh H, Klattk G D, Furdyna J K and Keesom P H 1980 Phys. Rev. B22, 3331
Nakanishi K 1981 Phys. Rev. B23, 3514
Nambu S 1985 Prog. Th. Phys. 74, 446
Neel L 1947 Ann. Geophys. 5, 99
Nelson D R 1983 in: Phase transitions and critical phenomena eds. Domb C and Lebowitz J L (Academic Press)

Nemto K and Takayiama H 1985 J. Phys. C18, L529
Ngai K L 1979 Comments on Solid State Phys. 9, 127
Ngai K L 1980 Comments on Solid State Phys. 9, 141
Ngai K L, Rajagopal A K and Huang C Y 1984 J. Appl. Phys. 55, 1714
Nieuwenhuys G J and Mydosh J A 1977 Physica 86-88B, 880
Nieuwenhuys G J G J, Stocker H, Verbeek B H and Mydosh J A 1978 Solid State Commun. 27, 197
Nieuwenhuizen Th M 1985 Phys. Rev. B31, 7487
Nigam A K and Majumdar A K 1983 Phys.Rev. B27, 495
Nishimori N 1980 J. Phys. C13, 4071
Nishimori N 1981a Prog. Th. Phys. 66, 1169
Nishimori N 1981b J.Stat. Phys. 26, 839
Nishimori H and Stephen M J 1983 Phys. Rev. B27, 5644
Nobre F D and Sherrington D 1986 J. Phys. C19, L181
Nogues M, Saifi A, Hamedoun M, Dormann J L, Malamanche A, Fiorani D and Viticoli S 1982 J.Appl.Phys. 53, 7699
Nordblad P, Svedlindh P, Lundgren L and Sandlund L Phys. Rev. B 33, 645
Novak M A, Symko O G, Zheng D J and Oseroff S 1985 J. Appl. Phys. 57, 3418
Novak M A, Symko O G and Zheng D J 1986 Phys. Rev. B33, 343
Nozieres P 1982 J. de Phys. Lett.(Paris) 43, L543
Ocio M, Bouchiat H and Monod P 1985a J. de Phys. Lett. (Paris) 46, L647
Ocio M, Alba M and Hamman J 1985b J. de Phys. Lett (Paris) 46, L1101
Ogielski A T 1985a Phys. Rev. B32, 7384
Ogielski A T 1985bb in: Proc. the Fall meeting of the Material Research Society, Boston
Ogielski A T and Morgenstern I 1985a Phys. Rev. Lett. 54, 928
Ogielski A T and Morgenstern I 1985b J. Appl. Phys. 57, 3382
Ogielski A T and Stein D L 1985 Phys. Rev. Lett. 55, 1634
Oguchi T and Ueno Y 1977 J.Phys. Soc. Jap. 43, 764
Okuda K and Date M 1969 J. Phys. Soc. Jap. 27, 839
Omari R, Prejean J J and Souletie J 1983a J. Phys. (Paris), 44,

1069
Omari R, Prejean J J and Souletie J 1983b in: Lec. Notes in Phys. vol.192
Omari R, Prejean J J and Souletie J 1984 J. de Phys. (Paris) 45, 1809
Ono I 1976a J. Phys. Soc. Jap. 41, 345
Ono I 1976b J. Phys. Soc. Jap. 41, 2129
Orbach R 1980 J. Magn. Mag. Mat. 15-18, 706
Orland H 1985 J. de Phys. Lett. (Paris) 46, L763
Oseroff S B 1982 Phys. Rev. B25, 6584
Oseroff S B and Acker F 1980 Solid State Commun. 37, 19
Oseroff S B, Calvo R and Giriat W 1979 J. Appl. Phys. 50, 7738
Oseroff S B, Calvo R, Fisk Z and Acker F 1980a Phys. Lett. 80A, 311
Oseroff S B, Calvo R, Giriat W and Fisk Z 1980b Solid State Commun. 35, 539
Oseroff S B, Mesa M, Tovar M and Arce R 1983 Phys. Rev. B27, 566
Oseroff S and Gandra F G 1985 J. Appl. Phys.57, 3421
O'Shea M J and Sellmyer D J 1985 J. Appl. Phys. 57, 3470
Owen J C 1982 J. Phys. C15, L1071
Owen J C 1983 J. Phys. C16, 1129
Owen J, Browne M E, Arp V and Kip A F 1957 J. Phys. Chem. Sol. 2, 85
Paladin, Mezard M and DeDominicis C 1985 J. dePhys. (Paris) 46, L985
Palmer R G 1982 Adv. Phys. 31, 669
Palmer R G 1983 in: Lec. Notes in Phys. vol. 192
Palmer R G and Pond C G 1979 J. Phys. F9, 1451
Palmer R G, Stein D L, Abrahams E and Anderson P W 1984 Phys. Rev. Lett. 53, 958
Palmer R G 1985 Phys. Rev. Lett. 54, 365
Palmer R G and Bantilan F T 1985 J. Phys. C18, 171
Pelcovits R A 1979 Phys. Rev. B19, 465
Papadimitriou C H and Steiglitz K 1982 "Combinatorial Optimizaton" (Prentice Hall)
Pappa C, Hammann J and Jacoboni C 1984 J. Phys. C17, 1303

Pappa C, Hammann J, Jehanno J and Jacoboni C 1985a J. Phys. C18, 2817
Pappa C, Hammann J and Jacoboni C 1985b J. de Phys. 46, 637
Parga N, Parisi G and Virasoro M A 1984 J. de Phys. Lett. 45, L1063
Parga N and Parisi G 1986 Preprint
Parga N and Virasoro M A 1986 Preprint
Parisi G 1979a Phys. Rev. Lett. 43, 1754
Parisi G 1979b Phys. Lett. 73A, 203
Parisi G 1980a J. Phys. A13, 1101
Parisi G 1980b J. Phys.A13, L115
Parisi G 1980c J. Phys. A13, 1887
Parisi G 1980d Phil. Mag.B41, 677
Parisi G 1980e Phys. Rep. 67, 25
Parisi G 1981 in: Lec. Notes in Phys. vol. 149
Parisi 1983a Phys. Rev. Lett 50, 1946
Parisi G 1983b J. de Phys. Lett. (Paris) 44, L581
Parisi G and Toulouse G 1980 J. Phys. Lett. (Paris) 41, L361
Paulsen C, Hamida J A, Williamson S J and Maletta H 1984 J. Appl. Phys. 55, 1652
Pelcovits R A 1979 Phys. Rev. B19, 465
Pelcovits R A, Pytte E and Rudnick J 1978 Phys. Rev. Lett. 40,476 and 1982 Phys.Rev.Lett. 48, 1297
Peretto P 1984 Biol.Cybern. 50, 51
Personnaz L, Guyon I and Dreyfus G 1985 J. de Phys. Lett. (Paris) 46, L359
Personnaz L, Guyon I and Dreyfus G 1986a in: Disordered Systems and Biological Organization (Springer)
Personnaz L, Guyon I and Dreyfus G 1986b Physica D
Personnaz L, Guyon I, Dreyfus G and Toulouse G 1986c J.Stat.Phys. 43, 411
Personnaz L, Guyon I, Ronnet J C and Dreyfus G COGNITA 85
Phani M K, Lebowitz J L, Kalos M H and Tsai C C 1979 Phys. Rev. Lett. 42, 577
Phillips W A 1972 J. Low Temp. Phys. 7, 351
Phillips W A 1981 in: Amorphous Solids: Low Temperature Proper

-ties ed Phillips W A (Springer, Berlin)
Pickert S J, Rhyne J J and Alperin H A 1975 AIP Conf. Proc. 24, 117
Pickart S J, Hasanain S, Andrauskas D, Alperin H A, Spano M L and Rhyne J J 1985 J. Appl. Phys. 57, 3430
Pippard A B 1966 Classical Thermodynamics (Cambridge)
Plefka T 1982a J. Phys. A15, L251
Plefka T 1982b J. Phys. A15, 1971
Plefka T 1982c Phys. Lett. 90A, 262
Poon S J 1978 Phys. Lett. 68A, 1014
Pound R V, Candela D, Buchman S and Vetterling W T 1983 in: Quantun Fluids and Solids 1983, ed. Adams E D and Ihas G G (American Inst. Phys.)
Prejean J J and Souletie J 1980 J. de Phys. (Paris) 41, 1335
Prejean J J, Joliclerc M and Monod P 1980 J. de Phys. (Paris) 41, 427
Prelovsek P and Blinc R 1982 J. Phys. C15, L985
Provost J P and Vallee G 1983 Phys. Rev. Lett. 50, 598
Provost J P and Vallee G 1983a Phys. Lett. 95A, 183
Proykova Y and Rivier N 1981 J. Phys. C14, 1839
Pynn R and Fender B E F 1985 Phys. Today 38, no.1, 46
Pytte E 1978 Phys. Rev. B18, 5046
Pytte E 1980 in: Ordering in strongly fluctuating condensed matter systems, ed. Riste T (Plenum)
Pytte E and Rudnick J 1979 Phys. Rev. B19, 3603
Raghavan R and Levy P M 1985 J. Appl. Phys. 57, 3386
Rajan V T and Rieseborough P S 1983 Phys. Rev. B27, 532
Rajchenbach J, Ferre J, Pommier J, Knorr K and De Graff A M 1980 J. Magn. Mag. Mat. 15-18, 199
Rajchenbach J and Bontemps N 1983 J. de Phys. Lett. (Paris) 44, L799
Rajchenbach J and Bontemps N 1984 J. Appl. Phys. 55, 1649
Rammal R 1981 Ph.D. Thesis (unpublished)
Rammal R and Souletie J 1982 in:Magnetism of Metals and Alloys, ed. Cyrot M (North-Holland)
Rammal R, Toulouse G and Virasoro M 1986 Rev. Mod. Phys

Ramond P 1981 "Field Theory: A Modern Primer" (Benjamin 1981)
Randeria M, Sethna J P and Palmer R G 1985 Phys. Rev. Lett. 54 1321
Ranganathan R, Tholence J L, Krishnan R and Dancygier M 1985 J. Phys. C18, L1057
Rao K V and Chen H S 1981 in: Proc. Int. Conf. on Rapidly Quenched Metals, Sendai, Aug. 1981
Rao K V, Faehnle M, Figueroa E and Beckman O 1983 Phys. Rev. B27, 3104
Rao K V, Fahnle M, Figuerra E, Beckman O and Hedman L 1985 Phys. Rev. B27, 3104
Rapaport D C 1977 J. Phys. C10, L543
Razavi F and Schilling J S 1983 J. Phys. F13, L59
Razzaq W A, Kouvel J S and Claus H 1984 Phys. Rev. B30, 6480
Razzaq W A and Kouvel J S 1984 J. Appl. Phys. 55, 1623
Razzaq W A and Kouvel J S 1985 J. Appl. Phys. 57, 3467
Reed P 1978 J. Phys. C11, L976
Reed P 1979 J. Phys. C12, L475
Reed P, Moore M A and Bray A J 1978 J.Phys. C11, L139
Reger J D and Zippelius A 1985 Phys. Rev. B31, 5900
Renard J P, Mirandy J P and Varret F 1980 Solid State Commun. 35, 41
Rhyne J J 1985 IEEE Trans. Magn. 21, 1990
Rhyne J J and Glinka C J 1984 J. Appl. Phys. 55, 1691
Rhyne J J and Fish G E 1985 J. Appl. Phys. 57, 3407
Richards P M 1984 Phys. Rev. B30, 2955
Riess I, Fibich M and Ron A 1978 J. Phys. F8, 161
Rivier N 1979 Phil. Mag. 40, 859
Rivier N and Adkins K 1975 J. Phys. F5, 1745
Roberts S A 1981 J. Phys. C14, 3015
Roberts S A and Bray A J 1982 J. Phys. C15, 1527
Rodrigues R, Fernandez A, Isalgue A, Rodriguez J, Labarta A, Tejada J and Obradors X 1985 J.Phys. C18, L401
Roshen W A 1982 Phys. Rev. B26, 3939

Rots M, Hermans L and van Cauteren J 1984a Phys. Rev. B30, 3666
Rots M, van Caurten J and Hermans L 1984b J. Appl.Phys. 55, 1732
Rowe J M, Rush J J, Hinks D G and Susman S 1979 Phys. Rev. Lett. 43, 1158
Rowe J M, Rush J J and Susman S 1983 Phys. Rev. B28, 3506
Ruderman M A and Kittel C 1954 Phys. Rev. 96, 99
Sadovskii M V and Skryabin Y N 1979 Phys. Stat. Sol. (b) 95, 59
Sakata M, Matsubara F, Abe Y and Katsura S 1977 J.Phys. C10, 2887
Salamon M B 1979 Solid State Commun. 31, 781
Salamon M B and Herman R M 1978 Phys. Rev. Lett. 41, 1506
Salamon M B, Rao K V and Chen H S 1980 Phys. Rev. Lett. 44, 596
Salamon and Tholence J L 1982 J. Appl. Phys. 53, 7684
Salamon M B and Tholence J L 1983 J. Magn. Mag. Mat. 31-34, 1375
Samara G A 1984 Phys. Rev. Lett. 53, 298
Sanchez J P, Friedt J M, Horne R and Van Duyneveldt A J 1984 J.Phys. C17, 127
Saslow W M 1980 Phys. Rev. B22, 1174
Saslow W M 1981 Physica 108B, 769
Saslow W M 1982a Phys. Rev. Lett. 48, 505
Saslow W M 1982b Phys. Rev. B26, 1483
Saslow W M 1983 Phys. Rev. B27, 6873
Saslow W M 1984 Phys. Rev. B30, 461
Sastry B S and Shenoy S R 1978 J. de Phys. (Paris) 39, C6-891
Sastry B S and Shenoy S R 1979 Physica 97B, 205
Sato T and Miyako Y 1982 J. Phys. Soc. Jap. 51, 2143
Sato T, Nishioka T, Miyako Y, Takeda Y, Morimoto S and Ito A 1985 J.Phys.Soc.Jap. 54, 1989
Savit R 1980 Rev. Mod. Phys. 25, 453
Schaf J, Campbell I A, Le Dang K, Veillet P and Hamzic A 1983a J.Magn.Mag.Mat. 36, 310
Schaf J, Le Dang K and Veillet P 1983b J. Magn. Mag. Mat. 37, 297
Schenk A 1978 J.de Phys. (Paris) 39, C6-1478
Schlottmann P and Bennemann K H 1982 Phys. Rev. B25, 6771
Schofield C L and Cooper L N 1985 Contemp. Phys. 26, 125
Schowalter L J and Klein M W 1979 J. Phys. C12, L395
Schreckenberg M 1985 Z. Phys. B60, 483

Schultz S, Gullikson E M, Fredkin D R and Tovar M 1980 Phys. Rev. Lett. 45, 1508
Schuster H G 1981 Z. Phys. B45, 99
Schwartz M 1985 J. Phys. C18, L1145
Senoussi S 1980 J. Phys. F10, 2491
Senoussi S and Oner Y 1985 J. Appl. Phys. 57, 3465
Sethna J P, Nagel S R and Ramakrishnan T V 1984 Phys. Rev. Lett. 53, 2489
Sethna J P and Chow K S 1985 Phase Transitions 5, 317
Sethna J P 1986 Proceedings of the New York Academy of Sciences
Shapir Y 1984 J. Phys. C17, L997
Shapiro S M, Fincher C R, Palumbo A C and Parks R D 1980 Phys. Rev. Lett. 45, 474
Shapiro S M, Fincher C R, Palumbo A C and Parks R D 1981a J. Appl. Phys. 52, 1729
Shapiro S M, Fincher C R, Palumbo A C and Parks R D 1981 Phys. Rev. B24, 6661
Shapiro S M, Maletta H and Mezei F 1985 J. Appl. Phys. 57, 3485
Shastry B S and Shenoy S R 1978 J.de Phys.(Paris) 39, C6-891
Shastry B S and Shenoy S R 1979 Physica 97 B, 205
Sherleker G, Srivastava C M, Nigam A K and Chandra G 1986 Phys. Rev. B34, 498
Sherrington D 1975 J. Phys. C8, L208
Sherrington D 1977 J. Phys. C10, L7
Sherrington D 1978 Phys. Rev. Lett. 41, 1321
Sherrington D 1979 in: Ill Condensed Matter, ed: Balian R, Maynard R and Toulouse G (North-Holland)
Sherrington D 1979a J. Phys. C12, L929
Sherrington D 1980 Phys. Rev. B22, 5553
Sherrington D 1981 in: Lec. Notes in Phys. vol.149
Sherrington D 1983 in: Lec. Notes in Phys. vol.192
Sherrington D 1984 J. Phys. C17, L823
Sherrington D and Kirkpatrick 1975 Phys. Rev. Lett. 35, 1792
Sherrington D and Southern B 1975 J. Phys. F5, L49
Sherrington D and Fernandez J F 1977, Phys. Lett. 62A, 457
Sherrington D, Cragg D M, Elderfield D and Gabay M 1983a J. Phys.

Soc. Jap. 52S,
Sherrington D, Cragg D M and Elderfield D 1983b J. Magn. Mag. Mat. 31-34, 1417
Shih W Y, Ebner C and Stroud D 1984 Phys. Rev. 30, 134
Shimizu M 1981 Rep. Prog. Phys. 44, 329
Shirakura T and Katsura S 1983a Physica 117A, 281
Shirakura T and Katsura S 1983b J. Magn. Mag. Mat. 31-34, 1305
Shklovskii B I and Efros A L 1980 Sov. Phys. Semicond. 14, 487
Shtrikman S and Wohlfarth E P 1981 Phys. Lett. A85, 467
Shukla P and Singh S 1981a J. Phys. C14, L81
Shukla P and Singh S 1981b Phys. Rev. B23, 4661
Siarry P and Dreyfus G 1984 J. de Phys. Lett. (Paris) 45, L39
Sibani P 1985 J. Phys. A18, 2301
Sibani P and Hertz J A 1985 J. Phys. A18, 1255
Silvera I F 1980 Rev. Mod. Phys. 52, 393
Simpson M A, Smith T F and Gmelin E 1981 J. Phys. F11, 1655
Singh R R P and Chakraverty S 1986 Phys. Rev. Lett. 57, 245
Smith D A 1975 J. Phys. F5, 2148
Sokol P E and Sullivan N S 1985 in: Orientational Disorder in Crystals, vol.5, No.3, ed. Ron Weir
Sommers H J 1978 Z. Phys. B31, 301
Sommers H J 1979a Z. Phys. B33, 173
Sommers H J 1979b J. Magn. Mag. Mat. 13, 139 (1979)
Sommers H J 1981 J. Magn. Mag. Mat. 22, 267
Sommers H J 1982 J.de Phys. Lett. (Paris) 43, L719
Sommers H J 1983a Z. Phys. B50, 97
Sommers H J 1983b J. Phys.A16, 447
Sommers H J 1985 J. de Phys. Lett. (Paris) 46, L779
Sommers H J and Usadel K D 1982 Z. Phys. B47, 63
Sommers H J, DeDominicis C and Gabay M 1983 J. Phys. A16, L679
Sommers H J and Dupont W 1984 J. Phys. C17, 5785
Sommers H. J and Fischer K H 1985 Z. Phys. B58, 125
Sompolinsky H 1981a Phys. Rev. Lett. 47, 935
Sompolinsky H 1981b Phys. Rev. B23, 1371
Sompolinsky H and Zippelius A 1981 Phys. Rev. Lett. 47, 359
Sompolinsky H and Zippelius A 1982 Phys. Rev. B25, 6860

Sompolinsky H and Zippelius A 1983a in: Lec. Notes in Phys. vol.192, Springer-Verlag
Sompolinsky H and Zippelius A 1983b Phys. Rev. Lett. 50, 1297
Sompolinsky H, Kotliar G and Zippelius A 1984 Phys. Rev. Lett. 52, 392
Soukoulis C M 1978 Phys. Rev. B18, 3757
Soukoulis C M and Levin 1977 Phys. Rev. Lett. 39, 581
Soukoulis C M and Levin 1978 Phys. Rev. B18, 1439

Soukoulis C M, Grest G S and Levin K 1978 Phys. Rev. Lett. 41, 568
Soukoulis C M and Grest G S 1980 Phys. Rev. B21, 5119
Soukoulis C M, Levin K and Grest G S 1982a Phys. Rev. Lett. 48, 1756
Soukoulis C M, Grest G S and Levin K 1982b J. Appl. Phys. 53, 7679
Soukoulis C M, Levin K and Grest G S 1983 Phys. Rev. B28, 1495
Soukoulis C M, Grest G S and Levin K 1983a Phys. Rev. Lett. 50, 80
Soukoulis C M, Grest G S and Levin K 1983b Phys. Rev. B28, 1510
Soukoulis C M and Grest G S 1984 J. Appl. Phys. 55, 1661
Souletie J 1983 J.de Phys. (Paris) 44, 1095
Souletie J 1985 Ann. Phys. Fr. 10, 69
Souletie J and Tournier R 1969 J. Low Temp. Phys. 1, 95
Souletie J and Tholence J L 1985 Phys. Rev. B32, 516
Sourlas N 1984 J. de Phys. Lett. (Paris) 45, L969
Southern B W 1975 J. Phys. C8, L213
Southern B W 1976 J. Phys. C9, 4011
Southern B W and Young A P 1977a J. Phys. C10, L79
Southern B W and Young A P 1977b J. Phys. C10, 2179
Southern B W, Young A P and Pfeuty P 1979 J. Phys. C12, 683
Spano M L and Rhyne J J 1985 J. Appl. Phys. 57, 3303
Spano M L, Alperin H A, Rhyne J J, Pickert S J, Hasanain S and Andrauskas D 1985 J. Appl. Phys. 57, 3432
Stanley H E 1971 "Introduction to Phase Transition and Critical Phenomena" (Oxford)

Stauffer D 1985 "Introduction to Percolation theory" (Taylor and Francis)
Stauffer D and Binder K 1978 Z. Phys. B30, 313
Stauffer D and Binder K 1979 Z. Phys. B34, 97
Stearns M B 1984 J. Appl. Phys. 55, 1729
Steglich F 1976 Z. Phys. B23, 331
Sullivan N S 1976 J. de Phys. (Paris) 37, 209
Sullivan N S and Devoret M 1977 in: Proceedings of the International Conference on Quantum Crystals, Colorado State University, Fort Collins
Sullivan N S, Devoret M, Cowan B P and Urbina C 1978 Phys. Rev. B17, 5016
Sullivan N S, Devoret M and Vaissiere J M 1979 J. de Phys. Lett. (Paris)40, L559
Sullivan N S and Esteve D 1981 Physica 107B, 189
Sullivan N S, Esteve D and Devoret M 1982 J. Phys. C15, 4895
Sullivan N S 1983 in: Quantum fluids and solids-1983 ed. Adams E D and Ihas G G, American Institute of Physics
Sullivan N S, Devoret M and Esteve D 1984 Phys. Rev. B30, 4935
Suran G and Gerard P 1984 J. Appl. Phys. 55, 1682
Suto A 1981 Z. Phys. B44, 121
Suto A 1984 in: Lec. Notes in Phys. vol. 206
Suzuki M 1977 Prog. Th. Phys. 58, 1151
Suzuki M 1985 Prog. Th. Phys. 73
Suzuki M and Miyashita S 1981 Physica 106A, 344
Swendsen R H 1982 in: Real Space Renormalization, ed: Burkhardt T W and van Leeuwen J M (Springer-Verlag)
Taggert G B, Tahir-Kheli R A and Shiles E 1974 Physica 75, 234
Takayama H 1978 J. Phys. F8, 2417
Takano F 1980 Prog. Th. Phys. Suppl.69, 174
Takeda Y, Morimoto S, Ito A, Sato T and Miyako Y 1985 J. Phys. Soc. Jap. 54, 2000
Tanaka F and Edwards S F 1980a J. Phys. F10, 2769
Tanaka F and Edwards S F 1980b J. Phys. F10, 2779
Taniguchi T, Miyako Y and Tholence J L 1985 J. Phys. Soc. Jap. 54, 220

Tank D W and Hopfield J J 1986 Preprint

Tari A 1976 J. Phys. F6, 1313

Teitel S and Domany E 1986 Phys. Rev. Lett. 55, 2176

Terauchi H. Futamura T, Nishihata Y and Iida S 1984 J. Phys. Soc. Jap. 53, 483

Theumann A 1986 Phys. Rev. B33, 559

Theumann A and Gusmao M V 1984 Phys. Lett. 105A, 311

Tholence J L 1980 Solid State Commun. 35, 113

Tholence J L 1981a Physica 108B, 1287

Tholence J L 1981b in: Proc.Trends in Physics, ed.Dorobantu I A (European Physical Society)

Tholence J L 1985 Physica 126B, 157

Tholence J L and Tournier R 1974 J. de Phys. (Paris) 35, C4-229

Tholence J L, Holtzberg, Godfrin H, Loehneysen H V and Tournier R 1978 J.de Phys. Coll.(Paris) 39, C6-928

Tholence J L and Salamon M B 1983 J. Magn. Mag. Mat. 31-34, 1340

Tholence J L, Yeshurun Y, Kjems J K and Wanklyn B 1986 J. Magn. Mag. Mat. 54-57, 203

Thomas H 1980 in: Strongly Fluctuating Systems, ed: Ordering in Strongly Fluctuating Condensed Matter Systems, ed. Riste T (Plenum)

Thomson J O and Thompson J R 1981 J. Phys. F11, 247

Thouless D J 1986 Phys. Rev. Lett. 56, 1082

Thouless D J 1986a Phys. Rev. Lett. 57, 273

Thouless D J, de Almeida J R L and Kosterlitz J M 1980 J. Phys. C13, 3272

Thouless D J, Anderson P W and Palmer R G 1977 Phil.Mag.35, 593

Toulouse G 1977 Commun. Phys. 2, 115

Toulouse G 1979 Phys. Rep. 49, 267

Toulouse G 1979 in: Lec. Notes in Phys. vol.115

Toulouse G 1980a J.de Phys. Lett.(Paris) 41, L447

Toulouse G 1980b in:Recent developments in Gauge theory, eds. 't Hooft G, Itzykson C, Jaffe A, Lehman H, Mitter P K, Singer I M and Stora R (Plenum)

Toulouse G 1981 in: Lec. Notes in Phys. vol.149 (Springer)

Toulouse G 1982a in: Anderson Localization, ed. Nagaoka Y and

Fukuyama H (Springer-Verlag)
Toulouse G 1982b Physica 109 & 110B, 1912
Toulouse G 1983 in: Lec. Notes in Phys., vol. 192 (Springer)
Toulouse G and Vannimenus J 1980 Phys. Rep. 67, 47
Toulouse G and Gabay M 1981 J. de Phys. Lett. (Paris) 42, L103
Toulouse G, Dehaene S and Changeux J P 1986 Preprint
Tovar M, Fainstein C, Oseroff S and Schultz S 1985 J. Appl. Phys. 57, 3438
Trainor R J and McCollum D C 1974 Phys. Rev. B9, 2145
Trainor R J and McCollumm D C 1975 Phys.Rev.B11, 3581
Uemura Y J 1980 Solid State Commun. 36, 369
Uemura Y J 1981 Hyp. Int. 8, 739
Uemura Y J 1982 Ph.D. Thesis, University of Tokyo
Uemura Y J 1984 Hyp. Int. 17-19, 447
Uemvan Dongen J C M, van Dijk D and Mydosh J A 1981 Phys. Rev. B24, 5110 J and Yamazaki T 1982 Physica 109&110B, 1915
Uemura Y J and Yamazaki T 1983 J. Magn. Mag. Mat. 31-34, 1359
Uemura Y J, Yamazaki T, Hayano R S, Nakai R and Huang C Y 1980 Phys. Rev. Lett. 45, 583
Uemura Y J, HUang C Y, Clawson C W, Brewer J H, Kiefl R F, Spencer D P and deGraff A M 1981a Hyp. Int. 8, 757
Uemura Y J, Nishiyama K N, Yamazaki T and Nakai R 1981b Solid State Commun. 39, 461
Uemura Y J, Nishiyama K, Kadono R, Imazato J, Nagamine K, Yamazaki T and Ishikawa Y 1983 J. Magn. Mag. Mat. 31-34, 1379
Uemura Y J, Harshman D R, Senba M, Ansaldo E J and Murani A P 1984 Phys. Rev. B30, 1606
Uemura Y J, Shapiro S M and Wenger L E 1985 J. Appl. Phys. 57, 3401
Ueno Y 1980 Phys. Lett. 75A, 383
Ueno Y 1983a J. Magn. Mag. Mat. 31-34, 1299
Ueno Y 1983b J. Phys. Soc. Jap. 52S, 121
Ueno Y and Oguchi T 1976a J. Phys. Soc. Jap. 40, 1513
Ueno Y and Oguchi T 1976b J. Phys. Soc. Jap. 41, 1123
Ueno Y and Oguchi T 1980 Prog. Th. Phys. 63, 342

Usadel K D 1986 Preprint
Usadel K D, Bien K and Sommers H J 1983 Phys. Rev. B27, 6957
van der Klink and Rytz D 1983 Phys. Rev. B27, 4471
van Dongen J C M, van Dijk D and Mydosh J A 1981 Phys. Rev. B24, 5110
van Dongen J C M, Palstra T T M, Morgownick A T J, Mydosh J A, Geerken B M and Buschow K H J 1983 Phys. Rev. B27, 1887
van Duyneveldt A J and Mulder C A M 1982 Physica 114B, 82
van Enter A C D 1985 Preprint
van Enter A C D and van Hemmen J L 1983 J. Stat. Phys. 32, 141
van Enter A C D and van Hemmen J L 1984a Phys. Rev. B29, 355
van Enter A C D and van Hemmen J L 1984b Phys. Rev. B31, 306
van Enter A C D and Froelich J 1985 Commun. Math. Phys. 98, 425
van Enter A C D and vanHemmen J L 1985 J. Stat. Phys.
van Hemmen J L 1980 Z. Phys. B40, 55
van Hemmen 1982 Phys. Rev. Lett. 49, 409
van Hemmen 1983 in: Lec. Notes in Phys. vol. 192
van Hemmen J L 1986 Preprint
van Hemmen J L 1986a J. Phys. C19, L379
van Hemmen J L and Palmer R G 1979 J. Phys. A12, 563
van Hemmen J L and Morgenstern I 1982 J. Phys. C15, 4353
van Hemmen J L, van Enter A C D and Canisius J 1983 Z. Phys. B50, 311
van Hemmen J L and Suto A 1984 J. de Phys. (Paris) 45, 1277
van Hemmen J L and Suto A 1985 Z. Phys. B61, 263
van Kranendonk J 1983 "Solid Hydrogen", Plenum, New York
van Kranendonk J and van Vleck J H 1958 Rev. Mod. Phys. 30, 1
Vannimenus J and Toulouse G 1977 J.Phys. C10, L537
Vannimenus J, Toulouse G and Parisi G 1981 J. Phys. (Paris) 42, 565
Vannimenus J and Mezard M 1984 J. de Phys. Lett. (Paris) 45, L1145
Varret F, Hamzic A and Campbell I A 1982 Phys. Rev. B26, 5285
Velu E, Renard J P and Mirandy J P 1981 J. de Phys. Lett. (Paris) 42, L237
Venkataraman G and Sahoo 1985 Contemp.Phys. 26, 579

Vernon S P and Halawith B N and Stearns M B 1985 J. Appl. Phys. 57, 3441

Viana L and Bray A J 1983 J. Phys.C16, 6817

Villain J 1977a J. Phys.C10,

Villain J 1977b J. Phys.C10,

Villain J 1978 J. Phys. C11,

Villain J 1979 Z. Phys. B33, 31

Villain J 1980 J. Magn. Mag. Mat. 15-18, 105

Villain J 1985 in: Scaling Phenomena in Disorered Systems, ed. Pynn R (Plenum)

Villain J, Bidaux R, Carton J P, Conte R and de Seze L 1980 J. de Phys. (Paris) 41, 1263

Violet C E and Borg R J 1966 Phys. Rev. 149, 540

Violet C E and Borg R J 1967 Phys. Rev. 162, 608

Violet C E and Borg R J 1983 Bull. Am. Phys. Soc. 28, 543

Viticoli S, Fiorani D, Nogues M and Dormann J L 1982 Phys. Rev. B26, 11

Volkmann U G, Boehmer R, Loidl A, Knorr K, Hoechli U T, Haussuehl S 1986 Phys. Rev. Lett. 56, 1716

Volovik G E and Dzyaloshinskii I E 1978 Sov. Phys. JETP 48, 555

von Loehneysen, van den Berg H and Lecomte G V 1985 Phys. Rev. B31, 2920

von Molner S, McGuire T R, Gambino R J and Barbara B 1982 J. Appl. Phys. 53, 7666

Wada K and Takayama H 1980 Prog. Th. Phys. 64, 327

Walker L R and Walstedt R E 1977 Phys. Rev. Lett. 38, 514

Walker L R and Walstedt R E 1980 Phys.Rev. B22, 3816

Walker L R and Walstedt R E 1983 J. Magn. Mag. Mat. 31-34, 1289

Walstedt R E 1981 Phys. Rev. B24, 1524

Walstedt R E 1982 Physica 109 & 110B, 1924

Wlastedt R E and Walker L R 1981 Phys. Rev. Lett. 47, 1624

Walstedt R E and Walker L R 1982 J. Appl. Phys. 53, 7985

Wannier G H 1950 Phys. Rev. 79, 357

Washburn S, Yu I and Meyer H 1981 Phys. Rev. A25, 365

Washburn S, Calkins M, Meyer H and Harris A B 1982 J. Low Temp.Phys. 49, 101

Washburn S, Calkins M, Meyer H and Harris A B 1983 J. Low Temp. Phys. 53, 585
Wassermann E F and Herlach D M 1984 J. Appl. Phys. 55, 1709
Wendler R and Baberschke K 1983 Solid State Commun. 48, 91
Wenger L E and Keesom 1975 Phys. Rev. B11, 3497
Wenger L E and Mydosh J A 1982 Phys. Rev. Lett. 49, 239
Wenger L E and Mydosh J A 1984 J. Appl. Phys. 55, 1717
Weisbuch G and Fogelman-Soulie F 1985 J. de Phys. Lett. (Paris) 46, L623 State Phys. Series Supplement 15, Academic Press
Whittle G L and Campbell S J 1983 J. Magn. Mag. Mat. 31-34, 1377
Wiedenmann A and Burlet P 1978 J. de Phys. (Paris) 39, C6-720
Wiedenmann A, Burlet P, Scheur H and Convert P 1981a Solid State Commun. 38, 129
Wiedenmann A, Burlet P, Scheur H and Gunsser W 1981b Solid Stae Commun. 39, 801
Wiedenmann A, Gunsser W, Rossat-Mignod J and Evrard M O 1983 J. Magn. Mag. Mat. 31-34, 1442
Wiedenmann A, Hamedouns M and Rossat-Mignod J 1985 J. Phys. C18, 2549
Window B 1969 J. Phys. C2, 2380
Window B 1970 J. Phys. C3, 922
Window B, Longworth G and Johnson C E 1970 J. Phys. C3, 2156
Wohlfarth E P 1977 Phyica 86-88B, 873
Wohlfarth E P 1979 Phys. Lett. 40A, 489
Wohlfarth E P 1980 J. Phys. F10, L241
Wohlfarth E P 1977 Phys. Lett. 61A, 143
Wolff W F and Zittartz J 1982 Z. Phys. B49, 229
Wolff W F and Zittartz J 1983 in: Lec. Notes in Phys. vol. 192
Wolff W F and Zittartz J 1985 Z. Phys. B60, 185
Wolff W F, Hoever P and Zittartz J 1981 Z. Phys. B42, 259
Wong P, Yoshizawa H and Shapiro S M 1985 J. Appl. Phys. 57, 3462
Yu F Y 1982 Rev. Mod. Phys. 54, 235
Yu F Y 1984 J. Appl.Phys. 55, 2421
Wu W, Mozurkewich G and Orbach R 1985 Phys. Rev. B31, 4557
Xu L B, Yang Q and Zhang J B 1985 in: Festkorperprobleme 25

Yamada I 1983 J. Magn. Mag. Mat. 31-34, 645

Yeshurun Y, Salamon M B, Rao K V and Chen H S 1980 Phys. Rev. Lett. 45, 1366

Yeshurun Y, Salamon M B, Rao K V and Chen H S 1981a Phys.Rev.B 24, 1536

Yeshurun Y, Rao K V, Salamon M B and Chen H S 1981b Solid State Commun. 38, 371

Yeshurun Y, Ketelsen L J P and Salamon M B 1982 Phys. Rev. B26, 1491

Yeshurun Y, Felner I and Wanklyn 1984 Phys. Rev. Lett. 53, 620

Yeshurun Y and Sompolinsky H 1985 Phys. Rev. B31, 3191

Yeshurun Y and Sompolinsky H 1986 Phys. Rev. Lett. 56, 984

Yoshida K 1957 Phys. Rev. 106, 893

Yoshizawa H, Shapiro S M, Hasanain S K and Guertin R P 1983 Phys. Rev. B27, 448

Young A P 1979 J. Appl. Phys. 50, 1691

Young A P 1983a in: Lec. Notes in Phys., vol. 192

Young A P 1983b Phys. Rev. Lett. 50, 917

Young A P 1983c Phys. Rev. Lett.51, 1206

Young A P 1984 J. Phys. C18, L517

Young A P 1985 J. Appl. Phys. 57, 3361

Young A P and Stinchcombe R B 1976 J. Phys. C9, 4419

Young A P and Kirkpatrick S 1982 Phys. Rev. B25, 440

Young A P, Bray A J and Moore M A 1984 J. Phys. C17, L149

Yu I, Washburn S and Meyer H 1981 Solid State Commun. 40, 693

Yu I, Washburn S and Meyer H 1983a J. Low Temp. Phys. 51, 369

Yu I, Washburn S, Calkins M and Meyer H 1983b J. Low Temp. Phys. 51, 401

Zastre E, Roshko R M and Williams G 1985 J. Appl. Phys. 57, 3447

Zeleny M, Screiber J and Kobe S 1985 J. Magn. Mag. Mat. 50, 27

Zhou W S, Meisel M W, Owers-Bradley J R, Halperin W P, Ochiai Y and Brittain J O 1983 Phys. Rev. B27, 3119

Zittartz J 1984 in: Lec. Notes in Phys. vol. 206

Zomack M, Baberschke K and Barnes S E 1983 Phys. Rev. B27, 4135

Zwanzig R 1985 Phys. Rev. Lett. 54, 364

Zweers H A and van den Berg G J 1975 J. Phys. F5, 555

ADDENDUM

Some interesting papers appeared after the completion of the manuscript. Besides, I missed some older papers while writing the original manuscript. A brief review of some of these works will be given in this addendum. Moreover, some specific points which were not very well explained in the text will be clarified here. Two recent reviews on spin glasses (Binder K and Young A P 1986 Rev. Mod. Phys., Mezard M, Parisi G and Virasoro M A 1986 "The Spin Glass Theory and Beyond", World Scientific), which discuss the replica approach in greater detail, should be considered as complementary to this book, in spite of considerable overlaps.

Chapter 2

The original work of Adkins K and Rivier N (J. de Phys. (Paris) 35, C4-237 (1974), J. Phys. F5, 1745 (1975), although unsuccessful, was the first attempt to interpret the SG transition as a phase transition where the local magnetization serves as an order parameter. Note that the Adkins-Rivier order parameter is quite different from the EA order parameter: the latter is a measure of the autocorrelation in time.

Chapter 3

It is not surprising that Ghatak (1979) observed no κ-dependence of the temperature corresponding to the maximum in $\chi(\kappa)$. In the spirit of the conventional Landau theories, Ghatak assumed that the coefficient of the q^2 term vanishes at a sharply defined temperature.

Chapters 4-10

The thermodynamic quantities computed from the iterative solutions of the mean-field equations

$$M_i = \tanh[\{H + \Sigma J_{ij}M_j\}/T]$$

by slow cooling from high temperature (Soukoulis C M, Levin K and Grest G S, Phys. Rev. Lett. 48, 1756 (1982), Phys. Rev. B28, 1495 (1983), Phys. Rev. B28, 1510 (1983), J. Appl. Phys. 53, 7679 (1982), J. Appl. Phys. 55, 1634 (1984), Grest and Soukoulis, Phys. Rev. B28, 2886 (1983), Phys. Rev. Lett. 50, 80 (1983), J. Appl. Phys. 55, 1661 (1984)), are not better than the corresponding Monte Carlo estimates (Reger J D, Binder K and Kinzel W 1984 Phys. Rev. B30, 4028). Besides, contrary to the usual expectations, the results become unphysical when appropriate care of the reaction field is taken into account (Ling D D, Bowman D R and Levin K 1983 Phys. Rev. B28, 262).

In order to stress the fact that the possibility (or impossibility) of ultrametric ground states of the short-ranged SG in d = 3 is far from established, I would request the reader to compare and contrast the works of Bray A J and Moore M A (J. Phys. A18, L683 (1985), J. Phys. C18, L699 (1985)), and Sourlas N (J. de Phys. Lett. (Paris) 45, L969 (1984), Europhys.Lett. 1, 189 (1986)) with those of Ogielsky.

Chapter 13

Recall that the order parameters for the m-vector SG were defined as $q^{\alpha\beta}_{\mu\nu} = q^{\alpha\beta}\delta_{\mu\nu}$ where α,β are the replica labels and μ,ν label the spin components. Goldbart P and Martin O (Phys. Rev. B34, 2032 (1986)) argue that a more precise choice should be

$$q^{\alpha\beta}_{\mu\nu} = q^{\alpha\beta}\,\Sigma\, R^{\alpha}_{\mu\rho}\; R^{\beta}_{\nu\sigma}\;\delta_{\rho\sigma}$$

where $\{R^{\alpha}_{\mu\nu}\}$ is a set of independent rotations for each replica. This observation, together with the fact that all rotations commute in two dimensions, suggests that the order parameter for the XY model should be given by

$$q_{\mu\nu}^{\alpha\beta} = s^{\alpha\beta}\ \delta_{\mu\nu} + a^{\alpha\beta}\ \epsilon_{\mu\nu}$$

where $s^{\alpha\beta}$ and $a^{\alpha\beta}$ are, respectively, symmetric and antisymmetric in the replica labels whereas $\epsilon_{\mu\nu}$ is totally antisymmetric in the labels denoting the spin components.

Chapters 15, 16 and 17

Let us compute the crossover exponent ϕ_2 associated with the Dzyaloshinski-Moriya (DM) and single-ion (SI) anisotropic exchange interactions in an m-vector SG. The nonlinear susceptibility χ_{nl} follows the scaling form (Pfeuty P and Aharony A in: Annals of the Israel Phys. Soc. 2, 910 (1978), Kotliar G, Preprint, 1986, Klein L and Aharony A (Private Communication))

$$\chi_{nl}^{-1} = (\tilde{t}^{-\gamma}h^2)^{-1}\ f((h^2/\tilde{t}^{\phi_1}),(d^2/\tilde{t}^{\phi_2}))$$

where $\tilde{t}$ is the reduced temperature, h is the reduced field and d is a dimensionless measure of the strength of the anisotropy. Note that the transition line in a pure vector SG is given by $\tilde{t} \sim h^{2/\phi_1}$. Recently, it has been observed (Kotliar 1986) that (a) ϕ_2(DM) is larger than ϕ_2(SI) and (b) $\phi_2(SI) = \gamma_s$, the exponent associated with the nonlinear susceptibility, to all orders in $\epsilon = 6 - d$. The latter result is consistent with the earlier prediction of Pfeuty and Aharony (1978).

In chapters 16 and 17 we mentioned that a special cancellation could explain why $\tilde{t} \sim H^{2/3}$ is the crossover line instead of $\tilde{t} \sim H$. We also mentioned that in the renormali-zation group (RG) terminology the existence of two dangerously irrelevant variables leads to the violation of simple scaling laws in the MFT. I would like to stress that this RG analysis by Fisher and Sompolinsky (1985) predicts $\tilde{t} \sim H^{2/(\beta+\gamma)}$ for both Ising and Heisenberg SG below six dimensions.

Extending some interesting ideas proposed by Anderson (1978) Shapir (1984) speculated that there exists an intermediate critical dimension d_{icd} such that for $d_{lcd} < d < d_{icd}$ the short-ranged Ising SG exhibit an entropy-dominated transition. Shapir suggested $d_{icd} = 4$.

The high temperature series expansion (Singh R R P and Chakravarty S 1986, Proc. 31st ann. conf. on Magn. Mag. Mat.) yields $\nu = 1.3 \pm 0.2$ and $\eta = -0.25 \pm 0.2$ for short-ranged SG in d=3. These values of the exponents are in excellent agreement with the corresponding MC estimates.

Chapter 19

Dotsenko V S (J. Phys. C18, 6023 (1985)) suggested that the relaxation in SG is a physical realization of a model of relaxation in a system with a fractal free energy surface, thereby making a connection between the hierarchical relaxation (see Chapter 19) and ultradiffusion (see chapter 27).

Chapter 20

The phase boundaries and the corresponding exponents for the generalized ±J model at T = 0 have been determined by expansion in powers of the concentrations of the positive and negative bonds (Aharony A and Binder K (J. Phys. C13, 4091 (1980)); the nonlinear susceptibility diverges for a critical concentration of the nonzero bonds. Various investigations of the ordering in dilute Ising models on triangular as well as f.c.c. lattices with nearest-neighbor antiferromagnetic exchange interaction (Grest G S and Gabl E G, Phys. Rev. Lett. 43, 1182 (1979), Anderico C Z, Fernandez J F and Streit T S J, Phys. Rev. B26, 3824 (1982)) suggest the existence of a SG phase for moderate dilution. The pioneering work on the antiferromagnetic Ising model on a f.c.c. lattice by Phani et al. (1980) has been extended by several other authors (e.g., Polgreen T L, Phys. Rev.

B29, 1468 (1984)); one of the most interesting features is that the high degeneracy of the ground state is lifted by small next-nearest-neighbor interaction (Mackenzie N D and Young A P, J. Phys. C14, 3927 (1981)).

Almost all the works on SG have been carried on the models with random scalar exchange interaction, the examples being the m-vector SG, the Potts glass, quadrupolar glass, etc. The Hamiltonian for a Gauge SG is given by (Goldbart P, Martin O and Fradkin E, Phys. Rev. B34, 301 (1986))

$$\mathcal{H} = - J \Sigma \vec{S}_i \cdot R(ij) \cdot \vec{S}_j$$

where R(ij) are (random) matrices in a subgroup of G, G being the group of linear transformations that leave the spin-spin product invariant. The nature of the SG transition and that of the corresponding low-temperature phase corresponding to the group of orthogonal transformations are different from those corresponding to the group of unitary transformations.

Chapters 21,22 and 23

The term "cluster" has been used in the SG literature by different authors in different senses. A summary of such usage can be found in Jacobs A E (Phys. Rev. B32, 7430 (1985)).

The two main drawbacks of the percolation model of SG transition should be mentioned here. The clusters in the percolation model become ill-defined in the case of the undamped RKKY interaction. However, this difficulty does not arise if the damping of the RKKY interaction is properly taken into account. The second drawback, which is believed to be more serious, is that the role of frustration in the percolation model is not clear. It is now almost certain that the SG transition is not a percolation phenomenon of the usual kind.

Feigelman M V and Ioffe L B (J. de Phys. Lett. (Paris) 45, L475 (1984)) suggested that a macroscopic condensation into a delocalized mode (recall the nature of the eigenstates of the exchange matrix J) takes place only for $d > 4$.

Chapter 24

A replica Monte Carlo simulation technique has been introduced by Swendsen R H and Wang J S (1986 Preprint). It is claimed that this approach requires less computer time than the conventional MC simulation. The data are qualitatively consistent with the earlier results.

Ogielsky (1986 Preprint) has computed the domain wall free energy

$$\Delta F(T) = T \int_{T}^{\infty} dT' \, \{\Delta U_L(T')/T'^2\}$$

where $\Delta U_L(T) = U_L^A(T) - U_L^P(T)$ is the difference between the equilibrium average values of the energy of the ±J model for system length L in d=3 with antiperiodic and periodic boundary conditions, respectively, in the z-direction and periodic boundary conditions in the other two directions. Interestingly, $(d/dt)W_L(T)$, where W_L is the RMS value of ΔF, exhibits a sharp peak at $T_g = 1.2J$ which is in excellent agreement with the earlier estimates of the corresponding T_g. The peak becomes sharper with the increase of L and, possibly, diverges as $L \to \infty$.

Chapter 26

We have summarized some of the interesting results on the one-dimensional SG models in various different chapters in this book. For example, some one-dimensional long-ranged models have been introduced in chapter 1 and the corresponding properties have been summarized in chapter 14. We shall now present the results on few other one-dimensional models in this section. Orland H, DeDominicis C and Garel T (1981) studied an one-dimensional model where $J_{ij} = J \sin k|x_i - x_j| \exp(-\gamma|x_i - x_j|)$, with $\gamma \ll k$ and c is the concentration of the sites occupied by the Ising spins. Below the transition temperature $T_g = (2Jc/k)$ the stable phase is non-homogeneous consisting of soliton-like coherent spin clusters.

The numerically exact study as well as MC simulation of the one dimensional model

$$J_{ij} = J_0 \cos(\alpha|x_i - x_j|)/|x_i - x_j|$$

(Ariosa D, Droz M and Malaspinas A, Helv. Phys. Acta, 55, 29 (1982)) suggest qualitative similarities of the thermodynamic properties with those of the SK model. In order to have a well-defined thermodynamic limit the exchange constant J_0 must scale with the number of spins N appropriately. Suppose, $J_0 = \tilde{J}_0/f(N)$ such that the free energy is extensive. Unfortunately, no convincing form of the function f(N) for this model is available.

One dimensional short-ranged SG models are trivial in the sense that there is no SG transition at any nonzero temperature. However, both analytical and numerical computations are easier in one dimension than in any higher dimension and hence might throw light on the nature of the ordering in the higher dimensional models. Gardner E and Derrida B (J. Stat. Phys. 39, 367 (1985)) (also see Chen H H and Ma S K, J. Stat. Phys. 29, 717 (1982)) showed that the zero-temperature magnetization $M \sim Ch^x$ where C and x depend on the probability distribution of the nearest-neighbor exchange interactions. Moreover, there is an exponentially large number of metastable states at T = 0 for magnetizations smaller than $M_{max} = 0.446042..$ but none for magnetizations larger than M_{max} (Derrida B and Gardner E 1986). On the other hand, Kaplan's (Phys. Rev. B24, 319 (1981)) treatment of the vector spin systems in d=1 showed the absence of metastable states. Another interesting feature of the short-ranged one-dimensional SG models is that there is a line of irreversibility on a finite time scale (this dynamical freezing effect is similar to that in d = 2); but the shape of the line is very different from that of the AT line (Reger J D and Binder K 1985 Z. Phys. B60, 137).

Stephen M J and Aharony A (J. Phys. C14, 1665 (1981)) predicted that the dipolar interaction

$$J_{ij} = J_0 \ (1 - d \cos \cos^2 \theta_{ij})/r_{ij}^d$$

can lead to SG ordering at sufficiently dilute systems. Some real materials are being seriously considered as candidates for such a phase (Private Communication from A Aharony).

There are not many rigorous results available for SG models. Readers interested in rigorous results should see van Enter A C D (in: Proc. les Houches 1984, in: Proceedings of the Groeningen Conference on Statistical Mechanics, ed. Hugenholtz N M and Winnink M, 1986, J. Stat. Phys. 41, 315 (1985)), Froehlich J and Imbrie J (Commun. Math. Phys. 96, 148 (1984)), Berretti A (J. Stat. Phys. 38, 483 (1985)), Froehlich J and Zegarlinski B (Europhys. Lett., 2, 53 (1986), Chayes et al.(1986)).

Chapter 27

Very recently, the rate of decrease of the transition temperature with the increasing pressure on $Rb_{1-x}(NH_4)_xH_2PO_4$ has been studied (Samara G A and Schmidt V H Phys. Rev. B34, 2035 (1986)).

A replica-symmetric solution of the random-link travelling salesman problem (Mezard M and Parisi G 1986 J. de Phys. (Paris) 47, 1285) agrees qualitatively with the corresponding numerical results. This apparent success of the replica-symmetric theory can be attributed to the fact that the energy, which is the analog of the optimum length of a tour, is approximated within 5% by the replica-symmetric SK solution in the case of the SK model of SG. See Baskaran G, Fu Y and Anderson P W (to be published) for a field theoretic representation of the travelling salesman problem.

For the effect of an external bias on the Huberman-Kerzberg model of ultradiffusion and for a generalization of the latter model to the two-dimensional ultradiffusion see Ceccatto and Riera (J.Phys.A19, L721 (1986)). Not very surprisingly, a transition from the power law to exponential decay of $P_0(t)$ is observed at a certain value of the biasing field. Kumar D and Shenoy S R (Phys. Rev. B34, 3547 (1986)) have solved the Ogielski-Stein model of ultradiffusion for arbitrary inhomogeneous branching under the constraint that the transition rate to a cluster of sites from a site outside is constant within each generation of the hierarchy. The subject of diffusion on fractal structures has developed into an interesting branch of statistical physics over the last five years. For the relation between ultradiffusion and diffusion on fractal structures see Havlin and Weissman (J.Phys.A, 1986).

Parisi (J.Phys. A19, L617 (1986)) has carried out computer simulation of a modified Hebb-Cooper rule with a cutoff A, i.e., $|J_{ij}| < A$, where A is a suitably chosen upper bound. The quantity of interest is the retrieval probability of the i-th pattern after i + j patterns have been stored. Parisi's work clearly demonstrates that this probability is the highest for the smallest j. Physically, it means that older patterns are automatically forgotten and the memory recalls the latest patterns most efficiently. However, such increase in the efficiency of retrieval with negligible confusion is achieved only at the cost of the storage capacity.

In the usual Hebb-Cooper prescription for learning the network can learn only when instructed to do so. A more natural learning mechanism (Parisi G 1986 J. Phys. A19, L675) is given by

$$dJ_{ij}(t)/dt = -\lambda J_{ij}(t) + S_i(t)S_j(t)$$

if the network is not in a state of confusion, and

$$dJ_{ij}(t)/dt = 0$$

if the network is in a confused state, where t is the time.

Computer simulation of Anderson's SG model of prebiotic evolution (Rokhsar D S, Anderson P W and Stein D L, J. Molecular Evolution, 23, 119 (1986)) not only showed "natural selection" among a wide diversity of the species that can grow but also demonstrated the competition between the different species and their adaptation to the changing environments. Very recently, lot of interesting work has been done on the Kauffman model of random Boolean networks (Derrida B and Pomeau Y 1986 Europhys.Lett. 1, 45, Derrida B and Stauffer D 1986 (preprint), Derrida B and Weisbuch 1986 (preprint), Derrida B and Flyvbjerg H (preprint)). A critical comparison of these works with Anderson's SG model of prebiological evolution and Hopfield's model of neural network might provide new insight into the challenging problems in life sciences.

Appendix B

Recent TRM data for AgMn SG alloys over a wide range of waiting times t_w have been successfully fitted to the form (Alba M, Ocio M and Hammann J, Europhys. Lett. 2, 45 (1986))

$$M_{TRM} = M_0 \lambda^{-\alpha} \exp[- \{\omega(\lambda/t_w^{\mu})\}^{1-n}]$$

with

$$\lambda = \{t_w/(1 - \mu)\}[\{1 + (t/t_w)\}^{1-\mu} - 1]$$

where $\mu < 1$. Note that if μ vanishes it would be possible to express M_{TRM} as the product of a power law decay and a stretched exponential decay in time.

Just like the earlier work of Souletie and Tholence (1984), the recent measurement of the a.c. susceptibility of rare-earth doped scandium SG (Wendler R, Pappa C, Eckert C and Baberschke K, to be published) could not conclusively prove whether power law is a better description of the $T_g(\omega)$ data than the Vogel-Fulcher law.

Recently, the scaling form for the ESR linewidth of the AgMn SG has been analyzed (Mahdjour H, Pappa C, Wendler R and Baberschke K, Z. Phys. B63, 351 (1986), J. Magn. Mag. Mat. 54-57, 179 (1986)) by a method somewhat different from that followed by Wu et al.(1985). Mahdjour et al. demonstrated that the excess width W_{ex}, which is obtained by subtracting the Korringa contribution from the total linewidth, depends only on χ_n/H^2, where χ_n is the nonlinear susceptibility.